Troubleshooting and Repairing Microwave Ovens

Other books by Homer L. Davidson

Troubleshooting and Repairing Solid-State TVs—3rd Edition
Troubleshooting and Repairing Compact Disc Players—3rd Edition
Troubleshooting and Repairing Camcorders—2nd Edition

Troubleshooting and Repairing Microwave Ovens

4th Edition

Homer L. Davidson

TAB Books
An imprint of McGraw-Hill

New York San Francisco Washington, D.C. Auckland Bogotá
Caracas Lisbon London Madrid Mexico City Milan
Montreal New Delhi San Juan Singapore
Sydney Tokyo Toronto

McGraw-Hill

A Division of The McGraw-Hill Companies

ISBN 0-07-015766-9 (hc.)   ISBN 0-07-015767-7 (pbk.)

McGraw-Hill books are available at special quantity discounts to use as premiums
and sales promotions, or for use in corporate training programs. For more information, please write to the Director of Special Sales, McGraw-Hill, 11 West 19th
Street, New York, NY 10011. Or contact your local bookstore.

Acquisitions editor: Roland Phelps
Editorial team: Lori Flaherty, Executive Editor
                 David McCandless, Supervising Editor
                 Anita McCormick, Book Editor
Production team: Katherine G. Brown, Director
                  Jennifer L. Dougherty, Layout
                  Jodi L. Tyler, Indexer
Design team: Jaclyn J. Boone, Designer
              Katherine Lukaszewicz, Associate Designer

EL1

Dedication

To my parents, Orpha and Chester, and to my wife's parents, Ethel and Richard.

Acknowledgments

Many manufacturers have contributed microwave oven data and circuits for this book, and to them I owe a great deal of thanks:

- Frigidaire Company
- General Electric Company
- Holaday Industries, Inc.
- Matsushita Services Company
- Narda Corporation
- Norelco Service, Inc.
- Samsung Electronic Corp.

Without the help of many others, this book would have been most difficult. A special thanks goes to Frank Besednjak, Manager—Technical Training of General Electric Company; Mr. Harry Foulds, National Service Manager of Matsushita Service Company (Panasonic); Mr. O.B. Walker, National Service Manager of Norelco Service, Inc. (North America Phillips Corporation); Mr. Ronald L. Hartle, National Service Manager of Samsung Electronic Corp. America, Inc., and Mr. A.N. Evans, Director of Communications of Tappan, a brand of Frigidaire Company.

Contents

3 Microwave oven circuits **77**

6 Servicing low-voltage circuits   *195*

11 Microwave leakage tests *391*

12 Microwave oven case histories *415*

Introduction

The microwave oven is here to stay. It is used every day in just about every home in America and all over the world. The microwave oven can be found in the kitchen, den, cottage, motel room, and motor homes. Who expected the microwave oven to be used as much as the color TV receiver? People who own microwave ovens cannot do without them. The oven is used in preparing breakfast, lunch, dinner, and sometimes in-between snacks. The microwave oven has become an integral part of our lifestyle.

Printed circuit boards were supposed to solve many wiring problems because small transistors were designed to never break down. Like the microwave oven circuits, printed circuit boards were designed for single and trustworthy operation—except that whenever you place mechanical and electronic components into a commercial unit, sooner or later you are going to encounter service problems. Like the TV chassis, microwave ovens are constructed better today and have less breakdowns. But when the oven quits, the customer wants it back yesterday.

Many ovens can be repaired by simply changing the ac line fuse. Of course, even replacing the fuse requires removing the back cover. These oven fuses are chemical types, in ceramic sleeves, so you cannot peek inside to see if they have blown open. Often, the monitor switch hangs up, line voltage overloads, or nearby lightning damages or opens the fuse. Whatever the breakdown, you should have a low-priced digital multimeter (DMM) to check for a defective fuse or poor interlock switch contacts.

Just about anyone can make basic repairs to the microwave oven using a few hand tools and the DMM found around the house or on the service bench. Defective fuses, interlocks, motors, fans, transformers, switches, and components in the low-voltage circuits can be checked with a little knowledge and the digital multimeter. Never place tools or hands into the oven while operating. Remember, these low-voltage parts operate from the ac power line. Treat the 120 volt ac circuits with extreme care. Leave the high-voltage problems up to the professional electronics technician. A Magnameter and leakage equipment should be used to locate the defective components in the HV circuits.

Always make sure to discharge that high voltage capacitor! Before replacing or attempting to service the microwave oven, discharge the HV capacitor. This capacitor can hold a charge for several weeks in some ovens, so discharge the high-voltage capacitor every time the oven is fired up. You can receive a shock of more than 3000 volts, which can injure or seriously damage a person. Remember, besides high-voltage, the amperage is very high in the magnetron circuits. Discharge the HV capacitor with a pair of long, insulated screwdrivers. Servicing microwave ovens can be extremely dangerous if you do not know what is going on.

The microwave oven has gone through many changes in the last ten years. Yesterday, all ovens were manually operated. Today, practically all ovens are operated with touch pads and control panel, like those found on a calculator or computer. The new microwave oven has many circuits on time, different foods, preheat, and defrost pads. Cooking begins and shuts off automatically for the selected time. Some ovens have variable-power cooking when touching the power level several times to select high, medium-high, medium-low, and low power. Multiple sequence cooking can be programmed for several cooking sequences, switching from one variable-power setting to another automatically. The electronic control brain provides a variety of cooking programs—and all are controlled with a microprocessor in the control panel.

In this fourth edition, valuable service information, data, and drawings of the General Electric, Norelco, Panasonic, and Samsung ovens are found scattered throughout the book. Each page has either a photo or drawing to illustrate the many oven components and circuits. Every component is shown, and you are instructed as to how to test, isolate, and replace each oven part. Although it is impossible to deal with every trouble found in microwave ovens, you will find that the great majority of problems are covered and resolved in this book.

The latest circuit diagrams, data, valuable service information, and drawings of General Electric, Norelco, Panasonic, Tappan, and Samsung ovens are scattered throughout the book. Fifteen different ovens with symptoms and troubleshooting methods are found in over 225 case histories. In my 25 years of experience in troubleshooting and repairing microwave ovens, I've encountered almost every problem found in such ovens, and these potential problems are all covered in this book.

Chapters 1 through 4 cover how the low-voltage circuits operate, with new required test equipment, and every kind of trouble. Chapter 5 shows how to troubleshoot those interlock switches with Chapters 6 and 7 discussing the high voltage circuits. How to locate, remove, and install the magnetron tube is explained in Chapter 8. How to service motors is explained in Chapter 9. How to locate and remove the defective control board circuits are discussed in Chapter 10. Chapter 11 explains how to tell if the microwave oven is cooking as it should. More than 250 actual oven case histories are found in Chapter 12. This chapter alone can help solve the many different microwave oven problems you may encounter every week. Some important do's and don'ts are listed in Chapter 13. Where to locate and find the microwave oven parts and manufacturers is given in Chapter 14.

The first, second, and third editions have gone through many printings. I have updated all the chapters and included new test instruments and oven service data in the Fourth Edition. The new General Electric tilt-down oven-door oven, Samsung, and Tappan latest ovens are found in this edition.

New methods of servicing the microwave ovens, new circuits, and required test instruments are found throughout the book. How to service the convection oven is found in Chapter 6, and Chapter 10 covers the latest control board troubleshooting data. You will find more than 450 illustrations in this book, and several microwave oven schematics are in the appendix. Just be sure to remember the danger involved in repairing microwave ovens.

1
CHAPTER

Some basic facts

Before long, nearly every home will have a microwave oven. Right now, microwave ovens are selling all over the United States, and nearly every major appliance manufacturer is building them. Once you own a microwave oven, you cannot do without it, just as you cannot do without a TV. Although microwave ovens might not require the same amount of service as a TV set, they do break down like any other electronic/mechanical device (figure 1-1).

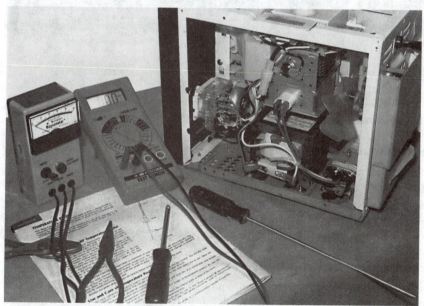

1-1 Like all mechanical-electronic devices, microwave ovens break down.

The microwave oven is an electrical appliance that is designed to warm or cook food within minutes or even seconds. Power is applied to the magnetron tube after a sequence of buttons, switches, and other components are energized. The magnetron

1

tube radiates RF energy that is tunneled into the oven to quickly cook the food that has been placed in the oven cavity.

Different types of ovens

Today, the microwave oven is found wherever food is prepared and served. Small, compact ovens (450 watts or less) are used in trailers, motor homes, and small apartments (figure 1-2). A standard tabletop microwave oven with 600 to 800 watts of power is found in most homes. Some of these tabletop ovens combine a separate ac heater coil for cooking and are referred to as convection ovens. Many large microwave ovens have a carousel or glass rotating dish that provides even heat in all areas. Large commercial microwave ovens are found in restaurants, hospitals, and nursing homes.

1-2 Microwave ovens can be used in the kitchen, cottage, or mobile home.

The compact oven

The small microwave oven is generally used for on-the-road cooking. It can be easily transported in the trunk of a car, trailer, or motor home (figure 1-3). These ovens perform only basic oven functions such as warming up food and drinks or light-meal cooking. Often, these units are less than 450 watts and pull less current than the larger models. They operate at the same frequency (2450 MHz) as most microwave ovens.

General Electric drop-door oven

General Electric has brought out a new Profile line with a new built-in combination/convection microwave oven. The JEBC200W model is white-on-white, and the JEBC200B model is black-on-black. These ovens feature an integrated exterior appearance, which eliminates the need for built-in trim kits when building them into matching cabinets.

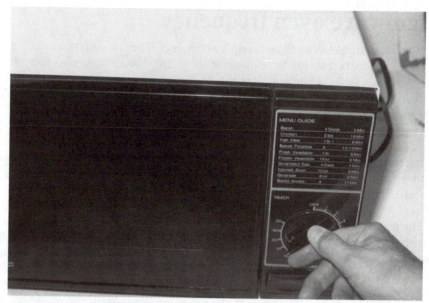

1-3 Small microwave ovens operate with less than 450 watts.

The new microwave ovens have a drop-down front door with a horizontal handle that matches the appearance of standard wall ovens to create a complementary appearance in the kitchen. You can purchase a regular microwave oven (without the convection oven) in models JEB100W (white-on-white) and JEB100B (black-on-black).

Samsung MW5510 and MW2500 models

The 800-watt Samsung microwave oven has a 1.0-cubic-foot auto-weight defrost system. With the automatic defrost feature, simply program the weight of the food, and the oven will automatically determine the correct cooking time. This MW5510 model also features a popcorn key, auto heat, one-touch cook, and ten power levels.

The Samsung touch-control microwave oven (MW2500) has a defrost feature for quick cooking right from the freezer. The oven has a two-stage cooking system and the glass tray can be removed for easy cleaning.

Sharp R3A54 12 instant action keys

The Sharp R3A54 microwave oven has three keys in four seperate lines to provide easy selection to twelve instant action keys. The first three buttons are popcorn, pizza, and beverage cooking keys. The other key buttons are baked potato, fresh vegetable, rice, dinner potato, pasta, muffin, main dish, vegetables, and roll muffin.

Good-bye guesswork! Popcorn, baked potato, and ten other favorite foods have their own designated keys. This oven also has an Express Defrost feature and a Minute Plus key that sets the oven on high for one minute per touch.

Microwave oven frequency

Microwaves are electromagnetic waves of energy. They are similar to light, radio, and heat waves. Microwaves have many of the same characteristics as light waves. They travel in a straight line and they can be generated, transmitted, reflected, and absorbed. In a microwave oven, the magnetron tube generates the microwaves. They are transmitted to the oven cavity, reflected by the sides of the oven area, and then absorbed within the food that is cooked in the oven cavity.

The microwave wavelength is relatively short compared to light. The frequency of a microwave oven is 2450 million cycles per second, with a wavelength of under 5 inches. There are three frequency bands allotted for microwave operation by the Federal Communication Commission. They are at 915 MHz, 2450 MHz, and the highest one, 5500 MHz. Today, almost all microwave ovens operate at 2450 MHz.

The magnetron tube generates the cooking energy in a microwave oven. Materials with a high moisture content (like most foods) absorb microwave energy. The food is made up of millions of molecules per cubic inch, which align themselves with the microwave energy when it enters the food. The microwaves change polarity every half cycle and oscillate back and forth at 2450 MHz, creating friction. The friction between the molecules converts the microwave energy to heat and in turn, cooks the food (figure 1-4).

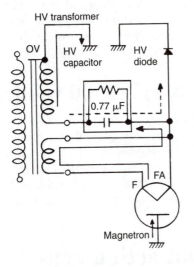

1-4
The magnetron supplies the RF
energy needed to cook food.

Microwaves can be reflected in the same manner as light. These short electromagnetic waves of RF energy pass through material such as glass, china, paper, and most plastics. Materials such as aluminum foil and stainless steel tend to reflect the microwaves, while ordinary steel might absorb some microwave energy. Metal material placed in the oven cavity should be used only as recommended in the cooking instructions.

The sides of the oven cavity are constructed of metal so as to remain cool while the food is cooking. Although the sides and bottom of the oven sometimes feel warm, this is caused by the transfer of heat from the food. The front door, which contains a metal plate with perforated holes, remains cool. The holes reflect the microwaves but allow light to enter so you can see inside the oven cavity. Remember: Water absorbs microwaves and begins to boil, while light passes through water.

Food prepared in a microwave oven should be cooked all the way through and not from the inside out. In some ovens, the food must be cooked for a few minutes, turned, and cooked again. Other ovens have a device that rotates the food, providing for even cooking throughout. With all food preparation and cooking, you should always follow the oven manufacturer's instructions.

Fundamental function of the magnetron

The cathode is located in the center of the magnetron. It is a filament that boils off electrons when it is hot. The cathode is connected to the negative side of the power supply, which has a potential of approximately 4000 volts with respect to the anode, which is connected to the positive side. The 4000-volt potential is produced by means of the high-voltage transformer and the doubler action of capacitor and diode, of which a detailed circuit description is given later in this book.

The electrons have negative charges, which means they are strongly repelled by the negative cathode and attracted to the positive anode. The electrons would travel straight from the cathode to the anode if the 4000-volt potential was the only force acting in the magnetron (figure 1-5). However, the magnetron is a type of diode with a magnetic field applied axially in the space between the cathode and anode by means of two permanent magnets (figure 1-6).

If a magnetic field of sufficient strength is applied between the cathode and the anode, an electron will travel in a path almost at right angles to its previous direction, resulting in a circular motion of travel to the anode (figure 1-7). Eventually, it reaches the anode.

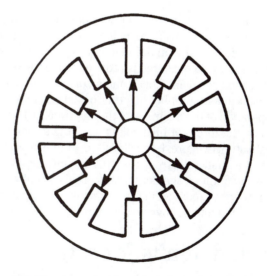

1-5 Electrons inside the magnetron travel from the cathode to anode.

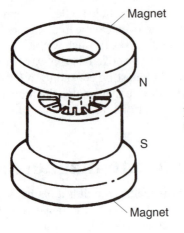

1-6
Heavy magnets are found within
the magnetron tube.

1-7
Electrons leave the cathode and
travel in a circular motion to
the anode element. Courtesy Matsushita
Electronic Corp. of America

This circular motion by the electrons induces alternating current in the cavities of the anode. When an electron is approaching one of the segments between two cavities, it induces a positive charge in the segment (figure 1-8). As the electron goes past and draws away, the positive charge is reduced while the electron is inducing a positive charge in the next segment. This inducing of alternate currents in the anode cavities can be thought of as a lumping together of the resonant circuit (figure 1-9).

In the actual operation of the magnetron, the electrons crowd together as they go around. Influenced by the forces of high voltage and the strong magnetic field, they form a spoke-wheel pattern (figure 1-10). This crowd of electrons, which has much stronger energy than a single electron, revolves around the anode and eventually reaches the cavities, resulting in the continuous oscillation of the resonant circuits. The high-frequency energy, produced in the resonant circuit (cavities), is then taken out by the antenna and fed into the oven cavity through the waveguide (figure 1-11).

Some do's and don'ts for safety

To prevent the risk of burns, fire, and electrical shock, the owner or operator should observe the following guidelines:

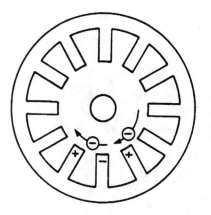

1-8
When electrons approach one of
the segments between cavities,
a positive charge is induced.
Courtesy Matsushita Electronic Corp. of America

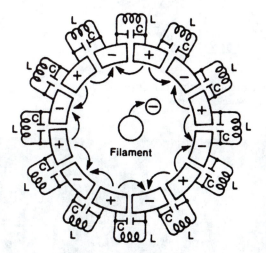

Filament

1-9
The resonant circuits are
lumped together. Courtesy Matsushita
Electronic Corp. of America

Follow the microwave oven instructions before attempting to cook. Don't forget that the oven is a high-voltage, high-current appliance. You must take extreme care at all times, but this does not mean that you have to be afraid to operate the microwave oven.

Make sure that the oven is installed properly. Most ovens include a three-prong plug. You can pick one up at your local hardware store or electrical dealer (figure 1-12). Do not under any circumstance cut or remove the third ground prong from the power plug. The small, flexible grounding wire with a spade lug should be screwed under the plate screw of the ac outlet. Use a voltmeter to determine if the center screw of the ac receptacle is grounded. Install a separate outlet from the fuse box for the microwave oven.

Do not use an ordinary two-wire extension cord to operate the oven. If you need an extension cord, contact a local electrician or the appliance dealer that sold you the oven. Make sure that the ac power outlet is never less than 105 volts or more than 125 volts for proper oven operation. The microwave oven is designed to operate from 115 to 120 Vac. Don't just plug the oven into an overloaded outlet where

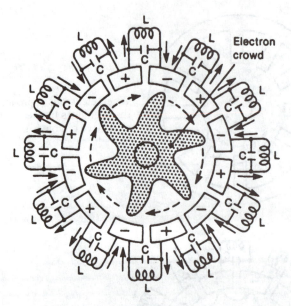

1-10 The strong magnetic force represents a spoked wheel pattern. Courtesy Matsushita Electronic Corp. of America

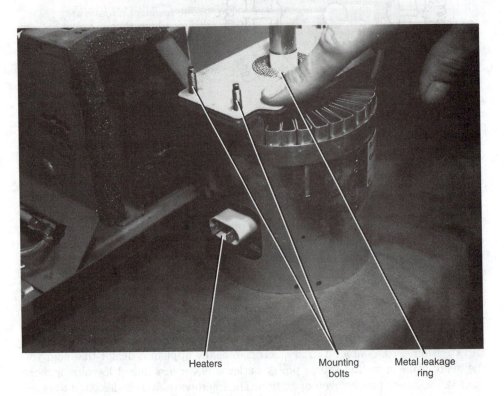

Heaters   Mounting bolts   Metal leakage ring

1-11 A magnetron tube is found in all microwave ovens.

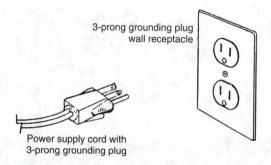

3-prong grounding plug
wall receptacle

Power supply cord with
3-prong grounding plug

1-12 Make sure the ac plug and three-prong plug are grounded.

several appliances are already tapped into it. If you do, the oven might operate erratically and never cook properly.

Install or locate the oven only in accordance with the installation instructions. Don't install the oven where the side or top opening might be blocked. The hot cooking air must escape, and fresh air must be pulled into the oven for normal operation. Don't operate the microwave oven if the power cord or plug is damaged. Repair the three-wire cord, and be sure that the ground wire is intact. Check the continuity from the metal cabinet to the ground wire terminal with the low range of an ohmmeter.

Supervise children who are operating the oven. Most microwave oven problems are related to improper oven operation. Don't use the oven outdoors. Do not let the cord hang over edges of a table or the cooking counter. Keep the ac cord away from heated surfaces or a nearby cooking stove. Don't use metallic cooking containers. Use only cooking utensils or accessories of the type recommended by the manufacturer or microwave cookbooks.

Don't use any type of material that might explode in the oven. Some sealed glass containers or plastic jars might build up pressure and explode. A regular paper sack with food inside might explode and cause a fire if steam or air holes are not punched in the top of the sack. A fire inside the oven might cause the plastic heat shields and front oven coverings to melt. This could result in a fairly expensive oven repair. Keep the oven cavity area spotlessly clean. Cooking fatty items such as bacon might in time cause grease to collect behind the plastic shelf supports on plastic microwave guide covers, resulting in excessive arcing and damage. Do not operate the oven with nothing in it.

Use liquid window cleaner or a very mild detergent with a soft, clean cloth over the oven's face and interior surfaces. Do not use any kind of commercial oven cleaner inside the oven area. Use a paper towel or sponge to blot up and remove spills while the oven is still warm. Pull the oven's power cord out if fire is discovered inside the oven cavity. Keep the oven door closed to help smother the fire. Clean out the top area or exhaust areas for signs of food pulled up by the fan (figure 1-13). Have the service person brush out these areas when the oven is serviced.

A collection of food particles might in time cause a persistent odor. Some odors can be eliminated by boiling a one-cup solution of several tablespoons of lemon juice dissolved in water in the oven cabinet.

1-13 Clean out bread crumbs, dust, and food particles from the exhaust fan area.

Don't change the operation of the oven while it is in operation. This is a good way to blow the fuse of a microwave oven. Touch the stop button and start the operation all over again.

Don't assume that the microwave oven is defective if lines are streaking across the face of a TV set in a nearby room. Most ovens cause some type of interference to nearby TV receivers.

When the microwave oven fails, make sure that the three-prong cable plug is installed in the ac outlet. Plug in a lamp or radio to make sure that the outlet is alive. Go over the operation instructions very carefully before calling the service person.

> **Caution:** It's possible to receive a dangerous shock or possibly be killed by trying to just change the 15-amp fuse. The highly charged capacitor must first be discharged before touching any electrical or electronic component (figure 1-14).

Don't remove the back cover of the microwave oven unless you know what you are doing. Keep the power cord pulled when the back cover is off.

Oven installation

Make sure that the oven is located on a firm table base to prevent vibration. The oven should be installed with at least a 2- or 4-inch clearance on all sides of the oven cabinet. Of course, this depends on where the oven vents are located. Leave at least one inch or more clearance at the top of the oven.

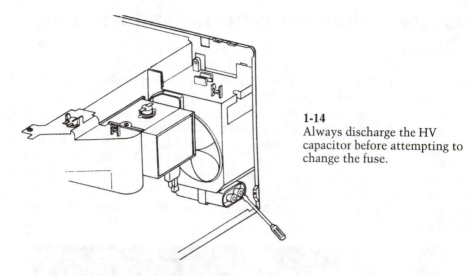

1-14
Always discharge the HV capacitor before attempting to change the fuse.

Try to avoid installing the oven near a heat source, heat duct, or cooking range. Keep the oven out of the sun where temperatures might exceed 110°F. A microwave oven will not operate if it is too hot. Avoid using the microwave oven when the humidity is very high. Position the oven away from radios and TVs; the microwave oven might interfere with the reception.

A microwave oven will operate successfully when adequate electric power is supplied to it. It's best to install a separate outlet for the oven. Because the oven is protected with a 15-amp fuse, additional appliances on the same circuit might overload the circuit and blow the house fuse (figure 1-15).

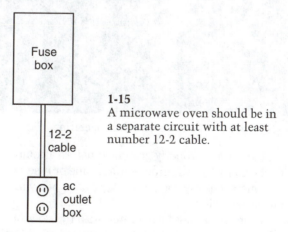

1-15
A microwave oven should be in a separate circuit with at least number 12-2 cable.

All ovens should be grounded for personal safety. A three-prong plug is used to supply power and ground the metal oven cabinet. Make sure that the third prong of the two-prong standard outlet is grounded. If installed by the appliance dealer, have the oven checked for proper grounds or have it checked by a local licensed electrician. While the electrician is present, have the power outlet checked for proper fuse protection and correct operation voltage (115 to 120 Vac).

Service technician cautions and warnings

The microwave oven is a high-voltage, high-current piece of equipment, so you must use extreme care while servicing it. Before taking off the back cover or outside wrap, remove your wristwatch. Make sure that the oven is unplugged at all times when replacing and testing components.

Never stick a tool or your hand inside the oven while the oven is operating (figure 1-16). Before checking any component or wiring in the oven, discharge the high-voltage capacitor—just like you discharge the CRT anode connection before working around the high-voltage section of a TV. The big difference between the high voltage of a TV receiver and a microwave oven is the amperage; you might be severely shocked or killed if the high-voltage capacitor is not discharged. Always discharge the capacitor with an insulated-handle screwdriver before working in the high-voltage area.

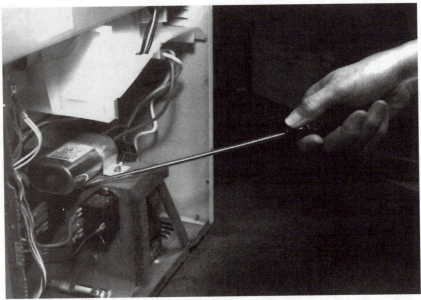

1-16 Do not stick tools inside the oven while operating.

Make sure that the oven is properly grounded before attempting to service it. If you're not certain, clip a flexible wire from a water pipe or fuse box to a metal screw on the back cover or metal chassis. Do not use a regular two-wire extension cord to operate the oven while servicing.

The microwave oven should not be operated with the door open. Do not for any reason defeat the interlock switches. Check and replace all defective monitors and latch switches if the oven will not shut off with the door open. The oven should never be operated if the door does not fit properly against the seal. Check for broken or damaged hinges. Visually inspect the seal gasket for possible cut or missing pieces. Check the gasket seal area for foreign matter. Make sure the oven door is mounted straight and fits snug against the oven. Readjust the door when play is felt between door and oven. Always remember to check all interlock switch functions before the oven is returned to the customer.

While servicing a microwave oven, it's best to have the service bench away from the customer and not to talk to anyone while working on the oven. Outside distractions might cause you to make a mistake that could result in an injury. Always keep your hands out of the oven operation area when the oven is operating.

After making any repairs or service adjustments, check the oven for excessive radiation. To ensure that the oven does not emit excessive radiation and to meet the Department of Health and Human Services guidelines, the oven must be checked for leakage with an approved radiation meter. Especially check for radiation around the door and vent areas. If by chance the radiation leakage is greater than 5 mW/cm^2 or someone has been hurt pertaining to microwave radiation leakage, report this to the oven manufacturer.

Although a few components can be substituted, most microwave oven parts should be replaced with the original part number. These can be obtained from the manufacturer or oven distributors. When replacing control panel circuits, do not touch any part on the board, as static electricity discharge might damage the control. Usually these controls come packed in a static-free wrap and carton.

Basic service precautions

When servicing any microwave oven, there are several basic safety precautions to follow:

Microwave oven door

Do not try to operate the oven with the door open, an improper seal, and/or damaged door hinges. If the oven has been dropped, inspect the door, seal, and hinges before you do any other servicing on it. Make sure that the door closes properly and no foreign matter is inside the door.

Extension cords

Always make sure that the oven is grounded with a three-prong-plug wall receptacle. Use only a three-wire extension cord between oven and wall outlet. It is best to have a competent electrician install the three-prong wall receptacle.

Unplug oven

Never reach into the oven component area while the oven is operating. When making ohmmeter or continuity tests, unplug the oven from the ac power line. If ac voltage measurements are to be made, clip on the voltmeter while the oven is off. Do not touch the voltmeter test probes while the oven is plugged in.

Shorting capacitor

Before checking any components in the oven, short out the two capacitor terminals. Always discharge the capacitor each time the oven is energized. Failure to discharge the capacitor could result in severe electrical shock.

Door interlock switches

Do not defeat or bypass the interlock door switches. Each interlock should be checked for proper operation.

High-voltage section

The power transformer, capacitor, diode, and magnetron tube can contain high voltage (HV) when the oven is operating. When the oven shuts off, discharge the HV capacitor before attempting to do any type of service. Be careful when taking high voltages in the HV section.

Magnetron tube

Check the magnetron level and secured bolts to eliminate hot spots or radio frequency (RF) arcing. Always replace the RF gasket found around the magnetron tube antenna. Do not use metal tools near the magnetron, as it can jerk them out of your hands.

Microwave leakage tests

Always make a microwave leakage test with the approved radiation meter after all repairs are made. This should be done before oven front and side cleanup procedures.

Checking the outlet receptacle

The microwave oven should have a separate outlet and be protected with a 20-amp fuse. The outlet should be polarized and grounded. An improper ac outlet might provide electrical shock for the operator. Check the ac outlet for correct ac voltage (117-120 Vac) (figure 1-17A). Determine if the left side of the outlet is grounded (figure 1-17B). If the ac receptacle is wired correctly, the right side of the outlet should be hot (figure 1-17C).

Pull the plug

Always remember to pull the plug or cord before attempting to service the oven (figure 1-18). Each time the oven is fired up, pull the plug and discharge the HV capacitor before putting your hands in the oven. Pull the plug before clipping test equipment to components in the oven. Pull the plug before taking continuity or resistance measurements of oven components. If by accident you are shocked by the HV capacitor and you end up with a metallic taste in your mouth, you will know how careless and how lucky you are.

Installing the ac cord

Try to replace the ac power cord with the original part number. This way, everything fits and is easily installed. If the new cord is to take some time or is not available from the manufacturer, check for heavy cord and a plug replacement at the local

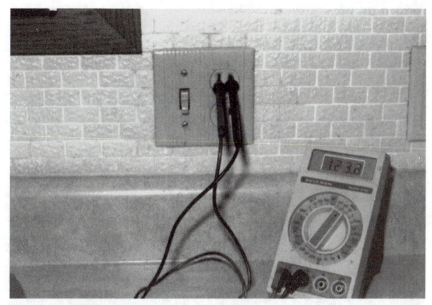

1-17A Check the ac plug for at least 120 Vac.

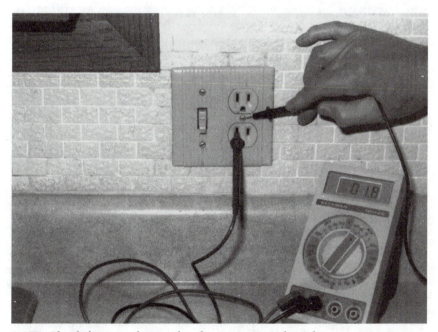

1-17B Check for a good ground and measure ac voltage between negative terminal and ground screw.

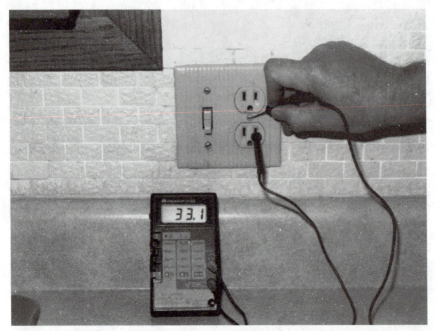

1-17C Check the ac plug for correct polarity.

1-18 Pull the ac cord before attempting to service inside components.

hardware store. You can replace the cord with a three-wire number-12 cable wire. Replace the ac plug with a heavy-duty, rubber, three-prong ac plug.

Make good, solid connections inside the oven area. Use crimp-on or eyelet connections. If these are not available, twist the copper braided wire into a small circle, to just fit over the bolt connections. Solder the entire eyelet or circle with a heavy-duty soldering iron for good, strong connections. Tie a knot in the cord, inside the oven, and wrap excess tape around the cord so it cannot be pulled out of the oven. Or use a cable clamp from the original cable, if it had one.

Checking defective power line cords

Periodically, inspect the microwave oven ac cord for breaks, check marks, frayed areas, and broken ac plug connections. The original ac cord has a molded ac plug with an extra ground wire. Sometimes, when the cord is yanked or the unit is dropped, the internal cable wires might break inside the molded plug without any sign of breakage. Cut off the cord three inches from the plug and install another heavy-duty three-prong ac plug.

Check the ground and both ac wires with the low ohm scale of the VOM or DMM. Make sure the green or ground wire has good continuity from the large center prong to the oven metal case, inside the oven. No resistance should be measured between these two areas (figure 1-19).

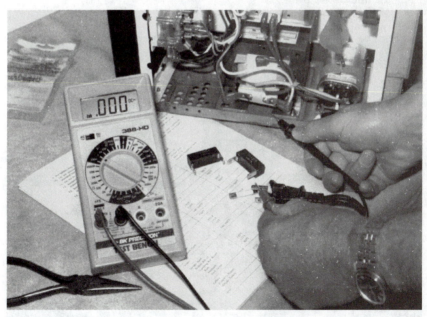

1-19 Check the ground plug from cord to metal base of oven.

Lightning protection

Many new microwave ovens have metal oxide varistors in the power line to protect the oven from lightning and power line surges (figure 1-20). These oxide varistors can be added to ovens that do not have this type of protection. Solder in the oxide varistors at the power cord fuse terminals inside the oven (figure 1-21). Even after you do that, a heavy line voltage surge or lightning a block away can cause the varistor to arc over and blow the oven fuse. Check the house fuse if the varistor and main fuse are blown.

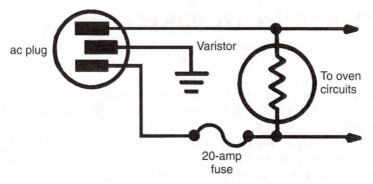

1-20 Check the oxide varistor to protect the touch panel control in the microwave oven.

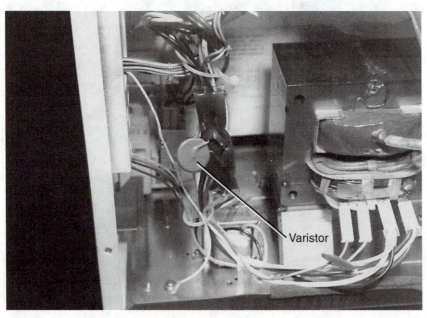

1-21 If the fuse keeps blowing, suspect a defective varistor.

Basic oven components

Some microwave ovens have different basic oven components than others. All microwave ovens operate basically the same, except that some ovens have many additional features. Here is a list and brief description of most of the components found in a microwave oven.

Antenna/antenna motor

The antenna and antenna motor are located below the removable glass floor in the GE JEBC200 built-in microwave oven. Remove the glass shelf (with a suction cup or putty knife) and lift the antenna off the antenna motor shaft.

To service the motor, disconnect the power and remove from the wall. Remove the antenna. Carefully place the oven on its back on a comforter or pad. Remove 2 screws to access panel on bottom of oven. Disconnect the motor lead and the 2 screws that secure the motor (figure 1-22).

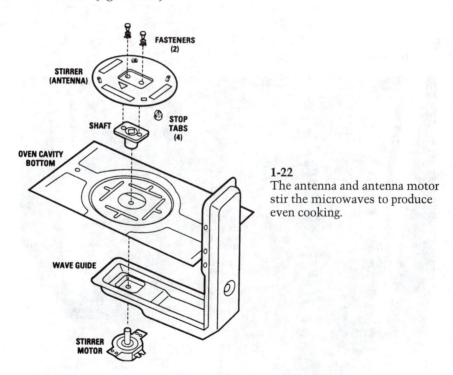

1-22
The antenna and antenna motor stir the microwaves to produce even cooking.

Bleed resistor

The bleed or bleeder resistor is around 10 megohms. It is found across the HV capacitor. When the oven is turned off, the bleed resistor discharges the capacitor within one minute (figure 1-23). Always discharge the HV capacitor before working inside the oven.

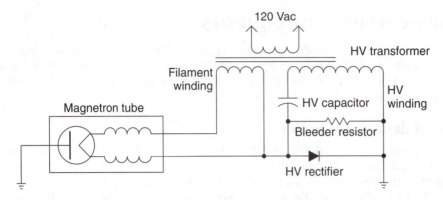

1-23 The bleeder resistor discharges the HV capacitor after the oven is shut off.

Blower motor

The blower motor draws cool air through the blower intake and is directed at the magnetron tube. Heat develops within the magnetron tube, and the tube needs to be kept cool with the blower fan blades. Most of the hot air is exhausted directly through the vents of the back plate (figure 1-24).

1-24 The fan blower brings cool air across the magnetron tube to keep it from overheating.

Buzzer or speaker

The buzzer or speaker is used in ovens with the power control circuits to indicate when the cooking process is finished. The audio device is located on the control circuit board (figure 1-25).

1-25 The piezo speaker or buzzer indicates when cooking is finished.

Capacitor

The oil-filled capacitor is located in the HV doubler circuit (figure 1-26). The ac voltage (2500 V) from the power transformer charges the capacitor through a silicon diode increasing to approximately 4000 peak volts. Always remember to discharge the two capacitor terminals before attempting to check or replace any component in the oven-operating area (figure 1-27).

Cavity light

The cavity light is located in the middle of the oven behind the top front grille. Disconnect the ac plug. Open oven door and remove two screws underneath grille lip. Remove screw holding light bulb housing (figure 1-28). Lift up on the door overhang and rotate the bulb housing towards the front of the oven. Unscrew and replace the 40 watt light bulb (WB02X4253).

Cook relay

The cook relay, when energized, closes the ac circuit so that the voltage is applied to the primary terminals of the power transformer. Sometimes this relay is referred to as the power relay.

Cook switch

Usually the cook switch is the last component to be punched after the oven settings. The cook switch completes the circuit to the power transformer. A relay might be energized by the cook switch, or the power transfer is completed through timer contacts, thermal cutout, and primary interlock switch.

1-26 The HV capacitor operates in the HV doubler circuit.

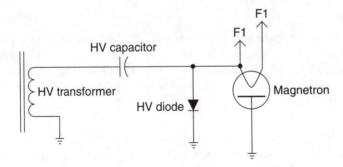

1-27 The HV capacitor is in series with the HV diode and common ground.

Control circuit

The control circuit might consist of an electronic control board mounted behind the front control panel. In some ovens, the control circuit is called the electronic controller. You might find that both the front panel and the control circuit board are replaced as one unit (figure 1-29).

Convection fan

The convection fan and heater are located at the rear of the GE JEB100 oven. There are two blades on the convection motor shaft, one moves the air across the heater (internal blade) and the outside fan is used to cool the fan motor (figure 1-30).

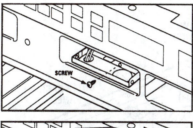

1-28
The cavity light provides light
inside the oven.

1-29 The control panel consists of key buttons, an LED or tube display, control components, and a ribbon cable.

Remove the convection fan motor by removing top panels and discharging the HV capacitor. Remove the screws that hold the fan blade in position. Remove the four screws that are located inside the heater element. Remove three screws at the other end of assembly. Now remove the two screws holding the motor to the motor bracket.

Damper door motor

The damper door motor and assembly is located next to the magnetron tube at the back right rear of the GE JEB200 oven. The damper door is closed during cooking

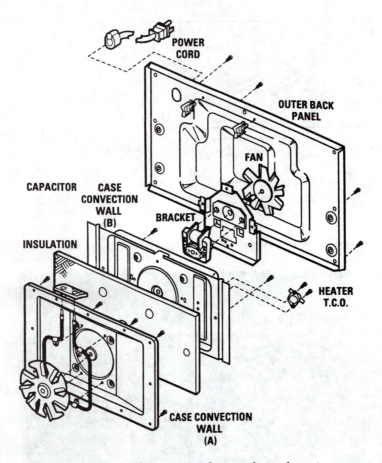

1-30 The convection fan removes hot air from the microwave oven.

functions because the magnetron fan motor is always running when the oven is on. This allows for more cooler air to be blown into the oven cavity (figure 1-31).

Defrost switch

The defrost switch is normally closed and allows the circuit to be completed for normal operation. The defrost circuit is made up of the defrost switch, defrost timer switch, defrost motor, and CAM. Placing the defrosting lever in the ON position opens the normally closed defrost switch and completes the circuit through the defrost timer switch.

Defrost timer switch

The defrost timer switch is opened and closed by the CAM attached to the motor shaft. Actually, the defrost timer switch is in parallel with the defrost switch (figure 1-32). The defrost and timer switches are similar to the vari-switch in other microwave ovens.

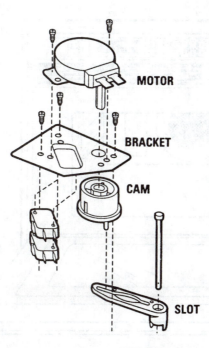

1-31
The damper door motor operates a door that opens during microwave cooking and closes during convection operations.

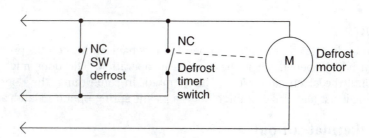

1-32 The defrost timer switch opens and closes the defrost switch.

Door—GE lay down

The lay down door in a GE JEBC200 consists of a handle, outer glass, sub-door assembly, choke cover, hinge assembly and latch pawls. Like several of the new ovens, the front door unlatches from the top and lays down towards the operator. The glass, handle, latch pawls, and door spring can be removed and replaced individually (figure 1-33).

Door seal

The primary door seal is sometimes called a choke. A choke cavity reflects the microwaves back into the oven cavity. The choke cavity is filled with a material called polypropylene, which is transparent to microwaves.

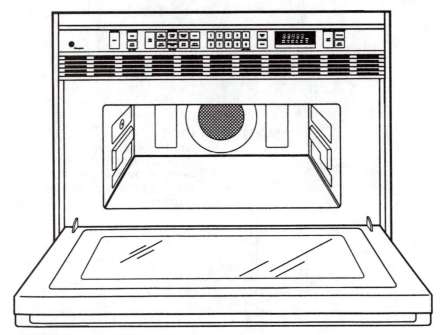

1-33 The front door of the JEBC200 microwave oven.

Door latch

The door latch is an electrically operated mechanical device that prevents the door from being opened when the oven is in operation. The door might remain latched until the electrical circuit is interrupted. In some units, the door latch is made of solid cast material or it might have a plastic spring-operated latch assembly.

Heater thermal cut-out

The heater TCO (thermal cut-out) is located on the back wall next to the convection motor. This safety device is responsible for not allowing runaway convection heater temperatures. If the oven gets too warm with convection heating unit on, the TCO will remove ac voltage from the heater coil.

Heating element

A separate heating element might be used for browning and cooking food in some microwave ovens. The heating element is located at the top of the oven cavity. The heating element might be controlled with the temperature control unit (figure 1-34).

Interlock switches

You can find from two to five interlock switches in the various microwave ovens. Primarily, the interlock switches are activated by closing the oven door (figure 1-35). A cook cycle cannot begin until the door is closed. No microwave energy is emitted while the door is open.

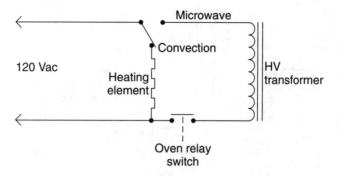

1-34 The microwave circuits are switched out with the convection oven cooking switch.

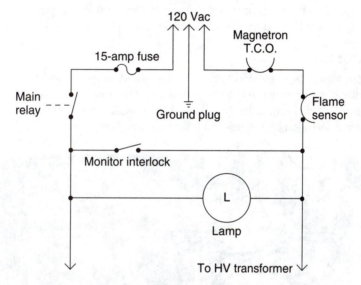

1-35 Several interlock switches are found throughout the oven.

Flame sensor

The flame sensor is located on the exhaust duct in this GE oven. The temperature rating is 248 degrees Fahrenheit. The flame sensor will open if the heat exceeds that limit. Check the flame sensor continuity with the ohmmeter range. The flame sensor is found in series with the ac line voltage (figure 1-36).

Fuse

Most ovens are protected with a 15-amp, chemically active fuse. Replace with the same type of fuse. You might find a fuse resistor circuit in other ovens (figure 1-37). The fuse resistor assembly senses an increase in current in the transformer secondary circuit and opens up the primary transformer circuit.

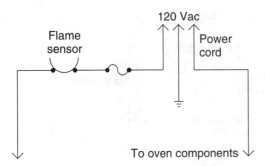

1-36
The flame sensor is located in series with the ac fuse.

Fluorescent tube

A fluorescent display shows the time and operation numbers, making them visible to the eye. The fluorescent tube is controlled by a microcomputer or control IC (integrated circuit) located on the control board.

Gas sensor

The gas sensor is a semiconductor device that detects vapor, aroma, moisture, and humidity coming from the food as it cooks. A gas sensor is now included in some microwave ovens. It automatically adjusts the cooking time for various foods. The gas sensor is controlled by the control board (figure 1-38).

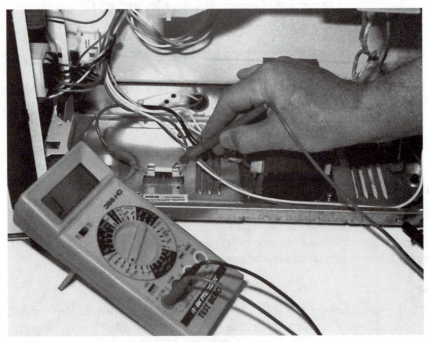

1-37 Check the 15-amp chemical fuse with an ohmmeter.

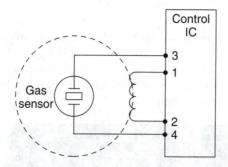

1-38
The gas sensor detects moisture, vapors, and humidity from the food in the oven.

Humidity sensor

The humidity sensor detects steam during the cooking period. In Panasonic ovens, the sensor is located close to the exhaust vents, where steam generated from the food is exhausted with circulating air (figure 1-39).

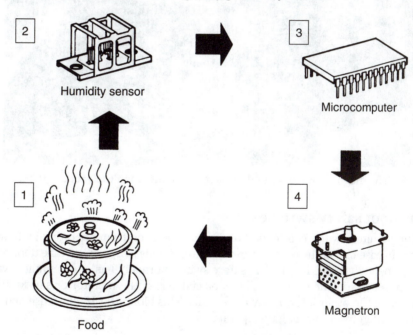

1-39 The humidity sensor detects steam in the cooking cycle.

Light switch

The light switch turns the oven light on so the operator can view the contents in the oven cavity. This switch might turn the light on when the door is opened or closed. Usually, no oven light indicates a blown fuse, a defective switch, or a defective light. You might find more than one oven light in some ovens.

Magnetron

The magnetron is a large vacuum tube in which the electrons flow from the heated cathode to a cylindrical anode surrounded by a magnetic field (figure 1-40). These electrons are attracted to the positive anode, which has a high dc voltage. The magnetron tube oscillator has a very high frequency of 2450 MHz RF energy. The radiated RF energy cooks the food placed in the oven cavity.

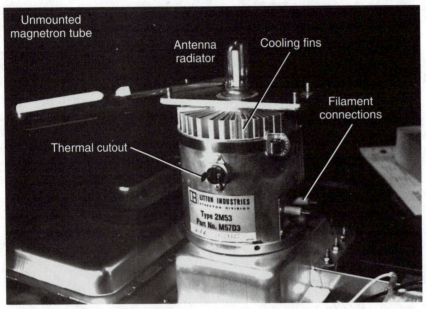

1-40 The magnetron has several different elements than the vacuum tube.

Monitor or safety switches

The monitor switch is intended to prevent the oven from operating by blowing the fuse in case one of the other interlock switches fails to open when the door of the oven is opened (figure 1-41). This safety switch normally has open contacts when the door is closed. The entire ac line is placed across the monitor or safety switch when one of the other interlock switches fails. This blows the 15-amp fuse and prevents radiation from reaching the operator.

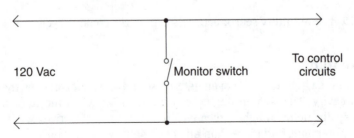

1-41 The monitor switch blows the fuse when the oven does not shut off after the door is opened.

Oven lights

Most oven lights come on when the door is opened or when the oven begins to cook. The oven light might be located in the circuit after the triac device or control relay (figure 1-42). The oven light is located outside of the oven proper and shines through perforated holes.

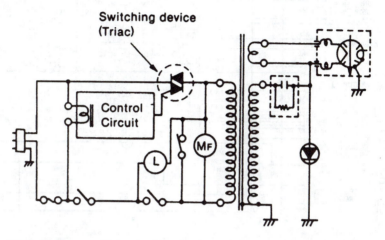

1-42 The oven light is found after the control relay or triac assembly.

Oven cavity repairs

When the inside of an oven receives heavy gouges or scratch marks, chips, and flakes inside the oven area, touch them up with a white MWO cavity paint spray. A specially formulated microwave oven paint is used to cover these areas. The ivory white aerosol oven touch-up paint is supplied by Panasonic #ANE0887L10AP. If the paint cannot be found locally, it must be shipped by UPS Ground Service, due to chemical content. Check electronics and microwave oven manufacturer wholesale or parts depots that might stock the touch-up spray paint.

Oven sensor

The oven sensor is a negative coefficient thermistor—that is, when the temperature goes up, the resistance goes down. The sensor in a GE JEB100 oven mounts on one side of the oven. Measure the resistance of the sensor around 60,000 ohms in this oven (figure 1-43).

Power relay

When the food is placed in the oven and the door is closed, the primary and secondary interlock (power relay) switches are closed (figure 1-44). Now when the start pad is touched, the main relay and the power control relay are energized by the touch control circuit. Both relay switches are in series with the HV transformer. If either relay will not function or has bad switch points, the 120-volt ac voltage is not found at the primary winding of high voltage transformer.

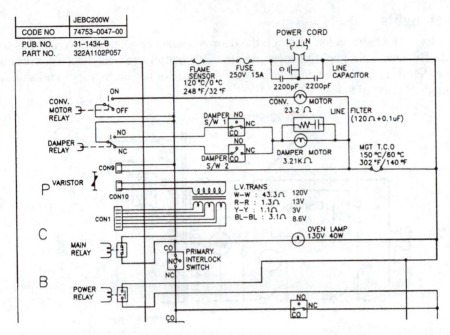

1-43 The oven sensor is connected to the touch panel circuits in the GE JEB100 oven.

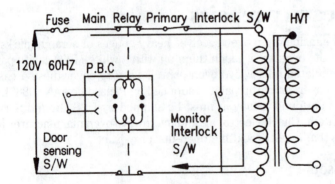

1-44 The power relay is in series with the bottom leg of primary winding of HV transformer.

Probe jack

The probe jack is located inside the oven and provides an easy method of plugging in the temperature probe. In some ovens, when the cooling probe is out of the jack, make sure to turn the cooking control off, or the oven will not operate. Always remove the temperature probe when the oven is empty. Do not let the probe touch any metal sides when in operation.

Rectifier

A special silicon diode is a solid-state device that allows current flow in one direction but prevents current flow in the opposite direction. The rectifier and capacitor provide an HV doubler circuit. The diode acts as a rectifier by changing alternating ac current into pulsating dc (figure 1-45).

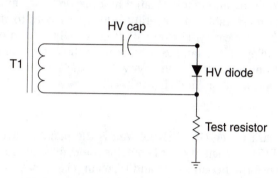

1-45 The silicon diode rectifies the ac current into a pulsating dc voltage.

Relays

The cook or power relay provides ac voltage to the power transformer circuits. In some ovens, the coil of the cook relay might be energized by the control unit (figure 1-46). A surge current relay, used in conjunction with a low-value resistor, might be used as a current limiting device. The contacts of the select relay complete the circuit to the fan motor and oven light.

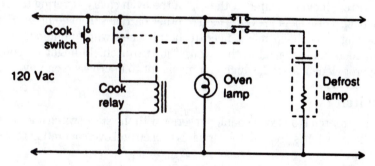

1-46 The power relay or cook switch provides power or ac voltage to HV transformer.

Resistors

A resistor is an electronic component that limits the flow of current or provides a voltage drop. The bleeder resistor across the HV capacitor lets voltage bleed off to chassis ground, through the HV diode. A fuse resistor in some ovens acts as a sensing device to an increase of current in the transformer secondary circuit and opens the transformer primary circuit. You might find a surge resistor located in the ac input circuit of some ovens to prevent damage to the oven circuits when power line outage or lightning occurs. A test resistor is used to measure the current of the magnetron tube.

Rotary antenna

Usually, the rotating antenna is located in a plastic tower in the middle of the cradle. The antenna assembly has several blades that rotate to stir the RF energy over the food in the oven cavity. The rotary antenna might be rotated with a separate motor or with a pulley on the fan blower motor. In the early ovens, the moving antenna was found on top of the oven cavity. Now the antenna assembly might be found under the bottom floor shelf (figure 1-47).

Smart board

The smart board in the GE JEB100 oven is the panel controls parts mounted upon a separate PCB. The smart board contains the convection motor relay, damper relay, varistor, transformer connector 1 and 10, main relay, power relay, and heater relay. All of these components are connected to the various oven circuits (figure 1-48).

Stirrer motor

A stirrer motor is located at the top of the oven cavity. The motor rotates or spreads the RF energy for even cooking in the oven (figure 1-49). Some ovens turn the food instead of using a stirrer motor. The stirrer motor might incorporate a separate ac motor, or the fan might be rotated by a pulley-belt arrangement on the oven fan.

Stirrer shield

The stirrer shield is located at the top of the oven cavity covering the RF waveguide opening. This shield prevents food or other particles from going into the waveguide opening. The shield might need to be replaced when bumped, cracked, or broken by food being placed in the oven. The stirrer shield might be constructed from polypropylene material, which is transparent to microwave energy.

Stop switch

The stop switch is a closed contact in series with the start switch and power relay. When the stop switch is pressed, the circuit is broken and oven operations are stopped.

Temperature fuse

The temperature fuse is in series with the ac oven fuse (20 amp). A thermal or temperature fuse can be found in some Canadian models. This fuse can be checked with the ohmmeter (figure 1-50).

Thermal cutout

The thermal cutout is designed to prevent damage to the magnetron tube if the magnetron becomes overheated. Under normal operation, the thermal cutout remains closed (figure 1-51). In some microwave ovens, an oven thermal cutout device

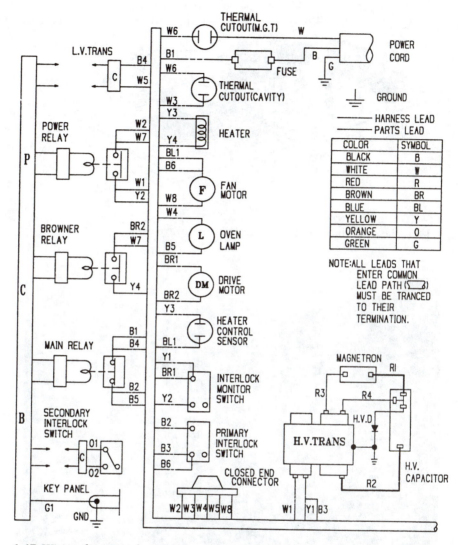

1-47 Wiring diagram of a Samsung MG 5920T microwave oven. Courtesy Samsung Electronics America, Inc.

is used in conjunction with the fan motor to cool the oven. The contacts of the oven thermal cutout are open at normal oven temperatures.

Thermistor

You might find that a thermistor is used with the temperature control unit. The resistance of the thermistor in the temperature probe changes with the food temperature determining the time of cooking. The temperature in the oven cavity is detected through the resistance of the thermistor. The thermistor is a negative temperature coefficient component. A temperature probe is not found in all microwave ovens.

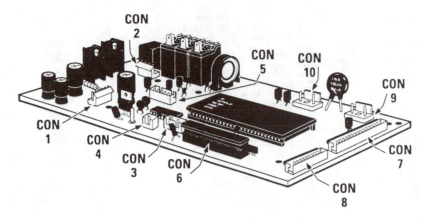

1-48 The smart board of the GE JEB100 microwave oven.

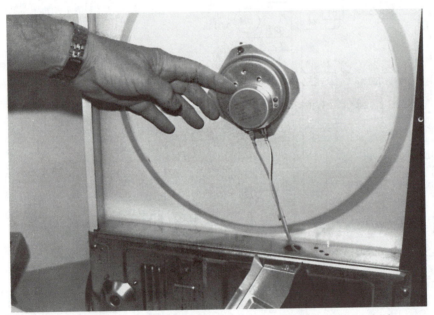

1-49 The stirrer motor is located at the top and turntable motor at the bottom of oven cavity.

Thermostat

A thermostat is found in some convection ovens. This thermal device senses the temperature of the convection air passing through the convection air guide. Different degrees of temperature might be selected by rotating the convection control shaft (90-465°F). Normally, the thermostat contacts are closed (figure 1-52). The thermostat is in series with the convection heating element.

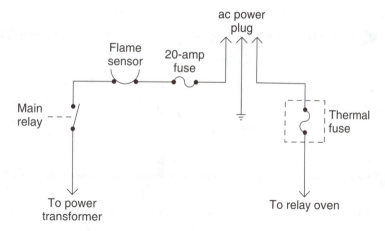

1-50 The thermal fuse is in series with the 15-amp oven fuse.

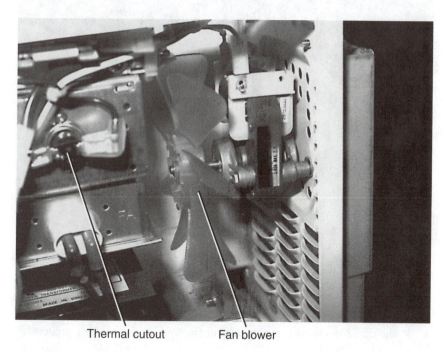

Thermal cutout   Fan blower

1-51 The thermal cutout upon magnetron shuts down the oven if magnetron runs too hot.

Timer

You might find more than one timing device in the microwave oven. One timer might be used for from zero to five minutes of cooking time, while the other timer has a longer cooking operation. The timer switch contacts are mechanically opened or closed by turning the dial knob located on the timer motor shaft (figure 1-53).

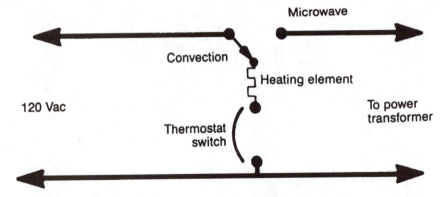

1-52 The thermostat or convection switch cuts out the ac voltage upon heater element.

1-53 The timer assembly turns on and off the oven in the cooking process.

Timer bell

The bell striker is mechanically driven by the timer motor. It rings once at the end of the cooking cycle, indicating that the food is ready. A defective timer assembly can prevent the bell from ringing.

Timer motor

The timer motor is energized through the timer contacts. When the timer reaches the 0 point on the scale, the timer switch opens the motor circuit, and the cooking cycle stops. The rotating timer motor determines how long the oven remains in the cooking mode. In some ovens, the fan motor remains in operation until the contact of the oven thermal output opens up.

Touch control panel

The touch control panel in the GE JEB200 contains key pads, control panel, LED display, ribbon cable and pin connectors. In other ovens, the touch panel may include the number display, varistor, main relay, secondary interlock relay, speaker, small power transformer, function key switches, touch control components, and flat ribbon cable (figure 1-54). In most cases, the control panel is either exchanged or sent in for repair.

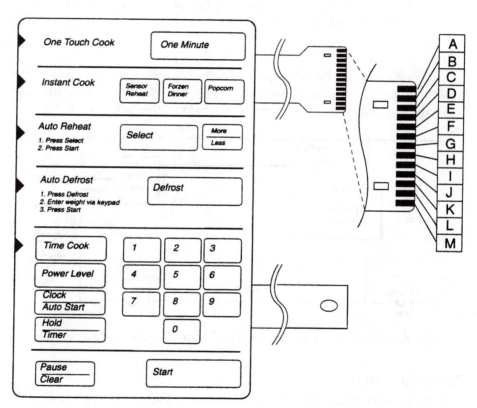

1-54 The touch panel control in a Samsung MG 5920T. Courtesy Samsung Electronics America, Inc.

Transformer

The HV power transformer is a step-up voltage device operating from the 120 (Vac) power line (figure 1-55). The stepped-up voltage of the secondary winding is approximately 2500 Vac and provides voltage to the voltage doubling circuit. A separate step-down secondary winding of the power transformer provides filament voltage to the magnetron tube.

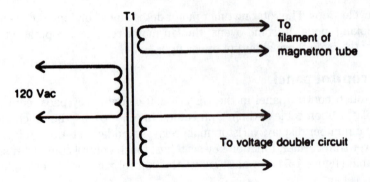

1-55 The HV transformer provides a filament or heater voltage and HV voltage for the voltage doubler circuits.

Triac

You might find a triac module in some microwave ovens in place of a power relay. Usually the triac module is controlled by the electronic controller circuit (figure 1-56). A defective triac might prevent the oven from going into the cooking cycle.

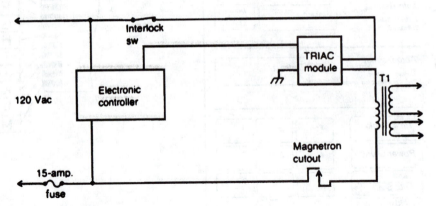

1-56 A defective tube might let the oven run after it has been shut off.

Turntable motor

In some ovens, a turntable device is used to rotate the food for even cooking. The turntable is rotated by a separate turntable motor assembly located at the bottom of the oven. A glass tray holds the food and rotates on several small rollers (figure 1-57).

Vari-motor

The vari-motor assembly consists of a vari-motor, vari-switch, CAM roller, gears, and mounting bracket. The motor rotates the vari-CAM and activates the vari-switch on and off intermittently, supplying power to the power transformer. The repetition rate can be changed by turning the mode selector.

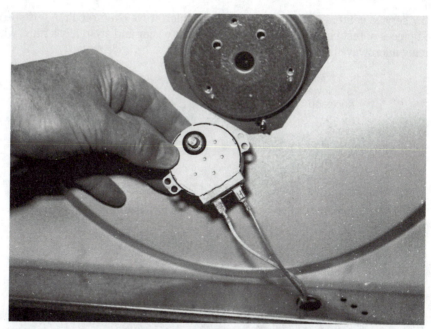

1-57 The turntable motor is at the bottom of the oven cavity.

Vari-switch

The vari-switch is part of the vari-motor assembly and operated by the CAM roller. If the variable cooking mode selector is set at warm, defrost, simmer, or roast position, ac power is supplied to the power transformer intermittently within a 30-second time base. Only a few microwave ovens contain a vari-motor and vari-switch assembly.

Varistor

You might find a varistor located in the ac input power circuit to protect the oven components from excessive power surges or lightning damage. The varistor will prevent the excessive voltage from entering these circuits. The varistor arcs over and opens the power fuse, preventing damage to the oven circuits. This same type of varistor components is found in the power line circuits of the latest TV receivers.

Waveguide

The magnetron tube is connected to the waveguide assembly. RF energy from the magnetron tube's antenna radiates power into the waveguide and oven (figure 1-58). The waveguide channels the RF power from the magnetron into the oven. A waveguide cover, mounted at the top of the oven cavity, prevents food particles from entering the waveguide assembly from the oven area. Like the TV receiver or any electronic device, various components can break down and cause many different symptoms within the microwave oven. With proper test equipment and tools, each defective component can be quickly replaced or repaired. Although most appliances and TV service centers have the basic tools for microwave oven repair, a few special added pieces of test equipment make the task a lot easier and are really a must when servicing microwave ovens.

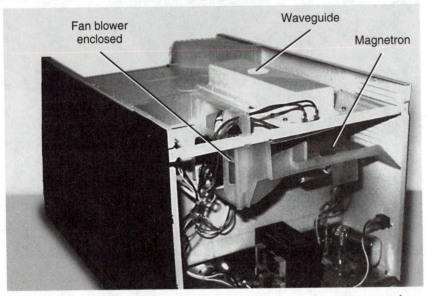

1-58 The waveguide assembly guides the RF energy to the oven cavity from magnetron.

The block diagram

The ac input circuit consists of interlocks, fuse, oven lamp, drive motor, fan motor, main and power relays. Many components found on the touch panel control parts are in the low voltage or ac input circuits. The control panel can control relays and triac assemblies in the primary winding of high voltage transformer. The HV transformer supplies high voltage to the magnetron tube, which is built up with a voltage doubled capacitor and HV diode. The heater terminals are at high voltage potential while the anode of magnetron is grounded (figure 1-59).

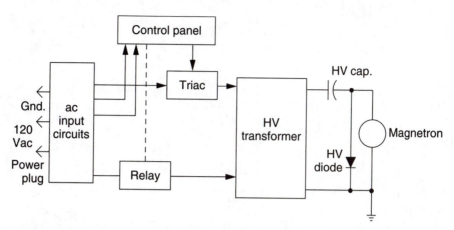

1-59 The block diagram of most microwave ovens with a control panel.

2
CHAPTER

Required tools and
test equipment

You need only a few basic tools to service minor problems with microwave ovens. Most are found around the house or in the average garage. A VOM and vacuum-tube voltmeter (VTVM) are needed to check continuity and voltage in the various components. For accurate high-voltage (HV) and current readings, a commercial meter is handy. Most of the test equipment you'll need can be found in an ordinary TV shop. Some additional test equipment and tools are described later in this chapter.

After an oven has been repaired or tested, a microwave leakage test must be made with a survey instrument. These tests are made to ensure that no oven leakage is present to cause possible injury to the owner. Several leakage testers recommended by the industry are the Narda 8100, Narda 8200, Holaday 1500, and Simpson Model 380m (figure 2-1). They can be obtained through many electronics distributors.

List of tools and test equipment

Many of the basic tools you'll need are found in the shop or on a TV repair bench. Only a few test instruments are required.

1. Basic hand tools
2. York and star screwdrivers
3. VOM and DMM
4. Microwave leakage tester
5. Triac and SCR tester
6. Magnameter
7. Capacitor tester
8. Fuse saver
9. Test bulb

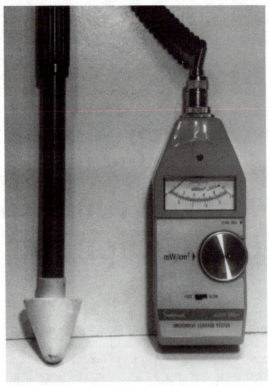

2-1 Always check the microwave oven for leakage with a certified leakage tester.

The microwave oven can be repaired with only a few basic hand tools and a DMM. For quick and efficient service, add a Magnameter, triac tester, and fuse saver. After making repairs, make a microwave leakage test with a certified leakage tester.

Panasonic's microwave exposure warnings

Do not operate a microwave oven with the door open. Before activating the magnetron or other microwave source, make the following safety checks on all ovens to be serviced and make repairs as necessary:

1. Interlock operation.
2. Proper door closing.
3. Seal and sealing surfaces (arcing, wean, and other damage).
4. Damage to or loosening of hinges and latches.
5. Evidence of dropping or abuse.

Before turning on microwave power for any service test or inspection within the microwave generating compartments, check the magnetron, waveguide, and cavity for proper alignment, integrity, and connections. Before releasing the oven to the owner, any defective or misadjusted components in the interlock, monitor, door seal, and transmission systems should be repaired, replaced, or adjusted according to the procedures described in the service manual. A microwave leakage check to verify compliance with the federal performance standard should be performed on each oven prior to release to the owner. Do not attempt to test the oven if the door glass is broken.

Attaching test equipment

Most microwave oven manufacturers warn service technicians not to use ordinary voltmeters to take voltage tests in the ovens. The Magnameter is the only meter you should use for voltage and current tests. Do not use a voltmeter unless it will test more than 4000 Vdc. Do not try to measure the filament voltage with a small voltmeter, as the filament winding has the highest dc voltage potential (figure 2-2).

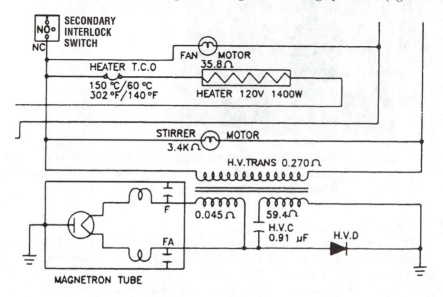

2-2 Never measure the filament or heater voltage with heater winding connected to HV winding.

When making voltage tests, clip all meter test leads into the circuit. If temporary parts are to be added or substituted, do not hold them in your hands. Clip parts into the circuit with clip-on test leads. Keep your hands out of the oven when operating. Before opening up the microwave oven, remove your wrist watch whenever you are working close to or replacing the magnetron. People wearing pacemakers must consult their physicians before attempting to service microwave ovens. It's wise to place an 8-ounce paper cup of water in the oven during testing operations.

Basic hand tools

Although the average service technician probably has the basic hand tools necessary for microwave oven repair, a list and description are given (figure 2-3). A set of screwdrivers, socket wrenches, pliers, test lamp, and homemade cables are required. The VOM and DMM help to locate open or shorted components. The Magnameter can

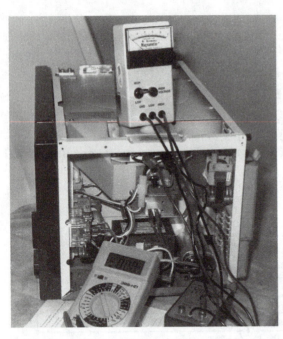

2-3
Basic hand tools and the VOM or DMM are required to service the microwave oven. To speed up the service process, use a Magnameter and circuit saver test instruments.

quickly locate a defective HV diode, capacitor, magnetron, and no high voltage in the high-voltage oven circuits. Another gadget, the Circuit Saver, replaces the fuse and services overloaded circuits in the oven.

Screwdrivers

Every home and shop has a lot of screwdrivers. You will need two Phillips screwdrivers, one large and one small. The larger Phillips screwdriver is needed to remove those stubborn screws that hold on the back cover. The nickel-plated metal screws are often tightened with a power screwdriver and are sometimes difficult to remove.

Two regular-size screwdrivers will come in handy to remove the various components and pry up a stubborn component from the metal chassis. Select two long, insulated screwdrivers for discharging the HV capacitor. The two screwdriver blades are jammed against the capacitor terminals and sandwiched together to discharge it. A couple of small, short screwdrivers come in handy when working in tight corners to remove or tighten those difficult-to-get-at metal screws. Some ovens require a special Torx or LHSTix screwdriver for removing power cords and other components.

A nut driver set can speed up the removal and replacement of the various components in the oven. A set of socket wrenches also can come in handy to remove those large power transformers and the turntable motor assembly. A small crescent wrench will do if a set of sockets are not available. To remove front knobs and shafts, keep a set of Allen wrenches handy.

A pair of regular pliers can be used in many ways for servicing a microwave oven. Long-nose pliers are handy when soldering cables or wires to the various parts. Select a good pair of long-nose and side-cutter pliers to make connections and remove wiring and cables from the replaced components.

Choose a regular 100-watt soldering iron or 150-watt soldering gun for most terminal connections. The larger cables and wires require more heat than those found in the TV chassis. You might find most connecting wires are crimped in ovens, but many have to be soldered. A small, low-wattage iron is handy when working around the control boards.

Special screw or nut drivers

You might find a few finished screws that hold components or attach the top cover to the microwave oven. In new ovens, yorx and star nut and screwdrivers are required to remove the finished screws. Often these screws were installed with a power nut or screwdriver and are difficult to remove. Make sure that you obtain the correct size of star or yorx screwdrivers so you won't rough up or destroy the head on the screw. Most manufacturers list the correct sizes for their ovens in the service literature.

Eyelet crimp tool

A multipurpose hand tool is comfortable to use. It cuts, strips, crimps, and serves as a small bolt cutter. The heavy-duty crimp tool crimps insulated/noninsulated and ignition-type eyelets (figure 2-4). These tools are handy to crimp on a broken connected wire and make good solid connections. These soldered connections can be fastened to number 22-10 copper wire. The crimped-on connector can be bolted or screwed to make electronic or electrical connections.

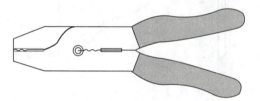

2-4 The crimp hand tool makes quick wire connections.

Homemade tools

Select three or four sets of alligator clip leads. They can be purchased at electronic supply houses. You can also make alligator clip leads yourself. Simply solder an alligator clip to a 12-inch flexible cable lead. The alligator clip ends should be of the rubber insulated type so they will not short out components when used in tight places. These cables come in handy to short around interlock switches or to temporarily clip components in the circuit (figure 2-5).

Take a long-bladed screwdriver (preferably a discarded Phillips) and solder a 2-foot cable with a large alligator clip at one end. A technician can use this tool to short the HV capacitor to chassis ground (figure 2-6). Clip the alligator clip to a good ground and place the blade of the screwdriver against both capacitor terminals. The grounding tool is also used to ground out any voltage at the filament terminals of the magnetron tube before the defective tube is removed.

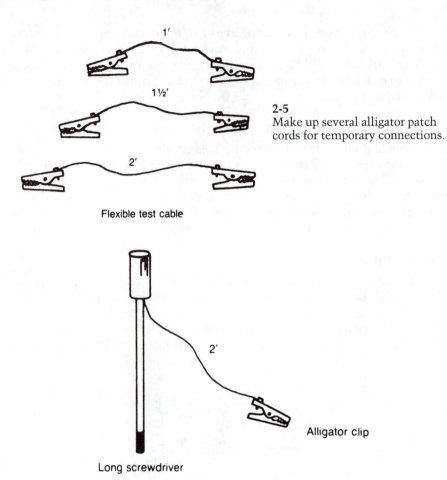

2-5
Make up several alligator patch
cords for temporary connections.

Flexible test cable

Alligator clip

Long screwdriver

2-6 Discharge the HV capacitor with a large screwdriver, alligator clip, and cable.

Another useful homemade tool is a fuse puller. Take a discarded screwdriver
with a long stem and bend a curl—a u-shaped end—so the tool will slip under the
fuse (figure 2-7). Apply pressure against the fuse holder or pull up on the fuse at one
end. Now the fuse can be removed. Sometimes chemical fuses are located inside the
oven wall or are mounted on the monitor switch assembly. They can sometimes be
difficult to remove.

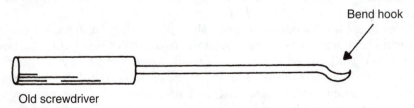

Bend hook

Old screwdriver

2-7 Pry out the fuse with a bent screwdriver tip.

Take a TV cheater cord and cut off the interlock portion to make a temporary ac connecting cable. Solder two regular insulated alligator clips to each end for easy attachment. This cable can be used to check the ac cable or to apply power directly to the power transformer leads to isolate the control circuit from the transformer circuits.

Pick up a pigtail light bulb socket for continuity and fuse blowing tests. These can be located in electronics departments or at your local electrical store (figure 2-8). Screw in a 100-watt bulb. Instead of blowing a chemical fuse each time a short in the oven or an interlock switch hangs up, the light will become bright with the full power line voltage across the pigtail test light. This troubleshooting method can save a lot of blown 15-amp fuses.

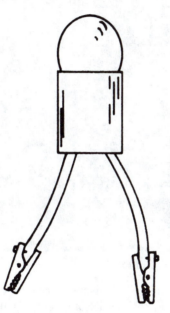

2-8
A pigtail light can be clipped across thermal cutouts, lamps, relay points, and transformer ac primary leads to determine if it is open, closed, or normal.

Volt-ohmmeter

The small, pocket-type VOM is a handy tester in checking continuity of a suspected fuse, solenoid winding, interlock switch contacts, and also for tracing wiring cables. You should have one of these VOMs in the toolbox when checking ovens in the home. A low ohm range from R×1 or R×1k is sufficient. All continuity measurements should be made after the power cord is pulled and the HV capacitor is discharged.

Only use the ac voltage range to check for ac power line voltage around the microwave oven. Do not use the dc range in the high-voltage circuits. These voltages are more than 2000 volts and will quickly damage or ruin the meter. Only use the small VOM for low-ohm and power line ac voltage measurements.

Digital VOM

The small digital VOM comes in handy when taking very low ohm measurements of transformer windings and filament continuity of the magnetron tube. Although the digital VOM is not a required test instrument, accurate low-ohm resistance tests (less than 1 ohm) might save a lot of valuable service time (figure 2-9). A low-ohm measurement is necessary because the primary winding of the power transformer might be less than 1 ohm and the secondary filament winding from 0 to 1 ohm. The resistance measurement of the filament or heaters of the magnetron tube might be from 0 to 1 ohm.

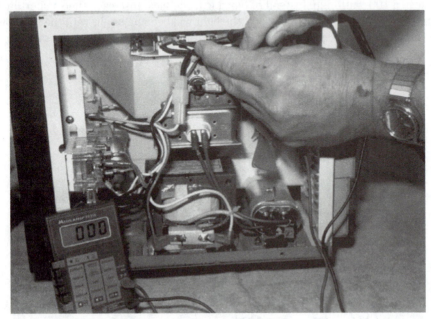

2-9 Check the fuse with the low ohm scale of DMM.

If a diode check is provided by the digital meter, this measurement can be used in checking diodes and transistors on the control board. Some technicians and oven manufacturers prefer to replace the control board instead of servicing it. The diode check is not accurate when testing the HV diode in the magnetron tube circuit. The milliampere scale can be used when the meter reads over 300 milliamps and a grounding test resistor is located in the HV diode circuit. Primarily, you should use the digital VOM for continuity and low ohmmeter measurements.

DMM diode tests

The digital multimeter (DMM) is ideal for taking low ohm-continuity tests on resistors, diodes, transistors, and IC components on the control board. A good diode shows a measurement in one direction only (figure 2-10). A low resistance measurement in both directions indicates a leaky diode. Some technicians and oven manufacturers prefer to replace the control board instead of servicing it. The diode check is not accurate if you test the HV diode in the magnetron circuit with a low-voltage powered DMM. Use a DMM with a 6- or 9-volt battery to provide any type of indication with the HV diode/rectifier.

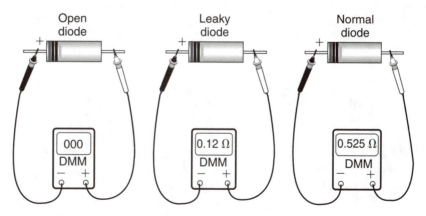

2-10 DMM diode tests.

DMM transistor tests

Transistor junction tests can be made with a DMM that has a diode test. Always place the red probes to the base terminal of an npn transistor and the black probe to the base with a pnp type (figure 2-11). You might find one or two elements of the transistor open or leaky. A normal transistor can show a measurement from base to emitter and base to the collector terminals. A low measurement in both directions indicates a leaky transistor. Check for improper voltages and low ohm measurements to ground on each IC terminal to locate leaky or shorted IC components.

VTVM

Most TV shops have a VTVM for radio and TV servicing. The VTVM can be used to check for continuity and make voltage measurements in the microwave oven. The various resistance and voltage ranges are ideal for checking the HV transformer, relays, magnetron tube, and to make voltage measurements on the control board.

The VTVM can also be used to check the HV circuits of the magnetron tube. A special TV HV probe must be used with the VTVM for these measurements. Do not try to make HV measurements without the probe. If the VTVM has a polarity switch for voltage measurements, it makes an ideal HV instrument.

Remember: the high voltage around the magnetron is negative in respect to the oven chassis. Use the negative dc voltage range. When taking HV measurements with the VTVM, keep the instrument off the oven, unless insulated from the metal cabinet. A piece of masonite, a book, or a piece of plastic will do.

Keep small meters out of the oven

The small VOM or DMM test meters should never be used to check voltages within the oven. This applies to meters that might be able to measure up to 3.5 kV. The VOM or digital multimeter (DMM) can be used to make continuity, resistance,

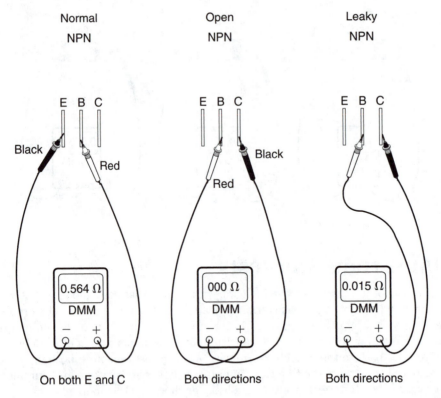

2-11 Transistor tests with DMM.

and diode tests inside the oven when the ac plug is pulled from the wall outlet. Always discharge the high-voltage capacitor before taking continuity or resistance measurements within the oven.

The VOM or DMM is ideal when taking continuity and resistance measurements of the HV transformer windings, filament windings, and small transformers found in the panel control circuits. Continuity tests of the chemical fuse, cavity thermal cutout, magnetron thermal cutout, oven cutout, and interlock switch contacts can be made with the VOM or DMM (figure 2-12).

The diode test of the DMM will not check out the HV diode. The higher resistors above 9–10 megohms can be checked by either meter, if the resistance goes that high. Do not attempt to check the HV capacitor or for HV on the magnetron filaments with either the VOM or DMM. You might damage it beyond repair.

Damaged small meters

Small VOM or DMM meters that will not measure up to 3.5 kV can be severely damaged while trying to measure the high voltage in the microwave oven. Do not attempt to use an analog meter (VOM), even if it measures up to 3500 volts. Use the Magnameter. It's safe and is designed to measure the high voltage in the microwave oven.

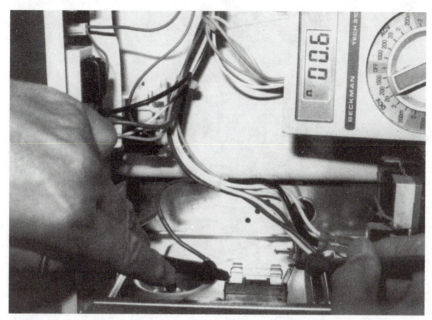

2-12 Keep the small meters out of oven for high-voltage tests.

Remember that oven voltage varies from 1.5 kV to 4.8 kV. Most VOM or DMM test meters will not go that high in voltage tests. When checking the HV diode, you might accidentally leave off the ground wire or fail to reconnect it. Even a small tester that might measure up to 3.5 kV can be damaged when taking high-voltage tests. An analog meter is damaged when a large amount of raw ac voltage is applied to the meter. These meters are damaged beyond repair and they might explode in your face. Even factory repair does not help to restore it. Be safe. Use only a VTVM with an HV probe or a Magnameter for HV tests within the microwave oven.

The Magnameter

One of the latest voltage and current meters, called the Magnameter (model 20-226), is available from Electronic Systems, of Rockford, Illinois. The Magnameter is a specialized instrument used to speed up and simplify microwave oven testing. With the Magnameter, you can quickly take voltage and current measurements without possible danger of HV shock and injury to the magnetron circuits (figure 2-13). The meter enables the technician to make both HV and plate current measurements of the magnetron and HV circuit with one setup. You don't have to handle or change test leads during the voltage and current checks.

The meter hookup is identical with positive and negative ground ovens. Most domestic ovens have a positive ground in the HV circuit. The meter is also well insulated with 10 kV rated test lead wires. A high impact plastic meter case prevents shock to the operator. No metal knobs or switches are exposed, preventing possible shock. The meter can be held in your hands or set on top of the metal oven while taking critical

2-13 The Magnameter can determine if magnetron, HV capacitor, and diode are defective.

voltage and current readings. The meter case is constructed of extremely tough thermoplastic material that resists breakage from all but the most severe abuse.

Besides the meter indication, a neon warning light flashes when high voltage is present—whether the switch is in the high (HV) or in low (current) position. Both readings can be obtained with just flipping the high/low toggle switch. First, take the HV reading, then make a current measurement.

The meter hand is rated from 0 to 10 kV. You can consult the oven manufacturer's service information for acceptable HV limits. Most domestic ovens operate between 1.5 and 3 kV, while commercial ovens operate from 2 to 4 kV applied to the magnetron tube. The toggle switch must be flipped to high to measure the high voltage in kilowatts.

Flip the toggle switch to low for direct magnetron plate current measurement. In some microwave ovens (especially the early ones), a 10-ohm plate resistor is found between the HV diode and chassis ground (figure 2-14). Using this 10-ohm resistor is a quick method of taking current readings without possible shock to the technician.

When no test resistor is found in the magnetron circuits, simply remove the ground side of the silicon diode (+) and connect the 10-ohm test resistor, provided with the meter, between diode and chassis ground. Actually, you are inserting the test resistor in series with the HV diode and common ground terminal. Connect the green meter lead to the 10-ohm resistor for current measurement. Then you simply multiply the meter reading 100 to get the actual current reading of the magnetron tube. For instance, if the meter hand was at 2, then 100 equals 200 mils of current. The current of most domestic microwave ovens ranges between 160 to 400 milliamperes, while the commercial ovens can have a plate current from 200 to 750 milliamperes.

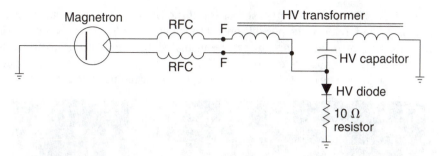

2-14 Connect a 10-ohm resistor in series with the HV diode and oven ground for voltage and current tests.

Very seldom does the manufacturer include the operating current of the microwave oven in the service literature. It's wise to take a voltage and current reading on every oven and mark them on the service literature for future reference.

Connecting the meter

After removing the back cover of the microwave oven, discharge the HV capacitor. Double check. Make sure the power plug is disconnected. Connect the black lead of the meter to the common chassis ground. These insulated alligator clips are strong and will stay in place. Clip the red lead to the HV capacitor or HV diode of the HV side (figure 2-15). If in doubt, trace the HV capacitor lead going to the filament circuit of the magnetron tube.

2-15 Connect the Magnameter to HV diode, capacitor, and ground.

Check for a 10-ohm, 10-watt test resistor. If the HV silicon rectifier goes directly to chassis ground, remove the positive end of the diode. Usually, this lead is grounded with a metal screw or nut. Next, insert the insulated 10-ohm test resistor provided and clip between the positive end of diode (end just removed) and common ground (figure 2-16). Most new ovens do not have a test resistor mounted in the HV circuit.

2-16 A 10-ohm resistor is in series with the HV diode and ground for tests with Magnameter.

Double check the meter connection. Make sure each clip is snug and tight. Each cable end has a colored rubber alligator clip for easy identification. (The red lead goes to the high-voltage side, the black lead to ground, and the green lead to the top side of the HV diode.) The meter should be connected as shown in figure 2-17.

Taking voltage and current measurements

After the cables from the meter have been attached to the oven circuits, start the cooking cycle. Set the meter toggle switch to high. Plug in the power cord. Place a liter of cool water in the oven cavity, turn the oven to maximum cook or heat, and push the cook button. Then check the high-voltage reading shown on the meter.

Change the toggle switch to the low position. Read the low voltage (which is 100), and this will give you the milliampere current reading. From these two readings, you can quickly determine if the HV circuits or the magnetron tube are defective. Return the toggle switch to the high position. Pull the power cord. Discharge the HV capacitor with the discharge switch on the side of the Magnameter. The meter

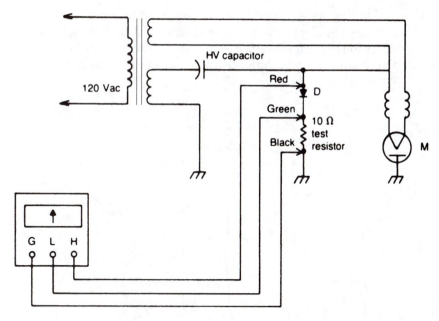

2-17 Connect the Magnameter as shown in drawing.

hand should return to zero and the warning light should be out, indicating that the HV capacitor is discharged. To make sure, you might want to discharge the capacitor with two screwdrivers.

Disconnect the meter and proceed to service the oven. If no high voltage is seen on the meter in the high position, suspect a leaky diode, HV capacitor, transformer, or no ac voltage supplied to the transformer primary winding (figure 2-18). You might find a grounded magnetron tube with no high voltage and no current reading. A shorted transformer and leaky diode can produce a low high-voltage reading (often below 1 kV).

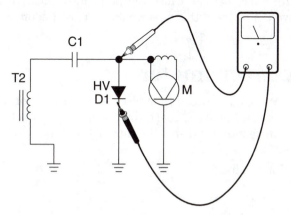

2-18 Suspect a power transformer, HV capacitor, and diode with no high voltage across diode.

An excessive high-voltage reading could be caused by a defective magnetron, open filament, open filament winding, or a high-voltage wire. You might have a low- or no-current reading with an excessive high-voltage reading from 3 to 5 kV. A leaky or shorted magnetron tube can produce a high-current measurement from 350 to 600 mils (figure 2-19).

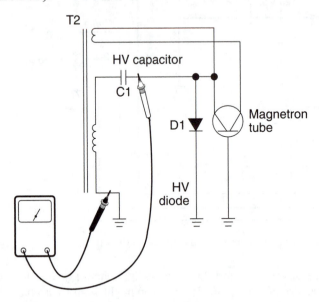

2-19 Suspect a transformer, HV capacitor, and magnetron tube with excessive high voltage.

The Magnameter can quickly indicate problems in the HV circuit as well as outside the magnetron tube circuit. Erratic meter readings indicate an intermittent magnetron tube or poor filament connections. Low current readings might indicate a weak magnetron tube that can cause the problem of taking too long for foods to cook in the oven. A meter of this type would be a great addition to your present VOM or VTVM.

The new Magnameter

The original Magnameter was invented by Nick Parnello of Rockford, Illinois. This first meter was distributed and sold through General Cement Company. Now the Magnameter is manufactured and sold through distributors by Electronic Systems, Inc. Their address is:

Electronic Systems, Inc.
624 Cedar Street
Rockford, IL 61102

Remember, the Magnameter is a specialized test instrument used to speed up and simplify microwave oven repair (figure 2-20). The new meter has a yellow area up to 2 kV, a green area from 2 kV to 4.5 kV, and a red area from 4.5 kV to 10 kV. The

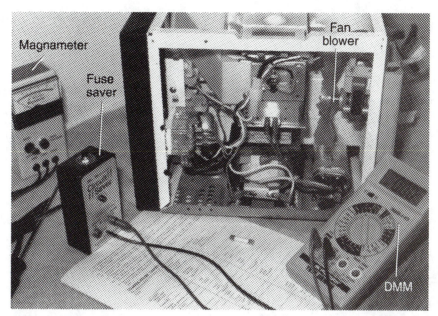

2-20 The normal oven will read in the green scale of the new Magnameter.

new meter does not have an HV capacitor discharge switch on the side; the original model does.

The Magnameter enables both high-voltage and plate-current (voltage across the plate resistor) measurements to be made with one setup. There is no handling or changing of test leads during voltage and current tests. The hookup is identical with reversed polarity (negative ground) ovens.

The new meter has superior insulation with special high-voltage test leads. The meter case is 18 inches thick and is made of high-impact plastic, with no metal knobs or switch actuators on the exterior of the unit. A red warning light comes on when high voltage is present, whether the switch is in high or low position.

Before you use the meter to check out the oven circuits, disconnect the oven from the wall outlet. Remove the outside cover of the microwave oven. Discharge the high-voltage capacitor. Double check the discharge tests. Connect the black lead to chassis ground. Connect the red lead to the high-voltage capacitor or rectifier (diode) on the high-voltage side (figure 2-21). Now you are ready to make high-voltage tests.

To make a current test (voltage across resistor), connect the green lead between plate resistor and magnetron tube plate lead wire. If the oven does not have a plate resistor in the circuit, disconnect the high-voltage diode ground wire. It is usually held in position with a metal screw to the oven chassis. Insert the 10-ohm resistor, which is included with the meter.

High voltage test

Now connect the oven to the ac outlet, and set the meter toggle switch to the high position. Insert water in a paper cup or container (6 oz.) to load the oven down. Turn the microwave oven on a high wattage setting. Now read the meter.

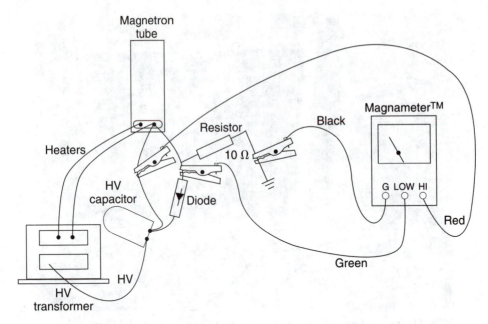

2-21 Check the pictorial diagram on how to connect the Magnameter.

If the meter hand appears in the yellow area, suspect a shorted magnetron, diode, high-voltage capacitor, shorted or open HV transformer, or an open HV fuse. When the reading is in the green area, the oven should be operating normally. If the meter reads in the red area, suspect an open magnetron, open filament, bad filament transformer, poor connections on the filament leads of magnetron, or an open HV wire.

Oven current tests

Rotate the switch to low position. Now read the low voltage, which represents current. A home oven should have a reading between 1.6 kV and 4.5 kV (equals 160–450 mA), while commercial ovens have a higher measurement. The voltage might vary from 2.0 kV to 7.0 kV (200–700 mA) of current. Any oven that reads above these figures indicates a leaky or shorted magnetron tube (Table 2-1).

**Table 2-1. Microwave oven
current and HV chart**

Oven	Mfg. current	Mfg. voltage
Litton	160–300 mA	1.9–2.5 kV
Sharp	200–450 mA	1.9–3.5 kV
Tappan	250–325 mA	1.9–2.2 kV
Whirlpool	240–300 mA	2.0–2.5 kV
Sanyo	200–270 mA	1.8–2.3 kV

To avoid possible shock, discharge the HV capacitor or wait at least 10 seconds after the warming lamp goes off to assure full discharge of capacitor. Periodically, check for a good ground (black) clip attached to a chassis bolt during all tests. Keep clear of meter and leads while the oven is operating. Each time it is connected to the oven, inspect the meter leads, which might show damage. Always discharge the high-voltage capacitor before connecting and removing the Magnameter.

Voltage tests (ac)

Regular ac voltage tests can be made throughout the oven to check the various components. Check the ac voltage at the fan blower to determine if defective fan has an open winding or connection. Take a voltage test across the lamp with no light (figure 2-22). Test the browner heater element with a voltage test and no heat with ac voltage test across heater terminals. Check ac voltage across stirrer or turntable motor with no rotation. Measure the ac voltage across the primary winding of power transformer to determine if all interlocks, relays, and triac assemblies are functioning.

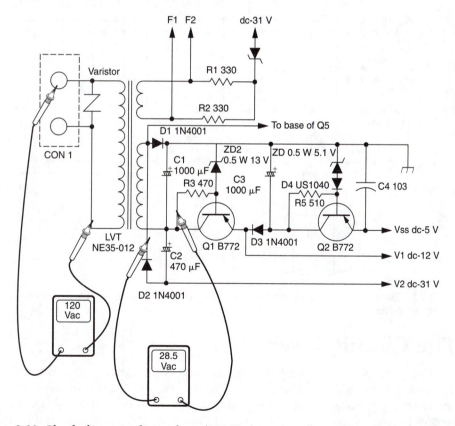

2-22 Check the power line voltage (120 Vac) across components to determine if voltage is present.

Resistance tests

Take low ohm resistance measurements with the ohmmeter set on the R×1 scale for most components in the microwave oven. Check for burned or corroded relay points with a low resistance measurement. Measure the resistance of the main or oven relay for open winding. Take a quick resistance measurement to see if an interlock is open or closed.

Check the continuity of a fan, stirrer, flame, door, and turntable motor to see if motor winding is open. Test the transformer primary and secondary windings for open or shorted windings. Remove the filament leads from transformer to magnetron heater or filament connections on magnetron for open filament. Check for open key buttons when a button is pushed and erratic or no operation is noted (figure 2-23).

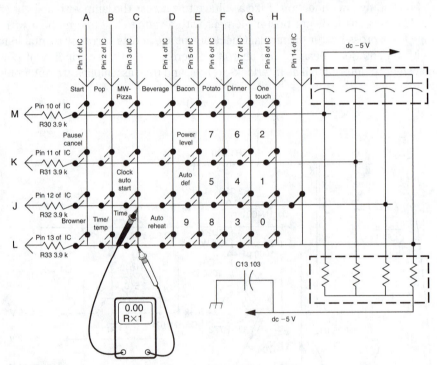

2-23 Check the resistance or continuity of oven components to determine if defective.

The Circuit Saver

The Circuit Saver is manufactured by the same firm that makes the Magnameter (Electronic Systems, Inc.). The Circuit Saver is designed to take the place of the blown oven fuse while testing and servicing the oven. Chemical fuses are quite expensive to use when you are testing an oven that is overloaded or has a leaky component. The Circuit Saver is ideal to use when the oven intermittently blows the main oven fuse (figure 2-24).

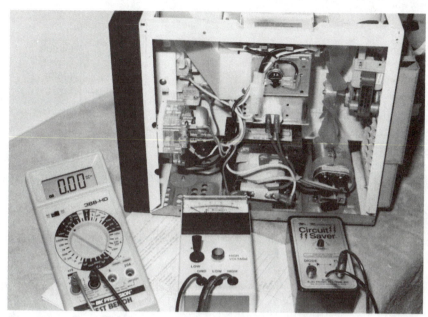

2-24 Plug in the Circuit Saver if the fuse keeps blowing.

This small tester can also be used to substitute an HV diode within the oven. A red high-voltage warning light is found on the front side of the meter. When using the Circuit Saver for diode substitution, if the high-voltage light does not appear after approximately 10 seconds, shut down the oven. Check for other shorted or over-loaded conditions in the oven.

Caution: Make sure that the high-voltage capacitor is discharged before connecting the circuit tester as a fuse.

Remove the blown fuse from the fuse holder in the microwave oven. Place the load (black lead) coming from point G of the Circuit Saver across one lead or clip of the fuse holder. Take the lead from point F (red) of the tester and place across the other end of the fuse holder. Then if the oven blows a fuse, just reset the circuit breaker of the tester.

Before substituting a diode, discharge the high-voltage capacitor. Remove the high-voltage wire from the old diode (at anode) of the magnetron. Connect lead D (red) wire from Circuit Saver to the high-voltage wire removed from the old diode in the microwave. Connect wire lead G (black) from the tester to the chassis of the microwave (figure 2-25). Diode substitution is now finished.

Check all leads of the tester before connecting to the oven. Make sure that no bare or broken wire areas are found. Always discharge the HV capacitor before and after each test. The Circuit Saver can save a lot of valuable service time when troubleshooting intermittent ovens.

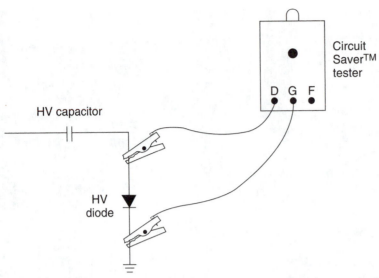

HV capacitor

Circuit Saver™ tester

D G F

HV diode

2-25 Connecting the circuit saver tester to check HV indication and the HV diode.

Triac/SCR diode tester

The Tri-Check 11 is a new test instrument designed by Electronic Systems, Inc. It can be used to test triacs, SCRs, and diodes in the microwave oven. This little test instrument is ideal to check out suspected triacs and high-voltage diodes (figure 2-26).

2-26 The new Triac-11 can test triacs, SCRs, and diodes.

Testing triacs

Disconnect the ac oven power cord. Disconnect all wires to the triac. Turn on the Tri-Check and the middle light should blink. Now connect the Tri-Check to the triac. Place black lead (T1) to MT1, clear wire (T2) to MT2, and red (G) lead to the gate terminal. Press the test button.

If the triac is normal, both lights should light when the test button is pressed and they should turn off when button is released. Both lights will always light with a shorted triac. No lights are on with an open triac. Always turn the tester off when not in use.

Testing Litton triac assemblies

Again, disconnect the ac power cord. Remove all wires and leads to the triac. You should mark the color code leads so they can be replaced with correct wires. The middle light should blink with Tri-Check turned on. Connect the tester to the triac assembly. Plug in optional test lead.

Connect the black (T1) terminal to marked MT1, and the clear cable (T2) to marked MT2 terminal. The red (G) lead is not used. Connect black terminal of tester to terminal marked blue and green to terminal marked green. Press the tester button. Both lights should light when the test button is pressed and turn off when button is released. Both lights will light with a shorted triac. The triac is open when no lights are on (figure 2-27).

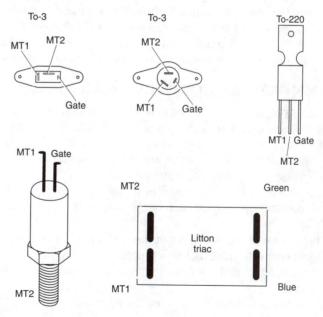

2-27 Triac outline and terminal connections.

Testing microwave diodes

Disconnect the ac power cord. Discharge HV capacitor. Disconnect all wires to HV diode. Turn on Tri-Check and middle light should blink. Connect tester to black

(T1) lead to the anode and clear (T2) to the cathode element of diode. Do not use the test button for diode tests. The diode light should light with a normal diode. Both lights will light with a shorted diode and no lights are on with an open HV diode.

Testing SCRs

Disconnect the ac power cord. Disconnect all wires to the SCR. Turn on Tri-Check and middle light will blink. Connect black lead (T1) to the cathode terminal, clear (T2) to anode, and red (G) to the gate terminal. Press the test button. The SCR light should light when the test button is pressed and turn off when button is released. Both lights will light with a shorted SCR. No lights are on with an open SCR.

Cooking test equipment

The power output of the magnetron can be measured by performing a water temperature rise test. A cooking-water test should be made after all repairs are finished. The cooking test might indicate a defective magnetron or HV component. When the oven takes too long to cook, suspect a faulty magnetron or related components and make the cooking test.

For making the cooking test, you'll need a 600-cc glass beaker or equivalent, a glass thermometer 100°C or 212°F, and two 1-liter beakers or equivalent.

You might find that each oven manufacturer has a somewhat different water test (each oven test procedure uses water, a container to hold the water, a thermometer, and a stopwatch). To ensure accurate water tests, the power line voltage should be 115 to 120 Vac. Here are several manufacturers' water test procedures.

Panasonic NN-9807 and MQ8897BW cooking tests

The power output of the magnetron can be determined by performing a simple water heating test as described in the following. Necessary equipment: two 1-liter beakers, a glass thermometer, and a wristwatch or stopwatch. Check the line voltage under a load. Low voltage will lower the magnetron output. Take the temperature readings and heating time as accurately as possible.

1. Fill each beaker with exactly one liter of tap water. Stir the water using the thermometer and record each beaker's temperature (recorded as T1A and T1B).
2. Place both beakers on the center of the cook plate. Set the oven for high power and heat it for exactly two minutes.
3. Stir the water again and read the temperature of each beaker (recorded as T2A and T2B).
4. Obtain the temperature rise by using the following formula:

$$\text{Average temperature rise} = \frac{\text{T2A} + \text{T2B} - (\text{T1A} + \text{T1B})}{2}$$

The normal temperature rise for these models should be 16 to 20°F (8.9 to 10°C) at the high power selection with the oven operating at the specified line voltage.

5. To obtain the power output, multiply the average temperature rise by 39 (or 70 if thermometer is in Celsius). Example: If you find average temperature rise of 18°F, the oven output is 18×39 = 702 watts.

Oven A

1. Pour 1000 ml of cool tap water into a large beaker and stir with the thermometer to measure the water temperature. Make a note of the initial temperature (TI). This temperature should be somewhere between 17°C and 27°C.
2. Place the beaker of water in the center of the oven and heat for 62 seconds at high power. Use the watch second hand instead of the oven timer.
3. When the time is up, stir the water with the thermometer and again measure the temperature. Record the final temperature (T2).
4. The output can be calculated with the following formula: wattage output (W) = (T2 – T1) 70 (Celsius).
5. The maximum temperature rise is 10°C; a minimum temperature rise is 8°C.

Oven B

1. Fill the measuring cup with 16 ounces (453 cc) of tap water. Measure the temperature of the water with the thermometer. Stir the temperature probe through the water until the temperature stabilizes. Record the water temperature.
2. Place the cup of water in the oven and set the cooking mode to full power. Allow the water to cook for 60 seconds while checking with a stopwatch, the second hand of a watch, or the digital readout countdown.
3. Remove the cup and measure the temperature. Stir the temperature probe through the water until the maximum temperature is recorded.
4. Subtract the cold water temperature from the hot water temperature. The normal result should be a 10 to 36°F (10.6 to 20°C) rise in temperature.

Oven C

1. Select a 1-pint (2-cup) pyrex measuring cup and a thermometer with a range of more than 180°F.
2. Fill the pyrex container with one pint of tap water. Use the thermometer and record the water temperature.
3. Place the filled container in the center of the oven cavity and set the timer at $2\frac{1}{2}$ to 3 minutes.
4. Time the oven at exactly two minutes with a watch or clock that has a second hand.
5. Remove the pyrex container and stir the water with the thermometer to check the rise in temperature. If the oven is operating successfully, the temperature rise should be approximately 55°C.

You should follow each manufacturer's cooking water test when working on a particular type of oven. All water tests indicate how the magnetron is functioning. If the water temperatures are accurately measured and tested for the required time period, the test results will indicate if the magnetron tube has low power output (a low rise in water temperature), which would extend the cooking time, or high power output (a high rise in water temperature), which would reduce cooking time.

One microwave oven technician I know has devised his own cooking water tests. Three new and different microwave ovens were used in the test. Each oven was checked against a beaker filled to a certain level and marked at the boiling point (figure 2-28). The time of each oven was recorded at 3 minutes. When the water began to boil, the time was recorded for each oven. A high and low line were drawn on the beaker. Most ovens with the same water level had very close water boiling times (2½ to 3 minutes).

2-28 Three cups are used to check the power and water levels.

When an oven takes longer than three minutes for water to start to boil, the technician knows trouble exists in the magnetron and HV section. In case the boiling point is more than 4 or 5 minutes, a voltage and current test of the magnetron tube will indicate a defective tube (figure 2-29). Of course, water boils at different temperatures in different parts of the country (depending on altitude), and this must be taken into account with this type of method. If the water boils in less than two minutes, the technician knows that the magnetron tube is defective and usually is running red hot.

If by chance you are on a house call to check a microwave oven and you forget to take a beaker or pyrex cup along, try a styrofoam cup. Today most kitchens have such cups available. These cups do not melt down, although they do run warm when the water begins to boil. Fill the cup with tap water and take the three minute boiling test. You should know how full to fill the foam cup. After several house calls or cooking tests, you know whether the oven is cooking accurately or if it is defective. If in doubt, bring the oven in and apply the manufacturer's water cooking test.

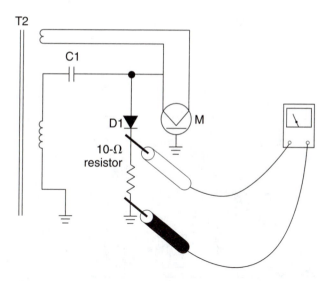

2-29 Check the magnetron current if the oven takes too long to cook food.

In case the cooking test takes a longer cooking time or the customer complains of burning up the food, check the magnetron tube. A magnetron tube pulling heavy current (400 to 600 mils) indicates a leaky or shorted tube. When the oven takes too long to cook, check for very low current readings and suspect a weak magnetron. Replace the magnetron if the correct HV voltage is present but there is an improper current measurement.

Leakage tests

Every microwave oven must have a leakage test taken after servicing the unit. All oven manufacturers require leakage procedures. Besides, most customers demand a leakage test. You might be asked to come to the home and only make a leakage test, since most people are scared to death of radiation. It's better to be safe than sorry, so take a leakage test for your own safety too.

There are several leakage test instruments recommended by the industry — Narda 8100, Narda 8200, Holaday 1500, and Simpson Model 380m (shown in figure 2-30). The price of a leakage tester can run close to $500. Two low-priced leakage testers are discussed at the end of this chapter.

The purpose of any radiation instrument is to check the radiation leakage around the microwave oven door, outer panels, vents, and door viewing window. After replacing a magnetron tube, check for leakage around the magnetron and waveguide assembly. Be careful. Avoid contacting any HV component with the radiation meter (figure 2-31).

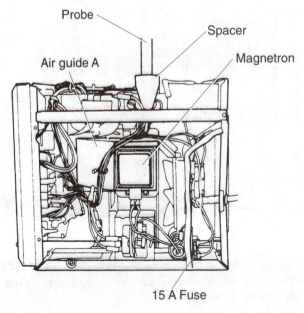

2-30 Check the door and oven areas for leakage with a certified tester.

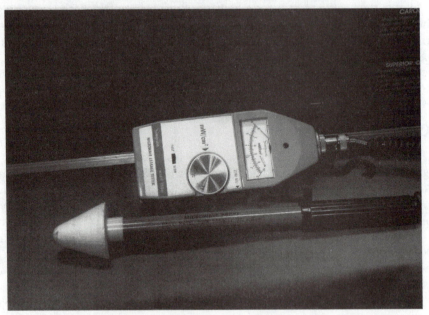

2-31 The Simpson model 380M microwave oven leakage tester.

The U.S. government has established a maximum of 5 mW/cm^2 leakage while in the customer's home. Many oven manufacturers request that the leakage should never be more than 2 mW/cm^2 (1 mW/cm^2 for Canada).

The power density of the microwave radiation emitted by a microwave oven shall not exceed 1 milliwatt per square centimeter at any point, 5 centimeters or more from the external surface of the oven, measured prior to acquisition by a purchase, and thereafter, 5 milliwatts per square centimeter at any point 5 centimeters or more from the extreme surface of the oven.

Low-priced leakage meters

The low-priced microwave oven leak detector is an economical leak detector used to check leakage around latches, gaskets, and around the control panels (figure 2-32). These testers instantly detect microwave leakage. They are intended to check for leakage by the operator and not used for certification of oven leakage.

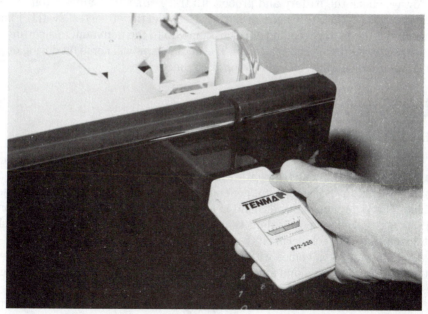

2-32 A low-priced oven leakage tester.

Just point the meter against the door gap area and watch the meter hand move. Some low-priced testers have a regulator meter hand, while others might have a red or green mark for the meter indication. If the oven shows any sign of leakage movement on the low-priced meter, have the oven checked with a certified leakage tester.

The Narda 8100 radiation monitor

The Narda instrument measures radiation leakage in milliwatts per square centimeter (mW/cm^2). A water load of 275 cc (approximately $1^1/_3$ cups of water) is placed in the oven and used as a load during leakage tests. The measuring probe should be used with a 2-inch cone spacer.

This tester can be operated on the internal rechargeable battery or from the power line. The battery can be charged from the power line. Switch the instrument to the 2450-MHz position. The fast and slow meter response switch should be set in the fast position. This tester has an available alarm system that warns the operator when high amounts of radiation might damage the meter. Set the alarm control to 50, which sounds the alarm when the meter reads 50% of full-scale deflection.

On ovens with an unknown leakage, use the high scale first and then switch to the low scale for low leakage. A test switch is used to check the battery and probe. The meter needle will not read above the minimum test mark on the meter if either scale is faulty. Set the meter to 0 with the 0 control.

Always check the battery and probe with the test switch before attempting to measure radiation leakage. Check both the battery and the probe test switch. Plug in the ac cord if the battery reading does not come up to the minimum test setting. Do not use the probe if the probe tests fail. The audio alarm will come on during operation if the probe becomes inoperative or disconnected.

The Simpson 380 series microwave leakage tester

The Simpson 380m is a portable, direct reading instrument designed for accurately measuring the amount of microwave leakage radiated by ovens, heaters, chargers, and other industrial equipment generating high power at microwave frequencies of 2450 MHz. The nonpolarized, hand-held, wide-range probe permits selection of five distinct ranges for power density measurements. This instrument has a slide switch, enabling the technician or operator to select the correct response time.

The Simpson 380m comes with a thermocouple probe, cone spacer, two 9-volt batteries, a carrying case, and an operating manual. When a new probe is required, the complete instrument must be returned to the factory for calibration. Probes are not interchangeable. Most radiation test equipment should be returned to the factory or factory authorized service centers once a year for overall check and calibration.

A six-position rotary-range switch is used to activate the tester, as well as to select the range for the radiation to be measured. The range selected is indicated by the marker. With the range switch set to either the 2.5 or 2 mW/cm^2 position, the lower (0 to 2.5) dial arc is read. The (0 to 10) dial arc is read when the range switch is set to the 10 or 100 mW/cm^2 position. Switch the range switch to the battery test position to check the condition of the batteries.

If the fast/slow switch (normally set to the fast position) that selects the response time of the meter remains at the set point (does not move toward zero), the probe is defective. Carefully read the manufacturer's operating manual and make sure the batteries are good before making any leakage tests. You will quickly get the hang of things after three or four leakage tests.

You can contact the following companies for more information about their microwave leakage detectors.

Holaday Industries, Inc.
14825 Martin Drive
Eden Prairie, MN 55344

Narda Microwave, Inc.
Plainview, NY 11803

Simpson Electric Co.
Elgin, IL 60120

The capacitor tester

The capacitor tester comes in handy to check capacitors in the control panel and HV capacitor (figure 2-33). Before making any capacitor tests, discharge the high-voltage capacitor. This capacity tester can check bypass and electrolytic capacitors for open, leaky, and correct capacity. Although the smaller capacity tester built into the DMM might not correctly check the large electrolytic capacitor, many do.

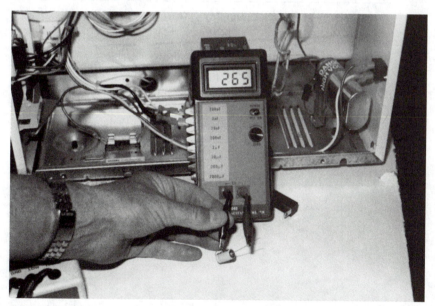

2-33 The capacitor tester can check small capacitors in the touch panel circuits.

Before checking any capacitors, discharge them. Set the function switch to the correct capacity range or position. If the value of the capacitor is not known, start at the highest range and reduce until a satisfactory reading is obtained. The open capacitor will have no measurement, and the leaky capacitor will have a low measurement. A capacitor that has lost its capacity or dried up will not have the same capacity as shown on its body.

Panasonic's confirm after repair guidelines

After repair or replacement of parts, make sure that the screws of the oven, etc., are neither loose nor missing. Microwaves might leak if screws are not properly tightened. Also make sure that all electrical connections are tight before inserting the plug into the wall outlet, and check for microwave energy leakage.

Samsung's operation guide

It is wise to know the cooking power output and level names in the microwave oven. The power level pad in the Samsung MW2500U oven has ten levels in cooking operation (Table 2-2). Ten power levels can be obtained by touching or pressing the power level pad one or more times. For instance, if you wanted to reheat food, press the power level button nine times, producing 470 watts of power. Touch the power level button once to operate the oven on high at 550 watts of power.

Table 2-2. The ten power levels in a Samsung MW2500U oven

Pad	Touch	Display	Output	Level name
	Twice	PL:10	90 W	Warm
	3 times	PL:20	140 W	Simmer
	4 times	PL:30	200 W	Defrost
	5 times	PL:40	250 W	Low
Power level	6 times	PL:50	310 W	Medium Low
	7 times	PL:60	360 W	Medium
	8 times	PL:70	420 W	Medium High
	9 times	PL:80	470 W	Reheat
	10 times	PL:90	530 W	Saute
	Once	PL:Hi	550 W	High

3
CHAPTER

Microwave oven circuits

Today's microwave ovens have four basic circuits. The low-voltage circuit consists of all components that operate from the power line. The high-voltage (HV) components are located between the HV power transformer to the magnetron circuits. All microwave ovens have the above circuits. Within the low-voltage stages, you can find several different circuits.

The first ovens were manually operated. Then automatic push-button cooking microwave ovens that operate with an electronic control panel went onto the market. Convection ovens can operate with either a manual or electronic control oven.

Problems not related to defects of oven

When you receive a complaint about a microwave oven, evaluate the problem carefully. It could prevent you from making an unnecessary house call or attempting to repair a working oven. Sometimes these symptoms can be caused by the owner not understanding how to operate their oven (Table 3-1).

Samsung MW5820T microwave oven circuit operation

When food is placed in the oven and the door is closed, the low-voltage transformer supplies voltage to the touch control unit. The primary interlock switch and secondary interlock switches are closed. Now the monitor interlock switch is open. This interlock monitor switch acts to blow the 15- or 20-amp fuse and stop magnetron oscillation if the door is opened during operation under abnormal conditions. Of course, the primary interlock switch does not open the circuit to stop oscillation of magnetron. The door key is caught by the door hook and the door sensing switch is closed to send the door-close signal to the touch control panel and circuits.

When the cooking cycle, power, and time are set by touching the function pads and the desired numerical pads, the cooking function word shown on the display window blinks to indicate the selected function. The time set also appears in the window. The touch control circuit stores the cooking data.

Table 3-1. Check the cause and remedy of problems not related to defects in oven

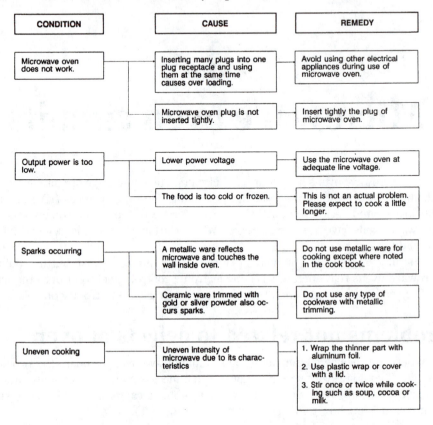

CONDITION	CAUSE	REMEDY
Microwave oven does not work.	Inserting many plugs into one plug receptacle and using them at the same time causes over loading.	Avoid using other electrical appliances during use of microwave oven.
	Microwave oven plug is not inserted tightly.	Insert tightly the plug of microwave oven.
Output power is too low.	Lower power voltage	Use the microwave oven at adequate line voltage.
	The food is too cold or frozen.	This is not an actual problem. Please expect to cook a little longer.
Sparks occurring	A metallic ware reflects microwave and touches the wall inside oven.	Do not use metallic ware for cooking except where noted in the cook book.
	Ceramic ware trimmed with gold or silver powder also occurs sparks.	Do not use any type of cookware with metallic trimming.
Uneven cooking	Uneven intensity of microwave due to its characteristics	1. Wrap the thinner part with aluminum foil. 2. Use plastic wrap or cover with a lid. 3. Stir once or twice while cooking such as soup, cocoa or milk.

After the start pad is touched, the main relay and power control relay are controlled by the touch control circuit (figure 3-1). And the oven lamp lights the inside of the oven by operation of the main relay in the touch control circuit. The switches and the main relay are shown by solid lines behind power control relay.

The fan motor rotates to cool the magnetron by blowing air coming from the intake on the back panel over the magnetron tube. After cooling the fins, the air is directed into the oven to blow out the vapor and odor in the oven through the rear vents. The drive or turntable motor rotates for even cooking. The speed rotation of glass turntable is less than 3 RPM.

When 120 volts ac is applied to the primary winding of the high-voltage transformer, 3.3 volts ac is generated from the filament or heater winding of the high voltage transformer. Usually, this 3.3 volts is supplied to the magnetron to heat the magnetron filaments through two noise-preventing choke coils.

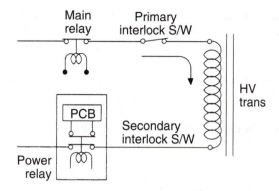

3-1 The main and power relays in Samsung's MW5820R oven. Courtesy Samsung Electronics America, Inc.

The high-voltage winding of power transformer generates 2120 volts ac and is fed to a voltage doubler HV capacitor. The HV diode rectifies the pulsating ac voltage which is then applied to the cathode or heater terminals. The anode terminal of magnetron is at ground potential.

The first half cycle of high voltage produced in the secondary winding of high-voltage transformer charges the high-voltage capacitor. The dotted lines indicate the current flow (figure 3-2). During operation of the second half cycle, the voltage produced by the transformer's secondary windings and the charge of the high-voltage capacitor are combined and applied to the magnetron. Then the magnetron starts to oscillate. The interference wave generated from the magnetron is suppressed by choke coils of 1.6 µH, a filter capacitor of 500, and the magnetron's shielded case so that TV and radio signals are not interfered with and to prevent black lines from appearing across the TV screen.

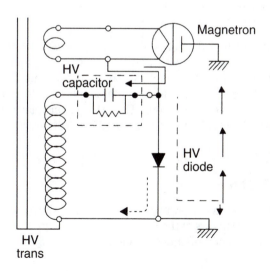

3-2
How the current flows in the high voltage circuit of a Samsung microwave oven. Courtesy Samsung Electronics America, Inc.

The power relay is intermittently turned on by the touch control circuit when the oven is set for other than for full power. The touch control circuit controls the on/off time of the power relay in order to vary the output power of the microwave oven from WARM to FULL power. The time of one complete on/off cycle of the power relay is 30 seconds. The cooling time is shown in the display and starts to count down.

When the door is opened during cooking, the primary switch opens to cut off the primary voltage of the high-voltage transformer to stop microwave oscillation. The door sensing switch closes to send the door open signal to the touch control circuit. The contacts of the main relay stay on, the power control relay turns off, and the display stops counting down.

The fan motor and turntable motor stop by the operation of the secondary interlock switch. Of course, the oven lamp stays on inside of the oven until the door is closed. Upon opening the door, the primary switch and the secondary interlock switch open, and the interlock monitor switch closes—short-circuiting the unpowered ac line. If both the primary interlock switch and the power relay do not function, the 15-amp fuse will blow due to the large current surge caused by the monitor switch activation (figure 3-3).

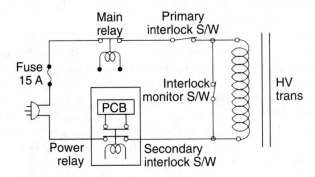

3-3 The low voltage components are in series with the power transformer and ac power line.

When the PAUSE/CLEAR pad is touched during cooking, the pad at once stops time cooking. Just touch the PAUSE/CLEAR pad twice and this cancels all programs stored in the touch control circuit except for the memory program. The time of day reappears on the display window. The oven lamp and all cooking indicators turn off. The fan motor stops. The turntable or drive motor stops. Then the power relay turns off to cut the primary voltage to the high-voltage transformer. Then the magnetron quits oscillating.

Dividing the circuits

You can easily troubleshoot and repair a microwave oven by dividing the oven into two different sections. The first section consists of the low-voltage circuits and the second, the high-voltage circuits. The basic components found in the low-voltage circuits are the fuse, switches, interlocks, timer, thermal cutouts, oven light, blower motor, turntable motor, fan motor, cook light, oven relay, triac, and primary winding of the high-voltage transformer (figure 3-4). Within the low-voltage circuits, you might find a heating element and power relay in the convection oven. The convection oven might have a humidity sensor, latch switch, heater fan motor, and thermal heater cutout.

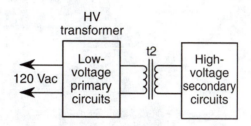

3-4 The ac circuit can be divided by the HV transformer.

All components in the primary section operate in the 120-volt ac circuits. The panel controller operates within the 120-volt circuits with a separate low-voltage power transformer that is used only for the controller circuits. Also, the panel control selects and controls the temperature probe and triac module.

The secondary circuit consists of all parts working in the high-voltage section. The high-voltage circuits consist of the secondary winding of the large high-voltage transformer, special filament transformer (found in some early models), filament or heater winding, high-voltage capacitor and diode, magnetron, and bleeder resistor.

You might find a bleeder resistor and voltage (current) resistor across the HV capacitor and a 10-ohm resistor in series with the HV diode to ground. A variable power switch might be located between the HV diode and cathode terminal of the magnetron. Besides the magnetron, a couple of choke coils might be found inside the magnetron cage. In early models, the HV diode was found inside the magnetron shielded area.

Different types of fuses

Some microwave oven manufacturers use a common household fuse to protect the oven circuits. But many manufacturers provide fuse protection with a long chemical-type fuse (figure 3-5). You might find another temperature-type fuse in series with the ac oven fuse. Usually, the oven fuse is located in the power line circuit before any switch operations.

Pull the power cord from the receptacle before replacing or testing the oven fuse. If the fuse is buried under wires and components, discharge the HV capacitor before removing the old fuse. Replace the monitor interlock switch if the fuse blows at once when the door is opened. The fuse might open with overloaded low- or high-voltage components. The oven fuse is located in the low-voltage circuits (figure 3-6).

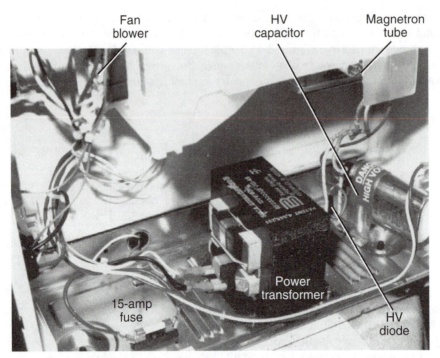

3-5 A 15- or 20-amp fuse protects oven components in the microwave ovens.

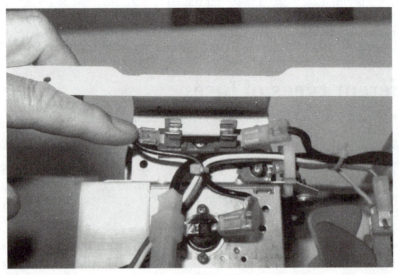

3-6 The oven fuse is located in one leg of the ac power line.

The schematic diagram

The schematic diagram of the microwave oven contains all of the components within the oven, with the exception of the touch panel control circuits (figure 3-7). The Samsung MW-2500 V microwave oven schematic shows the touch panel control, the main relay, the browner relay, and the power relay. Notice that the low-voltage transformer for the panel control circuits is found in the ac input line circuit, while the primary of the high-voltage transformer energizes when all ac input line circuits are closed. The 15-amp fuse, main relay, primary interlock switch, magnetron thermal cutout, thermal cavity cutout, and power relay are operating before ac voltage is applied to the primary winding of high voltage transformer. If any one of these components fail, no high voltage is applied to the magnetron circuits.

Child lock

The Samsung MW5820T oven has a special program for child safety. This function can protect your oven from certain danger caused by an unfamiliar person or child when you are absent. First, touch the PAUSE/CLEAR pad to lock the program. Touch and hold pads 5 and 7 at the same time until a tiny L appears (about 3 seconds) at the right of the display. Now, the oven will not operate and cooking cannot take place. Unlock the child lock with touch and hold 5, 7 at the same time until the tiny L disappears from the display. Now the oven is ready to be used.

The ac circuits

All components from the power line cord to the primary winding of the HV transformer might be included in the low-voltage circuit. Each component must operate to provide power line voltage (117 to 120 Vac) to the transformer (figure 3-8). The failure of the fuse, interlock switches, timer, cook switch, controller, triac, and oven relays might prevent the oven from operating. Other components in the low-voltage circuit might or might not prevent the oven from operating, but they provide the convenience of operation signals and indicators when troubleshooting.

Seventy-five percent of microwave oven problems are caused by open fuses and defective interlock switches. These two components must be operating correctly or no power line voltage is applied to the primary winding of the HV transformer. You might find from three to six interlock or latch switches in the low-voltage circuits of microwave ovens. Besides providing power to the oven circuits, the interlock switches also provide safe oven operation. Such low-voltage components as the blower, timer, turntable, and varimotors can be used for indication when checking the low-voltage circuits.

If a motor is not rotating, you might find no ac voltage at this point in the circuit or the motor might be defective. A dead oven light could quickly indicate that insufficient power is getting to the oven circuits. When the cook light or controller display does not light up, you should determine where the power line voltage stops in the low-voltage circuits.

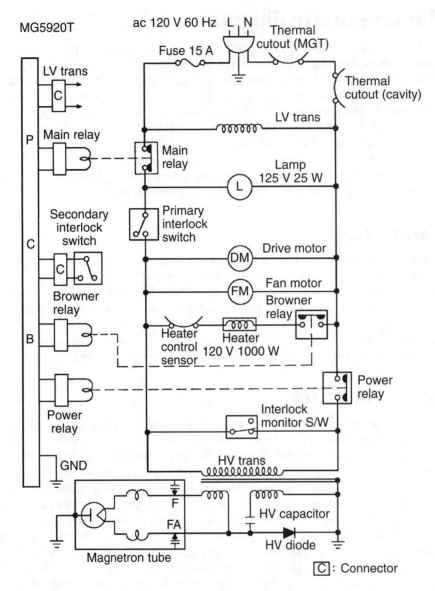

3-7 The microwave oven schematic diagram can help to locate the defective component. Courtesy Samsung Electronics America, Inc.

How an alternating electrostatic field is generated

The generation of an alternating electrostatic field can be explained by studying the generation of alternating current, which is widely known as the household power supply source. The alternating current (ac) takes place in a resonant circuit, which consists of a coil of wire and a capacitor. The coil and capacitor store energy respectively, but in different ways. When they are connected together, and connected to a source of energy, an alternating current is generated. You will see how this happens in the following sections.

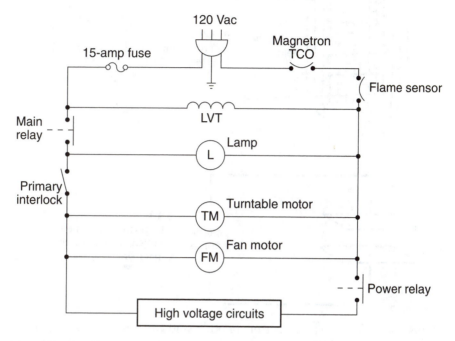

3-8 All components in series with the primary winding of the HV transformer must work for the oven to operate.

How the coil stores energy in a magnetic field

A coil of wire carrying an electric current generates a magnetic field around the coil. This field will have a north pole and a south pole at opposite ends of the coil, exactly like the permanent magnet does. If the direction of current through the coil is reversed, the direction of the magnetic field also will be reversed. If the voltage source supplying the electric current is cut off, the magnetic field around the coil collapses. This collapsing field generates a voltage in the coil, which for a short time keeps the current flowing in the same direction. In this way, the energy stored in the magnetic field is returned to the circuit. This ability of coil to store energy is called an inductance (figure 3-9).

How the capacitor stores electric charges

An ordinary capacitor is made of two metallic plates that are close together but separated by air, paper, oil, mica, or other nonconducting material. If the two plates are connected to the terminals of a battery or another source of energy, the plates of the capacitor will be charged with negative electrons on one plate and a positive charge on the other plate (figure 3-10).

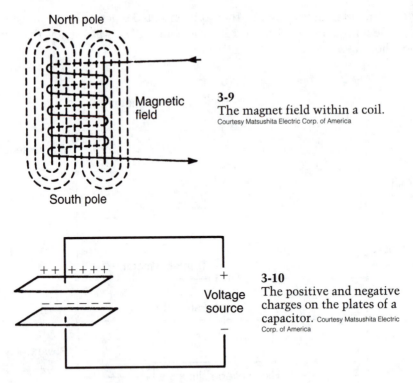

3-9
The magnet field within a coil.
Courtesy Matsushita Electric Corp. of America

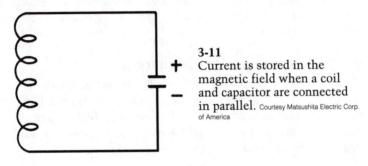

3-10
The positive and negative charges on the plates of a capacitor. Courtesy Matsushita Electric Corp. of America

When the connection to the battery or other voltage source is reversed, the electrons will travel through the external circuit until the polarity of the charges on the capacitor is reversed. This current through the external circuit is utilized to generate an alternating current.

When coil and capacitor are connected in parallel

When a coil and a capacitor are connected in parallel, energy is stored at one instant in the charge coil (figure 3-11). Then the current is stored in the magnetic field around the coil and starts flowing the electrons to the other end of the capacitor plate to charge the capacitor, but in the opposite polarity. This cycle will be repeated as long as energy is supplied to the circuit. This coil and capacitor operation is found in the HV section of the microwave oven.

3-11
Current is stored in the magnetic field when a coil and capacitor are connected in parallel. Courtesy Matsushita Electric Corp. of America

High-voltage circuits

Components operated from the HV power transformer can be considered part of the HV circuits (figure 3-12). The low ac power line voltage is applied to the primary winding of the HV transformer. High ac voltage from the secondary winding of the transformer forms a voltage-doubling network with the HV capacitor and diode.

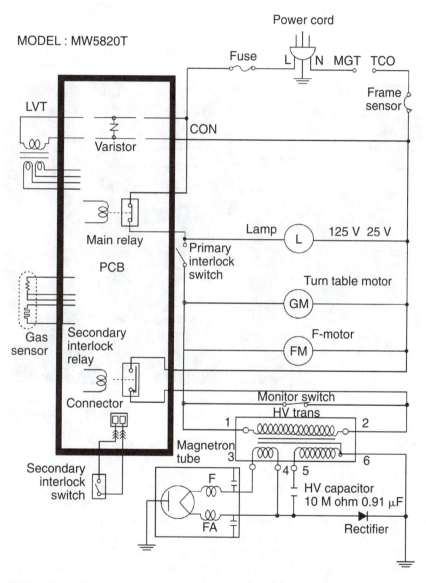

3-12 Schematic diagram of a Samsung MW5820T oven. Courtesy Samsung Electronics America, Inc.

Two separate voltages are fed to the magnetron tube. A low 3 Vac is applied to the heater or filament terminals of the magnetron. Also, high negative voltage is fed to one side of the heater terminal and positive voltage is fed to the grounded anode terminal (figure 3-13).

Caution: Always discharge the HV capacitor before attempting to connect test equipment or placing your hands in the oven.

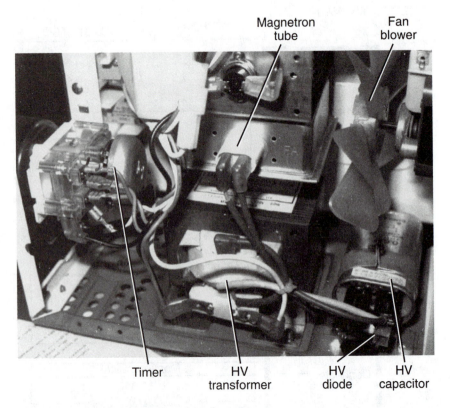

3-13 Critical components found in the microwave oven.

The HV circuits can be checked with low- and high-voltage test equipment. Although some oven manufacturers do not recommend HV tests (for reasons of safety), proper HV test equipment and safety precautions provide adequate measurements in the HV circuits. The voltage applied to the HV transformer will indicate power line voltage at the primary winding. Monitor the low voltage at the primary winding with an ac voltmeter or light bulb. An HV measurement at the diode and magnetron indicates that the transformer and voltage doubling circuits are functioning. A correct current measurement of the magnetron indicates that the tube is operating.

When servicing intermittent ovens, monitoring the power transformer input and high voltage at the magnetron is crucial. Intermittent current readings might indicate a defective magnetron. Intermittent operation of the HV circuits might indicate a defective component — which you can identify by taking HV and current measurements.

The controller circuits

The electronic control circuits provide an easy method of operating the oven (figure 3-14). Simply tap out or push the time, temperature, and cook buttons for easy oven operation. The electronic controller circuit automatically controls the oven circuits and turns the oven off after the cooking time has expired. A digital display window shows the time and the 64 microwave oven circuits' digital programmer control circuit memorizes the function you have tapped out.

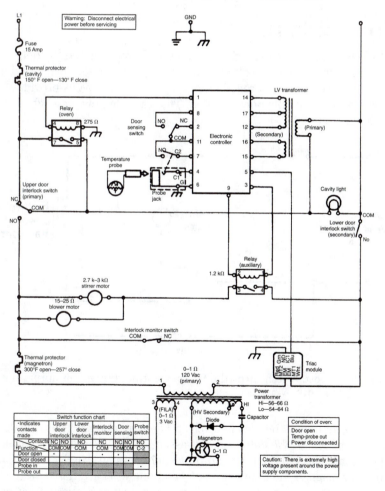

3-14 Electronic controller circuits in the Panasonic oven.
Courtesy Matsushita Electric Corp. of America

Most oven control circuits consist of a separate control board, digital clock display, and digital switch. Some ovens have a microcomputer controlling the pulse oscillator, buzzer, display, humidity sensor, temperature power relay control, clock pulse, door check, variable power relay control, power relay control, membrane switch keyboard, and a dc power supply powering these circuits.

Check the electronic controller if there is no cooking operation after the proper sequence buttons are tapped out. Measure the ac transformer voltage going in—and the control voltage going out—of the electronic control board. The electronic controller operates a triac module or oven relay. A dirty switch assembly or poor cable connection might produce erratic oven operation. Improper digital numbers might be caused by either a defective digital display or electronic control board. A control board should be exchanged or repaired if it is found to be defective.

Defrost circuits

In the early microwave oven circuits, a defrost motor rotated the defrost switch, causing the HV circuit to come off and on intermittently (figure 3-15). A defrost cycle with an electronically controlled circuit found in some ovens provides an on/off time operation of a variable power switch. This variable power switch is located in the HV circuit (figure 3-16).

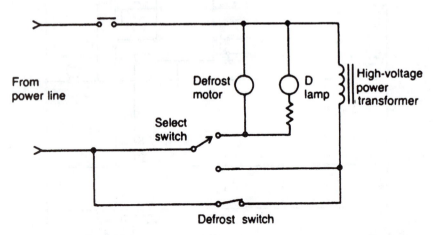

3-15 The defrost switch turns the HV transformer off and on during the defrost cycle in early ovens.

Check the setting of the select switch and defrost motor for possible defrost defects. Dirty contacts of the defrost motor assembly might produce intermittent defrost operations. Check the variable power switch for open switch connections when the defrost or magnetron circuits are not functioning. The variable switch controls might remain open, resulting in no cooking action when in the defrost or regular cooking modes.

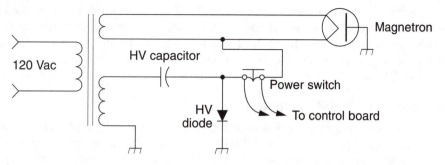

3-16 A variable power switch opens and closes the dc high voltage applied to the magnetron.

Quick defrost and automatic defrost

Some microwave ovens have an automatic defrost system in which you push the pad order of defrost and then place the food setting to whatever type of food is being cooked (figure 3-17). After selecting the defrost setting, touch the start pad and the oven begins the defrosting time. The total minutes of the defrost setting will then be seen on the display and the timer will begin to count down.

3-17 Touch the defrost pad to start defrosting of food.

In some ovens, the defrosting timer will stop after a few minutes and produce an audible sound. Open the door and take out the food, turn it over, and return the food to oven. Press the start switch and the defrosting stage will begin again. Of course, the original setting of the timer counts down until the original programming has been completed. Most defrost settings are given in the microwave oven instruction booklet.

Panasonic NN9807 cycle defrost

When defrost power and defrosting time are selected and the start pad is tapped:
1. The digital programmer circuit (DPC) divides the total defrosting time into eight equal periods consisting of four defrosting periods, each followed by a standing period (Table 3-2).

Table 3-2. The defrost on and off time in a Panasonic NN9807 oven
Courtesy Matsushita Electric Corp. of America

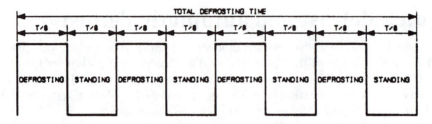

2. During defrosting power periods, power relay B is energized for 9 seconds and de-energized for 13 seconds by the DPC (Table 3-3).

Table 3-3. The on and off defrosting periods in a Panasonic oven
Courtesy Matsushita Electric Corp. of America

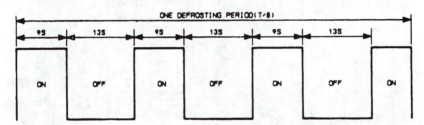

3. During standing periods, power relay B is always open, resulting in no microwave power.

The defrost time selected is converted into seconds by the DPC, but display will show selected time in minutes and seconds as programmed. The total number of seconds is divided into eight time periods. The remainder (seconds not equally divisible by 8) are added to the last standing time period.

Panasonic MN-9507 variable power cooking control

The coil of relay B is energized intermittently by the DPC when the oven is set at any power selection except for the high-power position. The DPC controls the on-off time of power relay B contacts in order to vary the output power of the microwave oven from warm to high power. One complete on-off cycle of power relay B is 22 seconds. The relations between indications on the control panel and the output of the microwave oven is shown in Table 3-4.

Table 3-4. Variable cooking control chart in a Panasonic NN-9507 microwave oven
Courtesy Matsushita Electric Corp. of America

Indication	Output power against high power	On, off time on power relay B (RYB)
High	22/22 (700W)	22S On — 22S On
Med	17/22 (70%)	17S On — 5S Off
M. low	13/22 (55%)	13S On — 9S Off
Low	8/22 (30%)	8S On — 14S Off
Warm	4/22 (15%)	4S On — 18S
Defrost	Approx. 110W	Please refer to description of cyclic defrost.

Auto touch control panel

Several ovens have an auto touch control system that is controlled within the control panel with an IC chip or processor. The microprocessor provides a variety of cooking methods with auto-touch control. Just touch the appropriate button on the front panel and the display will indicate the cooking time (figure 3-18).

You might hear a tone each time the control button is touched. Like all ovens at the end of the cooking cycle, an audible signal is heard. Follow the manufacturer's touch control instructions for easy operation.

Samsung auto start feature

The auto start feature programs the oven to start at any desired time of day in a Samsung MW5820T oven. It can be used to time the start of one, two, or three stage cook only. Touch the CLOCK/AUTO START pad for program desired cooking time. A zero (0) appears on the display. Now the START indicator light will blink. Set the desired cooking start time by touching the appropriate number pads.

3-18 The auto touch pads are controlled by a micro-processor on the control board.

For example, if you desire to start at 4:00 o'clock, touch pad numbers 4, 0, and 0 in sequence. The time between 00:00 and 00:59 is impossible to set.

Touch the START pad. Selected program disappears and time of day appears. AUTO START indicator lights up. Remember, any food cooked by AUTO START should be very cold or frozen before it is put into the oven. Most frozen foods should not stand at room temperature for more than two hours before cooking starts. Cook such foods as vegetables, fruit, smoked or frozen meals in the microwave. Avoid foods such as milk, eggs, cooked meats, poultry, or fish.

Temperature control circuits

You might find a separate temperature control circuit in manual-control ovens. The temperature probe plugs into a probe jack inside the oven cavity. The thermistor probe controls the temperature control circuits and in turn controls the oven circuits for correct cooking modes. In some ovens, the select switch must be turned away from the temperature probe switch position for the oven to operate.

In ovens with a DPC control, the temperature probe is operated from the control circuits. The thermistor temperature probe jack connects directly to the electronic control board (figure 3-19). When the temperature probe is out of the probe jack, the short circuit contacts provide regular oven operation. The temperature probe circuits are controlled by the controller circuit board. Always remove the temperature probe when not in use.

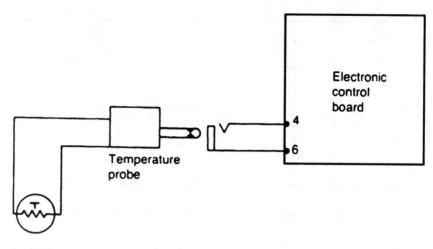

3-19 The temperature probe plugs into an oven jack connected to the electronic control board.

Panasonic temperature control circuits

The advantage of temperature controlled cooking is being able to eliminate the guesswork of cooking by monitoring the rise of the cooking temperature. The temperature control circuit controls the temperature of food from 115°F to 200°F (or 46°C to 93°C).

1. When the temperature setting on the front panel is set to the desired temperature, the probe jack is connected to the oven receptacle with the probe tip inserted into food, and the start button is pressed. Approximately 18 volts is generated in the tertiary winding of the HV transformer and is converted to 24 Vdc by diode D1. It is then smoothed by capacitor C13 (figure 3-20).

2. The 24 Vdc is then decreased to 15 Vdc by R20 and is applied to the sensor circuit. The zener diode stabilizes the 15 Vdc. This voltage forward-biases the base-emitter junction of transistors Q7 and Q9, causing them to conduct (figure 3-21). Conduction of Q9 allows Q10 to open so that 15 Vdc forward-biases the base-emitter junction of Q11 and it conducts. The 24 volts is then applied to the power relay coil, thereby energizing the power relay to close its contacts.

   These transistors can be considered switching devices. The circuit consists of resistive dc voltage dividers that cause transistors to be either in a conduction or nonconduction state. There are no signal voltages in the circuit. The temperature control circuits are controlled by the microcomputer system.

3. When the food temperature goes up, the resistance of the thermistor goes down. Therefore, the voltage between base and emitter of Q8 goes up (figure 3-22). When the base voltage of Q8 exceeds its emitter voltage, Q8 will conduct. This happens when food reaches a selected temperature.

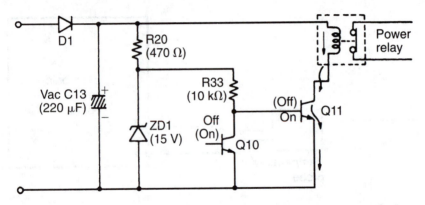

3-20 The low dc voltage source in the temperature probe circuits. Courtesy
Matsushita Electric Corp. of America

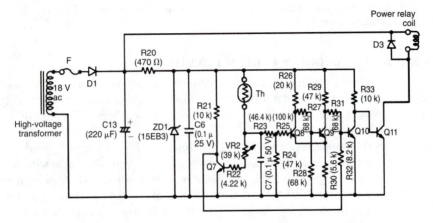

3-21 The temperature probe circuit of the Panasonic MQ8897B oven. Courtesy
Matsushita Electric Corp. of America

4. When Q8 conducts, it causes Q9 to cut off and Q10 to conduct.
5. When Q10 conducts, Q11 will cut off, causing the power relay to de-energize
   and open the power relay contacts. In this way, the temperature rise inside
   the food is controlled by the temperature circuit and the oven is automatically
   shut off when food temperature reaches the desired temperature. If a sensor
   probe component such as a thermistor or lead wire should go open, no bias is
   supplied to Q7, causing it to cut off. If Q7 cuts off, 15 Vdc forward-biases the
   base-emitter junction of Q10 through R21 and R32, allowing it to conduct and
   cause Q11 to cut off. If Q11 cuts off, it causes no power to be supplied to the
   power relay coil and the power relay shuts off the oven.

Construction of sensor probe

The sensor probe is constructed with a thermistor, stainless steel pipe, lead wire,
and jack assembly (figure 3-23). The probe is constructed so that no microwave

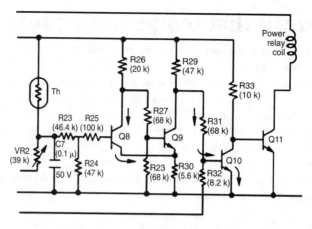

3-22 How Q8, Q9, Q10, and Q11 are biased during temperature probe cooking. Courtesy Matsushita Electric Corp. of America

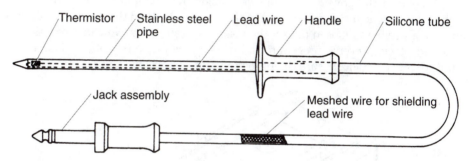

3-23 The various parts of a temperature probe.

leakage or water leakage to the thermistor occurs. This prevents the thermistor from being affected by such external conditions. To avoid possible water leakage into the thermistor portion, do not immerse the probe in water.

Temperature control probe

Not all ovens have a temperature control operation. For those that do, never leave the probe in the oven by itself (with nothing else in the oven). The temperature probe has its own probe plug that fits in a jack inside the oven cavity (usually, towards the oven control panel).

Prepare the meat to be cooked and controlled by the temperature probe. Insert the metal thermometer end into the food to be cooked. Push the probe end into the meat or food, about one-third the way down from the handle guard. Place the meat or food into the oven. Plug the temperature probe into the jack. Some ovens have the jack on the side or at the top of the oven cavity. Program the cooking temperature time and the oven will automatically shut off when the cooking cycle is finished. Most meat should be left in the oven for standing time before removing it and the temperature probe from the oven.

Take care of that temperature probe

Unplug the temperature probe when not in use. Keep the temperature probe in a safe place in the cabinets. Do not operate an empty oven with the temperature probe inside the area. Clean up the probe with soapy water, rinse, and wipe dry, like you would an electric cooking skillet. Never place the probe in an electric dishwasher.

Do not let the temperature probe touch the ceiling or sides of the oven cavity during cooking. Most temperature probes are designed for a specific oven and should not be interchanged. When servicing temperature probe circuits, do not take another temperature probe and try it in another oven. The temperature probe is a delicate instrument and should be handled with care. It should not be used when a steak is placed on a browning dish.

Tappan probe receptacle removal

Disconnect the power cord, remove outside oven wrapper, and discharge the high-voltage capacitor. Remove wires to the receptacle. Hold the body of receptacle and use a wrench to loosen nut on the inside of oven cavity. Now unscrew the nut from receptacle. Reverse procedure to install a new probe receptacle. Insert temperature probe and make sure the probe function is working (figure 3-24).

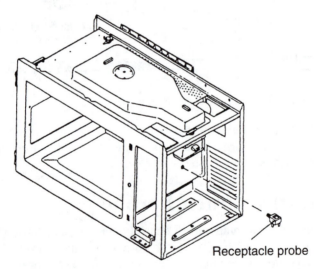

Receptacle probe

3-24 Removing the temperature probe receptacle in a Tappan oven. Courtesy Tappan, brand of Frigidaire Co.

Convection circuits

A convection oven consists of a microwave and separate heater control circuit. The separate heating element provides cooking or browning procedures. Power might be applied to the heating element from a separate select switch or power relay (figure 3-25). The heating element might be controlled by the DPC unit. In some ovens, a thermostat switch is located in series with the heating element to prevent overheating the oven cavity. Usually a heater fan or convection motor circulates the hot air through the oven cavity.

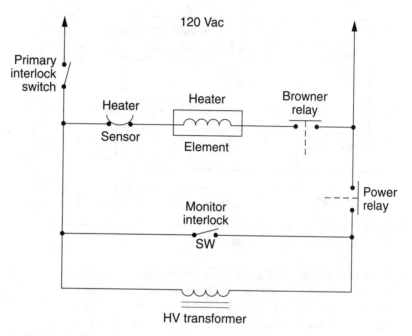

3-25 The heater element is operated by the browner relay control in the panel control board.

You might find a large, single heating element at the top of the oven for browning or cooking modes. In some early ovens, the heating element could be raised or lowered over the food to be cooked. Other convection circuits had a round heating element with a fan motor in the center of the element. These circular heating elements are generally located at the center and top area of the oven cavity.

Panasonic convection cooking control

The digital circuit controls the on/off time of the heater in order to control oven cavity temperature.

1. After the start pad is tapped with the desired convection program set, an 18 Vdc signal comes out of the DPC and is applied to the coil of power relay C.
2. When the contacts of power B close, a power source voltage (120 Vac) is applied to the heater and the heater turns on.
3. When the oven temperature reaches the set temperature, the DPC senses the temperature through the oven temperature sensor and stops supplying an 18 Vdc signal to the coil of power relay C, and the heater turns off.

4. After the heater turns off, the oven temperature will continue increasing for awhile and then decrease (figure 3-26). When the oven temperature drops below the set temperature, the DPC senses the signal and starts supplying an 18 Vdc signal to the coil of power relay again.

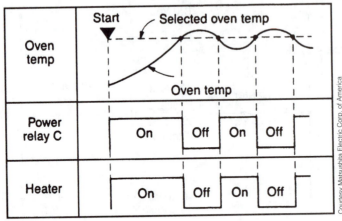

3-26 The on and off periods of convection cooking.

In the Panasonic NN-8907 oven, an upper and lower heating element is located. Here, power relay D and C closes the contacts for the heating elements (figure 3-27). The DPC controls four different power relays. Only the top element is used in the broil mode. The upper and lower heating elements alternately go on and off in accordance with the cooking temperature. Some of these ovens have a thermal cutout in series with the heating element to remove the element from the circuit.

The cooking cycle

Before attempting to service the microwave oven, make sure you know how to operate the oven controls. The manually operated oven is very easy to operate, while some electronically controlled ovens are more difficult. Always have the operation manual of the oven handy when testing or using the oven. Not only will the oven operation manual start you off in the right direction, but it also might help you locate the defective component. A typical normal cooking sequence of a manual operated oven might begin as follows:

1. Place the food to be cooked in the oven and close the door. When the door is closed, all interlock switches are closed except the monitor or safety switch. The monitor or safety switch opens with the door closed (figure 3-28). Follow the bold lines in the schematic diagram. Here the oven is in the idle or off condition until the timer is set.
2. Next turn the timer dial to the desired cooking time. With the door closed and the timer set, the fan blower motor will start to run. The air starts to circulate in the oven and cool the magnetron. The oven light will come on (figure 3-29). Notice that power is applied to the cook switch.

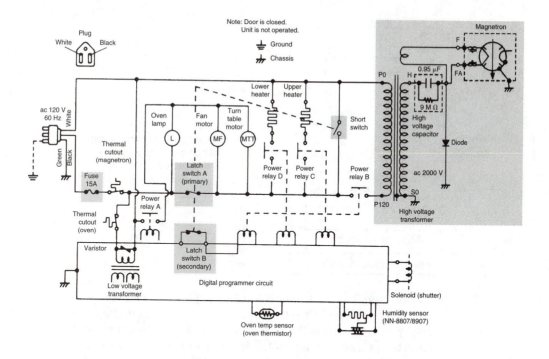

3-27 Power relays C and D control the heating elements of convection cooking. Courtesy Matsushita
Electric Corp. of America

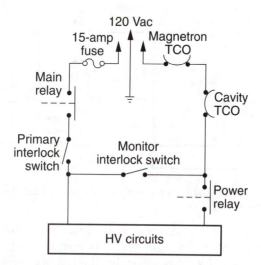

3-28 The monitor interlock switch is on
when oven is opened and off when the door
is closed.

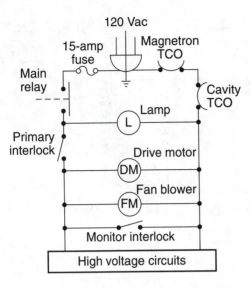

3-29
In most ovens, the oven lamp is on when the oven door is closed or opened.

3. Push the cook button. The cook indicator light will glow, indicating that the oven is in the cooking cycle. When the cook switch is depressed, the timer motor begins to operate (figure 3-30). Power line voltage is found across the cook light and the primary winding of the HV transformer. High ac voltage from the secondary of the power transformer feeds into a voltage-doubling diode and capacitor network with a high negative voltage applied to the heater terminals of the magnetron. Low ac voltage is applied to the heater or filament terminals. The magnetron begins to oscillate, providing RF energy to cook the food in the oven cavity.

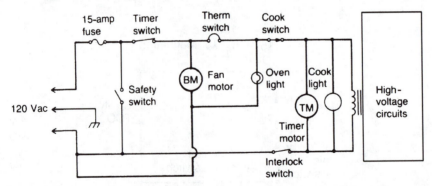

3-30 In this oven, the fan blower operates and oven light comes on when door is closed and the timer is set.

4. When the cooking cycle has finished, the timer control opens up, removing the power line voltage from the rest of the circuit. The timer returns to 0 on the dial. A timer bell sounds, indicating the end of the cooking cycle. After the cook cycle has ended, the food is removed from the oven. When the door opens, the monitor or safety switch closes contacts and the other

interlock switches open up. Remember, the safety switch contacts are closed with the door open (figure 3-31). This prevents the oven from operating and will blow the 15-amp fuse if any of the other interlock switches hang up. Knowing the cooking sequence with each component in operation is helpful in locating the defective component. Follow the schematic as each component begins to operate.

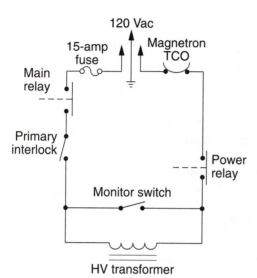

3-31
The monitor switch is open when the oven door is closed.

When the next cooking sequence ceases to function, suspect the corresponding component. A typical microwave oven cooking sequence with a digital electronic controller is given as follows:

1. When not cooking with the temperature probe, make sure the probe is unplugged and out of the oven.
2. Open the oven door and place food or water test in the oven cavity. With the oven door closed, the low-voltage transformer supplies ac voltage to the control circuit as the power cord is plugged in. When the door is closed, the contacts of the safety switch are open. The latch or interlock switch contacts are made to furnish power line voltage to the low-voltage circuit. The oven light turns on. In some ovens, the oven light is on when the door is opened or closed (figure 3-32).
3. Tap the cooking time and power controls of the digital control board. The power indicator light turns on to indicate that power has been set. The time appears in the digital display window. The control circuits also memorize the functions you have set.
4. Tap the start or cook control button. The coil of the power or oven relay is energized by the control board. In other ovens, the triac assembly is energized by the control board. Power line voltage (120 Vac) is applied to the primary winding of the HV transformer through the relay contacts or triac assembly (figure 3-33).

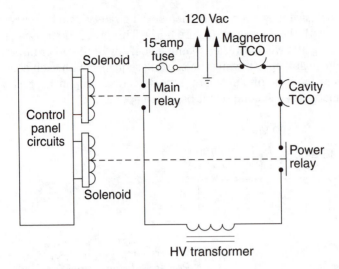

3-32 The main and power relays must operate to provide ac voltage to the HV transformer.

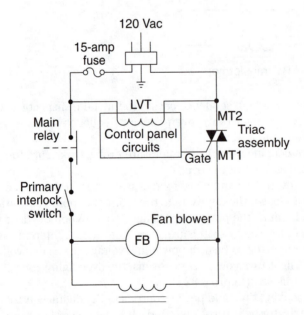

3-33 The low-voltage transformer is on at all times in this oven circuit.

The fan motor begins to operate, blowing air against the magnetron and exhausting vapor through the vent areas. In some ovens, the same fan motor might rotate the stirrer blade with a pulley-gear belt arrangement. In other ovens, you might find a separate stirrer motor. In some Sharp ovens, the turntable starts rotating.

A power indicator light might start blinking to indicate the oven is functioning with the oven lights on. The cooking time starts to count down in the digital display window. The power line voltage is applied to the primary winding of the HV transformer, filament, or heater voltage is applied to the magnetron with high ac voltage fed to the voltage-doubler network. The HV capacitor and diode combine to form the dc voltage-doubling circuit. A negative high voltage is applied to the heater circuit of the magnetron. The magnetron begins to oscillate with the RF energy being emitted through the waveguide assembly to the food in the oven cavity.

The stirrer motor or stirrer-fan motor, which is found in several ovens, starts spreading RF energy to prevent hot spots in the oven and to prevent uneven cooking. The lower motor has a pulley with a chain-type belt that rotates the blade, while the stirrer motor is a direct drive type. The rotary blade might be called a rotary antenna in the Amana and radiating antenna in the Hardwick ovens (figure 3-34).

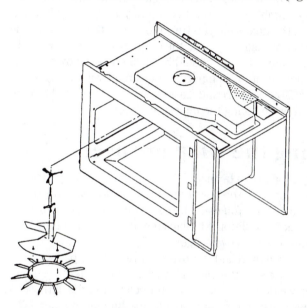

3-34 The rotating antenna at the top of a Tappan oven.
Courtesy Tappan, brand of Frigidaire Co.

In some of the ovens, including Sharp and Samsung, a turntable rotates the food for even cooking. The motor for the turntable has a reduction gearbox in the bottom of the oven. You might find a variable power switch or relay in the HV circuits. It is energized intermittently, often with a digital control circuit.

With this knowledge of operating the oven, how the cooking sequence functions, and closely following the circuit diagram, you might be able to quickly locate a defective component. First, isolate the service problem with the circuit diagram. Next, see how far into the suspected circuit various components are actually working. Take voltage and resistance measurements and locate the defective component. Then replace the defective part.

The wiring diagram

The wiring diagram of any microwave oven is somewhat different than the schematic diagram. While the schematic shows all the components tied together in the circuit, the wiring diagram illustrates how each component is tied to one another (figure 3-35). Notice that in Samsung's MG5920T wiring diagram the color-coded connecting wires have a color label. For instance, the two yellow (y) wires (y3 and y4) connect to the heater element. The drive motor has brown (BR1 and BR2) connected to motor terminals from the touch panel circuits.

New turntable motor

The new turntable motor might operate in any direction. Do not be alarmed when your old turntable motor rotates one way, and the new oven turns both directions. Of course, the next time you turn it on, the glass tray goes the other way. The plastic motor rotator might lift off so the entire oven can be cleaned up (figure 3-36).

The turntable and support should be cleaned as the oven is wiped up. Carefully pull up the plastic turntable support and clean off with soapy water. After cleanup, replace the turntable support and glass tray. Do not drop either one. Always remove the glass tray when servicing the microwave oven.

Removing the wrap or cover

Before any repairs can be made, the back cover or wrap must be removed. First, disconnect the oven from the power line plug. Do not leave the oven plugged in when removing the outer case. You could find several screws at the back along the top and side edges. Check for additional Phillips metal screws along the bottom edge of both sides of the outer case or cover (figure 3-37).

In older and commercial-type ovens, several back panels might need to be removed so you can get at the various working components. The wraparound metal cover might be welded to the metal bottom plate. Servicing these types of ovens often takes longer because the components are difficult to reach.

When all cover screws have been removed, slide the entire case back about one inch. You might have to lift up the back side of the case to free it from retaining clips on the cavity face plate. Set the outer cover out of the way so it will not get scratched up (figure 3-38).

Caution: Before touching or checking any component, discharge the HV capacitor.

After all oven repairs have been made with cook and leakage tests, replace the outer case. Make sure the power cord is pulled. Double check the top and side front edges. Be sure the edges are tucked inside before replacing the metal screws. If not,

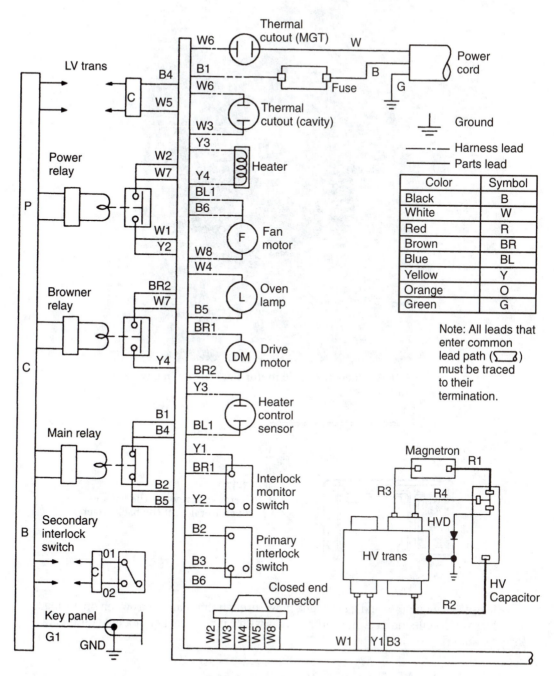

3-35 The wiring diagram shows how each component in the oven is connected. Courtesy Samsung Electronics America, Inc.

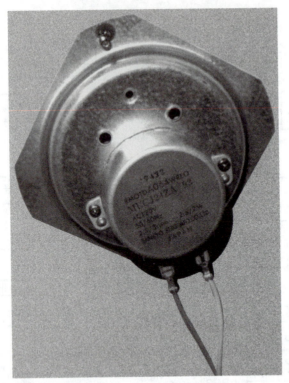

3-36 The new turntable motor can start up in any direction.

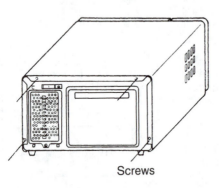

3-37
Remove screws at the back of oven to take off the back cover.

Screws

the case might vibrate and light might show through the seam areas when operating. Clean off the outside case and control areas with window cleaner or a mild detergent and water.

Exploded view of oven parts

The exploded view of any oven displays the various components and where they are located in the oven. The circled numbers of each part can be located in the parts list. Simply locate the defective component, look up the part in the exploded view, and compare that number with the part list (figure 3-39). Order out that new part for the oven. Replace all parts with the exact part numbers.

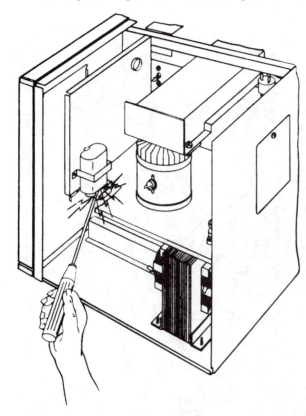

3-38
After removing the outside cover, discharge the HV capacitor. Courtesy Norelco Corp.

Clean up

Usually, the outside case of the microwave oven is a laminated plastic over metal, or all metal cabinet. Clean up the cabinet after all repairs are made. Do not use strong household cleaners on the oven finish. Clean the outside case with regular soapy water. Then wipe dry. Do not apply any type of polish (such as furniture polish) to the outside cabinet. Clean up the front door with mild soap and warm water.

Food spillovers in the oven can be cleaned up with mild soap and baking soda. Remove the glass tray and turntable support, if possible. Clean up the side walls with mild soap and hot water; rinse and wipe dry with a soft cloth. Be real careful when replacing the glass tray.

Extreme care must be exercised when cleaning up the touch control panel. You can easily scratch the plastic cover. Pull the ac plug or open the front door, so when a switch or button is accidentally moved or touched, the oven will not start up. Wipe the plastic panel with a moist, soft cloth. Do not use window spray, cleaner, or chemical sprays on the front area of the oven.

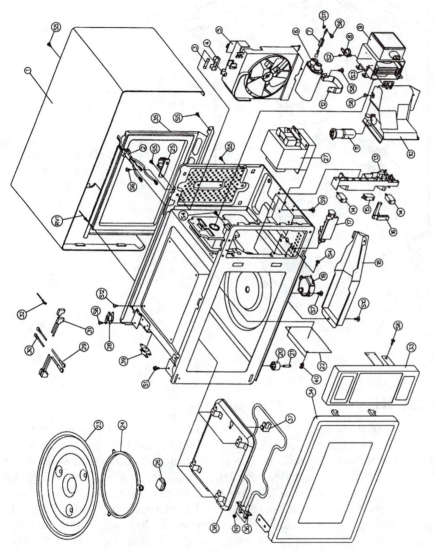

3-39 An exploded view of oven lists and numbers the parts for easy location of the part list.

Discharge

Technicians who regularly service microwave ovens should have a sign above the bench: Discharge the high-voltage capacitor! I cannot say this enough. This high-voltage capacitor can knock you down if touched during operation or afterwards. You can be seriously injured. A lot of amperage, current, and high voltage are stored in this capacitor, making it more dangerous than the high voltage in a TV set. So, remember to discharge that high-voltage capacitor after each time the oven is fired up.

4
CHAPTER

Basic troubleshooting procedures

Before attempting to service any microwave oven, you should know the trouble symptoms, have the correct test instruments, know certain test procedures, secure the oven schematic, and use a lot of common sense. The more you know about the trouble symptoms, the easier it will be to locate the defective component. You should ask the owner several questions about how the microwave oven operates:

1. What were you doing when the oven quit?
2. Is the oven dead or intermittent?
3. At what point in the cooking cycle did the oven stop operating?
4. Is there anything more that you can tell us about the oven?
5. How old is the microwave oven? (Date purchased.)
6. Is the oven under warranty? (Warranty registration card.)
7. Where was the oven purchased? (Name and location.)

In case the oven is under warranty, make sure that you see the warranty registration card or bill of sale. Most ovens have a one- or two-year labor warranty, two years on parts, and from five to seven years on the magnetron tube. Besides the trouble symptoms, you should list the model and serial number of the oven on the service repair tag. List the date and place where the oven was purchased. At this point, you should tell the customer if he or she is to pay labor, parts and labor, or if the entire repair is covered by the warranty. This data must be listed on the warranty repair order to collect payment on warranty service.

Owner's responsibility

Many of the troubles with microwave ovens are caused by the operator of the oven. Make sure the oven is plugged into the outlet receptacle. If the oven does not heat properly and is connected to an improper extension cord, check for proper ac voltage at the oven plug (117-120 Vac). If there's no voltage, check the house fuse. A separate outlet should be installed for the microwave oven with at least number-12 wire and proper grounding (figure 4-1).

4-1 Check the wall outlet with any small appliance in the house.

Know how the oven operates. Read the oven instructions and starting and cooking-test procedures several times. Improper oven operating results in another service call. Have the service technician or salesperson go over the instruction manual in detail. Do not be afraid of any buttons on the oven. Do not be afraid of radiation from the oven.

If the front glass accidentally gets cracked or the door will not close properly, pull the plug of the oven and call the electronics service technician. The oven technician should always check the oven for radiation leakage before it is returned to the owner. Have the technician check for radiation of the oven in your home if you are still in doubt.

Neon circuit tester

If you do not have a VOM or DMM handy, you can test the ac circuit with a neon circuit tester, also known as a tell-tale light. These testers can be picked up for a couple of bucks at hardware, electrical, and department stores. They are also available at Radio Shack (part number 22-102). The tester consists of nothing more than a Ne-2 neon bulb, limiting resistor (68 k), resistor, and two male plugs (figure 4-2).

If ac voltage is found at the wall outlet where the oven is plugged in, the microwave oven might have knocked out the main fuse in the fuse box. Quickly make a test with the neon or tell-tale light by plugging into the ac outlet. If the neon light is lit, you know that power line voltage is found at the wall outlet. Check for correct polarity in the outlet by touching one probe to the center screw on the outlet plate, or ground plug, and inside the spade terminals. Try both sides.

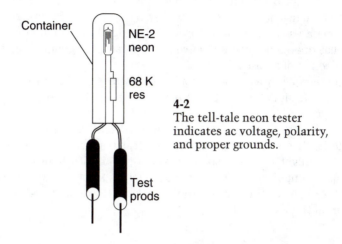

Container
NE-2 neon
68 K res
Test prods

4-2
The tell-tale neon tester indicates ac voltage, polarity, and proper grounds.

If neither side produces a light, suspect a poor ground. This tester will tell if the oven is properly grounded. Call the local electrician to check the outlet for a good ground. The microwave oven can be a shock hazard without a good center-prong ground. You should have a light indication (bright) from the hot side of the power line to the ground hole where the third prong fits into the wall outlet. If the light is weak, check for a poor ground. The negative side of the power line-to-ground prong will not show a light, which is normal. If the light is dim, a poor ground exists from ground outlet prong to the main fuse box.

Oven safety instructions

To prevent electric shock, fire, burns, and microwave radiation, play it safe when operating the microwave oven. Keep either food or a cup of water in the oven to make tests. Never leave the microwave oven empty and turn it on. Do not turn the oven on with the temperature probe lying on the oven cavity.

Make sure the microwave oven is properly grounded. Do not use extension cords, unless they have a three-wire cable and plugs. The wall outlet should be grounded at the outlet and main fuse box. Measure the resistance between the long prong on a three-wire plug to the metal cabinet with less than 1 ohm of resistance. Static shock might occur when you touch the metal part of the oven after walking on a kitchen or living room carpet; this static is normal.

Inspect the ac cord for cuts or frayed areas. Check the front-door oven seal. In older ovens, these rubber-like seals might crack or peel, making a poor seal. Look for a misaligned door, dents, a warped door, and damaged gaskets on the oven door.

Do not cook whole eggs, especially with the shells, in the oven. To avoid explosions, keep glass jars or sealed glass or crock containers out of the oven. Follow the manufacturer's instructions on cooking popcorn in the oven. Several ovens have key or pad operation for popcorn.

Keep the oven away from the rear wall so the fan can exhaust oven fumes and heat. Inspect the fan blower vents for a collection of dust and dirt. This might cause oven heating of the magnetron and keep the fan motor from exhausting hot air. Make sure the fan operates all the time the oven is cooking. If the fan motor comes on after the oven has operated for three to five minutes, suspect gummed-up bearings or excessive collection of grease and dirt.

Make sure the microwave oven is not too close to hot, open stove burners. Keep the ac cord away from hot areas. Leave the cord plugged in and not left out in the open for damage.

To prevent fire in the oven, do not cook food until it burns. Watch for combustible material that is placed in the oven (such as paper, plastic, and regular popcorn bags). You might find that some recycled paper with filings will burn more easily with mixed impurities. Keep all metal (such as twist-ties for paper or plastic bags) out of the oven. If a fire starts in the oven, keep the door closed and pull the ac power cord from the wall outlet.

Customer service checks

Before calling a service technician, the customer should make a few preliminary tests. When the door is open, does the oven light come on? If not, there might be a blown oven fuse or defective ac outlet. Try another appliance or lamp in the same outlet and see if both operate. If not, check the main fuse box. Proceed to a cook test if the outlet is normal.

Place a plastic or paper cup filled with water into the oven. Shut the door and see if the oven light goes out. Turn on or program the oven for two or three minutes. The oven light should come on and you should hear the fan blower operating. Listen closely at the rear oven vents for a rush of air. Do you hear a tone or buzzer sound when the oven turned off? Call for service if there is no oven light, fan operation, or cooking process.

Customer leakage radiation exposure

The microwave oven is a safe appliance when operated properly. The only area where radiation might occur is through and around the oven door. If the oven operates with the door open, pull the plug and call a service technician. An interlock might hang up and the safety switch might be defective. Do not defeat or tamper with the safety or monitor interlock switches just to make the oven operate.

Keep the area between the door and seal clean. Keep all food out of this area. Wipe up the area when food boils over or is spilled inside the oven cavity. Do not operate the oven with a damaged door, loose hinge, or bent-out-of-line door. Check to see if the oven door is level with the oven top area. Do not operate the oven with a damaged glass door.

Don't attempt to use the oven if the door will not latch or if you must hold it shut manually in order for the oven to operate. The front door needs alignment and the interlock switches must be adjusted. Keep the temperature probe out of the oven door area and seal. Make sure that the oven door closes snug and tight. Place a low-priced microwave oven radiation meter around the door area to check for signs of leakage. If in doubt, call a microwave oven technician.

Checking the ground wire

Make sure the oven is grounded. Most ovens are grounded through the 3-prong ac cord (figure 4-3). Do not operate the oven on a 2-wire extension cord. The microwave oven is designed to be used when grounded. Check the oven 3-prong plug to make sure the oven is grounded prior to repair. Check the resistance from the long middle ground tip of the ac cord to the metal chassis of the oven.

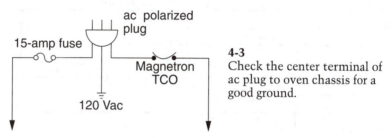

4-3
Check the center terminal of ac plug to oven chassis for a good ground.

Where to start

Refer to the list of trouble symptoms in the service literature, isolate the trouble, locate the defective component, and replace the new part (figure 4-4). Turn to the service diagnosis and trouble symptom charts in the service literature. Most oven procedures start at the beginning, and each stage is checked off as you go through the various test procedures. When the malfunction occurs, test the various components listed.

The oven symptoms are often listed as low-voltage, control, and high-voltage (HV) circuits. Any possible symptom or trouble that might occur with the power line voltage (117 or 120 Vac) is listed as being in the low-voltage circuits. The control circuits are either manual or electronic board control circuits. The manual circuits are controlled by the oven switch and timer, while the electronic control circuits are controlled by the electronic push-button control board. All HV circuits exist from the power transformer through the magnetron tube circuit.

GE oven operation

When the start pad is touched, the relay contacts are closed and Ry-1 and Ry-2 are turned on. 120 volts ac is supplied to the primary winding of the power transformer and

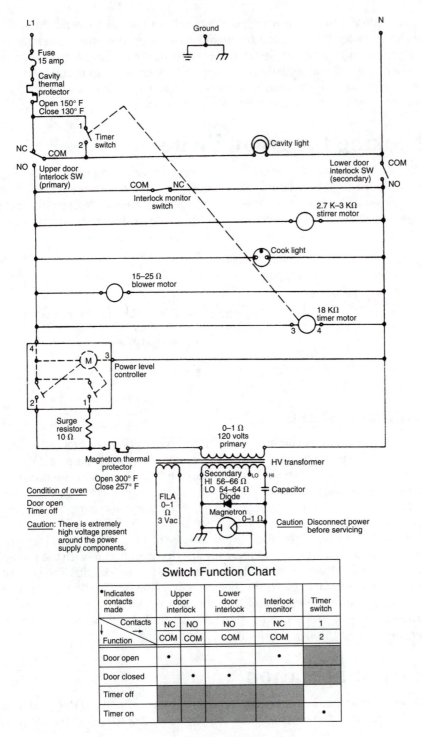

4-4 Schematic diagram of an early Norelco MCS8100 microwave oven.

is converted to about 3.1 volts ac output on the filament winding, and around 2000 volts ac on the high voltage winding. The filament winding voltage heats the magnetron filament and the HV is sent to a voltage doubler circuit (–4000 Vdc half phase).

The microwave energy produced by the magnetron is channeled through the wave guide into the cavity feed box, and then into the oven cavity where the food is cooked. Upon completion of the cooking time, the power transformer, oven lamp, and other parts are turned off and the generation of microwave energy is stopped. The oven reverts to the OFF condition.

When the door is opened during a cook cycle, the monitor switch, door sensing switch, secondary interlock switch and primary interlock deactivate the main relay (figure 4-5). Thus, the circuits to the cooling fan motor and high voltage components are turned off, with the oven lamp on and the digital readout displays the time remaining on the cook cycle when the door was opened.

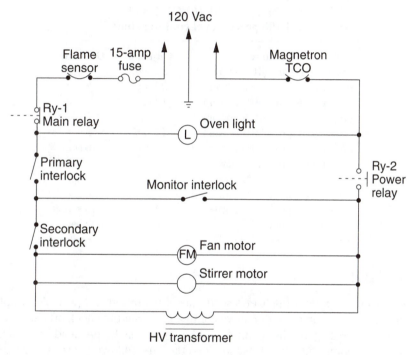

4-5 Components found working in the GE cooking operation.

The monitor switch electrically monitors the operation of primary interlock switch. It is mechanically associated with the door and functions in the following sequence: When the door opens, the primary interlock switch and secondary interlock switch open their contacts, then the monitor switch contacts are closed. When the door is closed, the monitor switch contacts open, and the contacts of the secondary interlock switch and primary interlock switch close.

If the primary interlock switch fails (contacts remained closed) when the door is opened, the closing of the monitor switch contacts will form a short circuit through the monitor fuse and primary interlock switch causing the monitor fuse to blow. Actually, the monitor switch is now across the ac power line. The fuse should only blow when the primary and monitor interlock hangs up or if a defective component is found in the oven.

GE variable power cooking

When a variable power cooking is programmed, 120 volts ac is supplied to the power transformer, intermittently through the contacts of relay (Ry-2), which is operated by the control unit within a 22-second time base. Microwave power operation is found in Table 4-1. Notice the on/off time ratio does not correspond with the percentage of microwave power, because approximately 3 seconds are needed for heating of the magnetron filament.

**Table 4-1. GE JEBC200 and
JEB100 power level cooking chart**

Vari-mode	On time	Off time
Power 10 (High) (100% power)	22 sec.	0 sec.
Power 9 (approx. 90% power)	20 sec.	2 sec.
Power 8 (approx. 80% power)	18 sec.	4 sec.
Power 7 (approx. 70% power)	16 sec.	6 sec.
Power 6 (approx. 60% power)	14 sec.	8 sec.
Power 5 (approx. 50% power)	12 sec.	10 sec.
Power 4 (approx. 40% power)	10 sec.	12 sec.
Power 3 (approx. 30% power)	8 sec.	14 sec.
Power 2 (approx. 20% power)	6 sec.	16 sec.
Power 1 (approx. 10% power)	4 sec.	18 sec.
Power 0 (0% power)	0 sec.	22 sec.

Isolation

Once you understand the oven's sequence of operation and possible symptoms, you can quickly eliminate the defective components. Simply check off each working component according to the symptom. When you reach the dead area, use the schematic to point out the possible defects. The following are several actual microwave oven problems, broken down with a partial schematic to demonstrate each trouble.

Dead condition

The oven door is closed, timer set for three minutes, and the oven cook switch is pushed, but nothing happens. The oven appears dead, even the oven lamp does not come on, and there is no fan rotation. What is the trouble? Let's take a breakdown of a simple microwave oven circuit (figure 4-6).

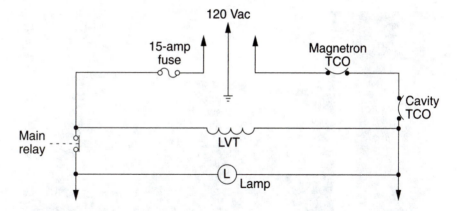

4-6 The components that must be closed in the Samsung MG5920T oven to let the oven lamp become an indicator.

First, remove the fuse and check continuity with the ohmmeter. Replace the fuse if it is open. Then see if the oven operates. The 15-amp fuse opens up when there is an overloaded condition in the oven. Maybe one of the interlock switches did not open up when the door was opened. Perhaps the magnetron or HV diode became leaky or shorted out. It's possible the power line might be overloaded, causing the fuse to open. Sometimes the fuse is blown when the operator rapidly changes the timer and push-button control board without shutting the oven door. Often, the open fuse symptom will show up the possible defective component in a four-hour cooking test.

The fuse tested good, but you replaced it anyway, so what could be the trouble? Glancing at the schematic, the timer switch must be defective if the blower motor doesn't start to operate. If the fan motor is operating and there is no oven light, the thermal cutout on the magnetron must be open. Pull the ac power cord. Take an ohmmeter reading across the thermal cutout contacts. When normal, you should have a shorted condition with no resistance measurement.

The oven light should come on when the timer switch contacts are closed or the timer is set for a few minutes on the dial. Check the timer switch contacts and thermal cutout assembly with the low ohmmeter range. Both components should show a shorted or closed condition. If the oven light now is on but the oven still does not function, suspect a defective interlock switch, cook switch, or cook relay (figure 4-7).

Oven light on but no cooking

When the cook switch is pushed, the oven relay should energize and apply ac to the primary winding of the power transformer (figure 4-8). Follow the double-lined wires in the schematic. We know that ac voltage is at point (1) to the blower motor and the oven light. The oven relay must be energized to have voltage at point (2). Listen very carefully; you can hear the cook relay working when the cook switch is pushed. A thud or jamming noise indicates the relay is operating (figure 4-9).

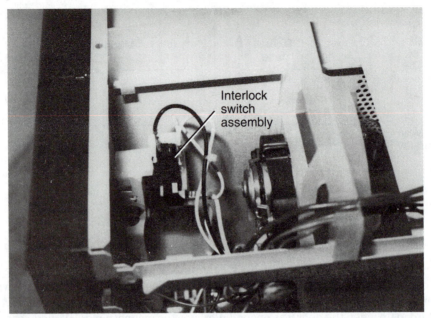

4-7 The door interlock switch can be checked with the low ohm scale of DMM.

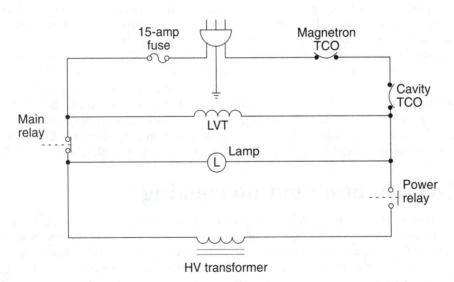

4-8 Suspect a defective power relay if you don't hear a thud or energized noise when the lamp is on with no cooking.

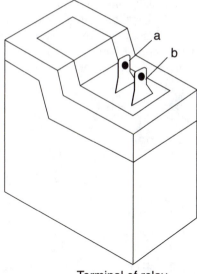

4-9
The oven relay is mounted on the control panel in some microwave ovens.

Terminal of relay

Because the cook relay is not energizing, suspect the cook switch contacts, interlock switch 2, or an open cook relay solenoid. Always check the interlock switches after the fuse for possible trouble. They cause more problems than any other component in the oven. The interlock switches are closed and opened each time the oven door is opened. So the most likely component is interlock switch 2. Locate switch 2 in the service literature (figure 4-10).

Interlock switches

4-10 Check the interlock switch with an ohmmeter while in the oven.

The contacts of interlock switch 2 are normally open when the door is opened. Once the door is closed, the contacts of interlock switch 2 are closed. With the door closed, clip the ohmmeter across the two switch contacts. You should have a shorted measurement (figure 4-11). If not, the switch is open. Now open the door and notice if the meter hand will show a shorted reading and open when the door is opened.

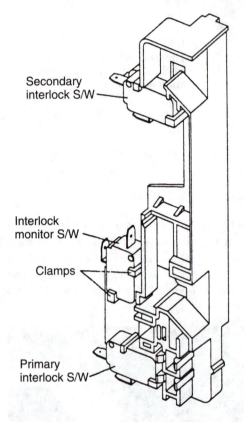

Secondary interlock S/W

Interlock monitor S/W

Clamps

Primary interlock S/W

4-11
Locate the interlock switch and check across the terminals with R×1 ohm scale of DMM.

In case the switch 2 is normal, check the cook switch and timer switch contacts for closed contacts. When the cook switch is pushed, the timer switch contacts will close with the timer set at a few minutes on the timer dial. Sometimes this timer switch contact is erratic in operation, causing the oven to work one minute and not the next. Intermittent or erratic oven conditions are a little more difficult to locate.

In case interlock switch 2 is normal, check the switch contacts with the ohmmeter. Make sure the ac power cord is disconnected. Clip the ohmmeter leads across the oven switch terminals. Push in on the oven switch. Most oven switch assemblies have momentary contacts. Contact is made only when the switch is pushed. You should measure a dead short across the normally open switch. Let up on the oven switch, and the meter reading should be open.

The only component left is the oven relay. There are two ways to check the oven relay. Measure the ac voltage (120 V) across the relay solenoid terminals or take a resistance reading. When ac voltage is across the solenoid winding with no movement of the relay, suspect an open winding. Double check with the ohmmeter.

With the dead condition test, you are using the oven lamp and fan motor as indicators to break down the operating circuits. It's possible to have a defective oven lamp, but in some ovens there are at least two different bulbs. Because the oven light is now on, you know that part of the circuit is operating to point (1). Also, with the fan blower rotating, you know that voltage is normal this far in the circuit. It must be said here that all fans and oven lights might not be wired the same in different microwave ovens. By using different components as indicators, you can quickly find the defective component.

In ovens with tab button or electronic control circuits, the electronic control unit must operate before the oven light comes on. The electronic control circuits might operate a power relay or a switching device (triac). The electronic control circuits and triac must function before the oven light comes on in a Panasonic oven (figure 4-12).

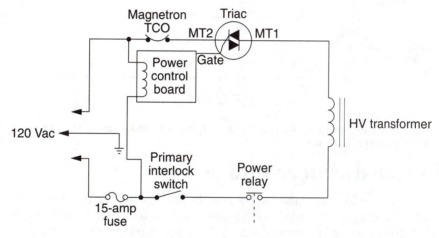

4-12 The control board provides gate voltage for the triac assembly.

Oven light on/relay energized/no cooking

You know the relay is energizing because you can hear it close; also, the oven light is on, indicating voltage at point 2. To determine if voltage is getting to the power transformer, measure the ac voltage across the primary terminals. These terminals are usually enclosed in plastic covering and should either be pulled back or you should use needle-type test probes that go into the wire insulation. It's best to clip the ac meter across the primary winding of the power transformer.

Because ac voltage is found at point 2 of the cook relay, suspect interlock switch 1 for being open or erratic in operation (figure 4-13). Both switch 1 and the monitor or safety switch should be replaced when one is found to be defective. Locate the interlock switch 1 close to the front door latch. Normally, the switch is closed when the oven door is shut and opens up with the door open. Clip the ohmmeter leads across the terminals. Open and close the door and check the ohmmeter readings. Interlock switch 1 and 2 and the 15-amp fuse produce more dead and intermittent operation than any other oven components.

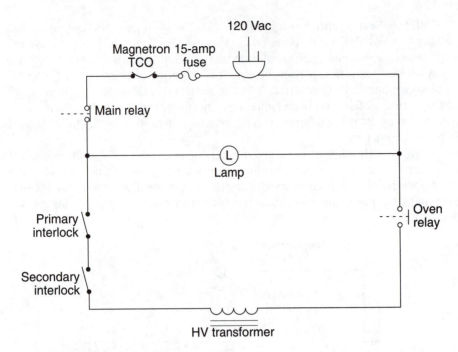

4-13 Check for open interlock switch if the lamp is on and no ac voltage is present at the primary of the HV transformer.

No transformer voltage

Clip the ac voltmeter or test lamp across the primary winding of the power transformer. If the lamp is on and 120 Vac is measured at the transformer, suspect a defective component in the HV circuits. Suspect poor power relay contacts or a defective triac and control circuit (figure 4-14). Does the fan blower come on? The fan blower is another indicator showing that ac voltage is applied this far in the circuit.

Transformer voltage but no cooking

In some ovens, you might find a defrost timer motor and indicating lamp that is wired across the primary winding of the power transformer (figure 4-15). When the defrost mechanism is operating, you can assume that ac voltage is applied to the large power transformer. In other ovens, the fan motor might be tied across the ac primary winding. Look for working components through the schematics to help locate the various troubles.

After a normal voltage measurement is found on the primary winding, with no operation of the magnetron, suspect a defective component in the HV section. Before taking any voltage or resistance measurements in the HV circuit, discharge the HV capacitor each time the oven is fired up. This HV capacitor should hold a charge unless a high-megohm resistor is found across the capacitor to leak off the voltage.

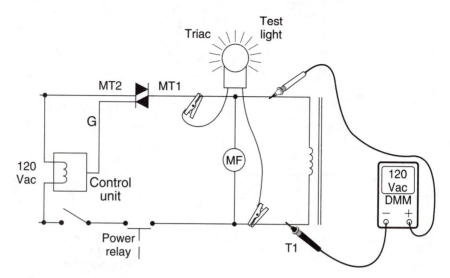

4-14 Suspect a defective power relay, triac, or control unit if you find no voltage at primary winding.

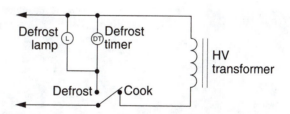

4-15 Check the defrost switch for bad contacts with no defrost or cooking of food.

The most likely defective component in the HV circuits are the HV diode and magnetron tube (figure 4-16). Correct HV measurement at the magnetron tube indicates the HV circuits are normal and the magnetron tube must be defective. No or low voltage at the magnetron tube might result from a leaky magnetron tube or any component in the HV circuit. Proper high voltage with no current measurement at the magnetron tube indicates a defective magnetron.

Measure the ac high voltage across the secondary with the Magnameter or an HV probe. Remember, this voltage might be from 2 kV to 3 kV across terminal numbers 3 and 4. A very low ac voltage measurement might be caused by a leaky capacitor, diode, or magnetron tube. The filament voltage might vary between 3.15 and 3.2 Vac (figure 4-17). Be careful because these filament leads are at a high dc voltage potential to chassis ground. Information about servicing the HV and low-voltage circuits is given in chapters 6 and 7.

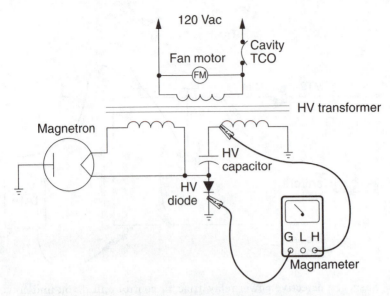

4-16 Check for high voltage across secondary winding of the HV transformer with Magnameter.

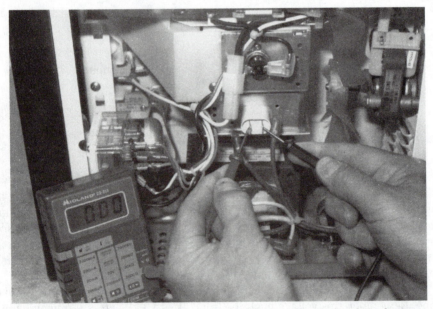

4-17 The magnetron heater or filament terminals will measure less than one ohm when good.

Typical troubleshooting charts

Table 4-2 and Table 4-3 show two different microwave oven troubleshooting charts. Table 4-2 is for a typical manual-type timer oven with possible defects. Table 4-3 is a typical troubleshooting chart for an electronic control board. Most troubleshooting charts provided by the manufacturer are about the same, but each one might have a different method of servicing. If a service manual is handy for the particular oven you are servicing, use it. You might want to take a few notes of a given trouble and list them right in the schematic for future reference.

Table 4-2. A typical manually operated oven with possible defects

Problem	Possible defective component	Test and replace
House fuse blows when power cord inserted	Shorted power cord & plug	Replace ac power cord
Fuse blows when open the door	Check interlock monitor and safety switch	Replace both safety and interlock switch
Fuse blows when oven turned on	Check interlock switch	Replace interlock switch See if shorted
	Check line varistor	See if line varistor arcing—Replace
	Check safety switch	Hang up safety switch Replace
Fuse blows when cook switch is pushed	Check HV capacitor Check HV diode Check magnetron Check power transformer	Test and replace the defective component
Oven dead with timer and cook button on	Check for blown fuse	Replace with 15-amp chemical fuse
	Check for power in home	Measure with HV voltmeter
	Check defective timer	Replace timer
	Check component in HV circuitry	Test all HV components
	Bad latch or interlock switch	Check continuity and replace
	Defective thermal cutout	Check continuity
Blower motor won't rotate	Bad plug-in connection Poor soldered cable	Check and reset ac plug Check voltage at motor terminals
Blower motor won't rotate	Defective motor	Check motors for dry or gummed up bearings
Oven light does not come on	Burned-out bulb	Replace bulb
	Defective timer contact	Check timer switch with ohmmeter
	Defective thermo cutout	Measure across terminal with ohmmeter, should have short reading
	Poor wiring to socket	Visually inspect wiring
Oven goes into cook cycle—timer does not operate	Defective timer motor Open wiring to timer	Replace timer assembly Check for ac voltage (120 V) at timer motor terminals
Timer erratic—sometimes stops	Defective timer assembly	Replace entire timer assembly

Table 4-2. continued

Problem	Possible defective component	Test and replace
Oven cooks—no cook light	Defective cook light	Replace cook light—some of these are new type bulbs
	Bad wiring	Check for 120 Vac across light bulb terminal assembly
Oven light on—No heat or cooking	Defective power transformer	Check voltage at primary 120 Vac
	Defective HV diode	Check if warm or leaky
	Defective HV capacitor	Open or shorted—replace if in doubt
	Defective magnetron	Measure HV and current of magnetron tube

**Table 4-3. A typical troubleshooting chart
for an oven with an electronic control board**

Problem	Possible defective component	Test and replace
Oven light out when door opened	Defective interlock switch	Check with ohmmeter with door open and closed
	Open or poor wiring	Visually inspect wiring
	Defective light	Replace bulb
Oven light does not light at all	Burned-out bulb	Replace—check maybe 2 bulbs in oven
	Poor wiring to light socket	Check and repair wiring
Oven ready with water test—components stop operating when door closed	Defective oven relay	Check solenoid with ohmmeter
	Defective triac	Check contacts
	Display stops and doesn't count down	Check as in chapter 6
		Check door-sensing switch
		Check wiring
		Check low-voltage transfer
		Replace defective electronic circuit control-board assembly
Cook indicator light on— No fan motor rotation	Check door interlock switch	Check continuity with ohmmeter while opening and closing door
	Check door alignment	Realign door
	Defective Auxiliary relay	Check relay
	Defective blower motor	
Cook indicator light on— No fan motor rotation	Broken wires to motor	Measure voltage at fan motor
		Lubricate bearings
		Check wires for poor connection or plug end

Problem	Possible defective component	Test and replace
Turnable motor doesn't rotate	Defective turntable motor	Measure ac voltage at motor terminals
		Measure voltage at motor terminal
		See if motor coupling binding
	Check wiring	Visually inspect wiring and voltage
		Replace motor if defective
Stirrer motor does not rotate	Check defective motor	Check motor continuity and voltage
	Belt off fan blade	Visually inspect belt and replace
First steps time out—Tone sounds—Oven shifts to memory time—Color begins to flash—Tone sounds three times	Check for improper oven grounding	Check continuity of grounds with ohmmeter and voltmeter
	Low line voltage (below 110 V)	Measure ac line voltage
		Notify electric company if low
	Defective electronic contact board assembly	Replace electronic contact board assembly
Oven operates, but low heat or heats slowly	Check line voltage (below 110 V)	Measure ac line voltage. Should be 115 to 120 Vac
	Defective power transformer	Check primary & secondary voltage
	Defective HV capacitor	
	Defective thermal cutout (intermittent)	Take winding resistance test
	Defective magnetron	Replace
	Defective cook relay	Place ac meter across thermal contacts when in question
	Defective triac	
	Defective control unit	Take high voltage and current measurements
		Check continuity of solenoid
		Check for leaky conditions—replace
Oven erratic cooking	Defective touch panel	Replace
	Defective triac module	Replace
	Defective electronic control-board assembly	Replace
Oven goes into cook cycle—Uneven heating or cooking	Stirrer blade does not rotate	Check stirrer motor or belt assembly
	Burned waveguide cover	Replace cover
	Turntable does not operate	Check ac voltage on motor terminals
	Defective magnetron	Replace broken coupling bushing
		Replace after voltage and current tests
Using temperature probe—Oven does not function	Switch not turned to probe	Check switch control setting
	Temperature probe open or shorted	Check temperature probe with ohmmeter

Table 4-3. continued

Problem	Possible defective component	Test and replace
Using temperature probe—Oven does not function	Defective probe jack	Check jack contacts with ohmmeter
	Defective electronic control-board assembly	Replace control board assembly
Push temperature pad center 140—Close door —Oven begins to operate—Oven beeps when pad is touched	Defective probe Defective probe jack	Check with ohmmeter Check with ohmmeter

Typical oven symptoms

All of the following symptoms are actual troubles that have been found while servicing microwave ovens. Each symptom is given with the actual breakdown of the component. This list is given to help you quickly locate a definite symptom with the possible defective component. These problems are not listed in any special order.

Dead/nothing works

1. Open 15-amp fuse.
2. Bad latch switch.
3. Replace open interlock switch.
4. Broken door latch spring, will not activate interlock switch.
5. Realign and adjust door.
6. Replace switch and monitor assembly.
7. Only oven light operation—control set to probe—no probe inserter.
8. Loose plug or power board.
9. Replace defective control board (figure 4-18).
10. Defective oven relay.
11. Bad wire connections broken.
12. Open thermal cutout.
13. Upper door switch.
14. Open low-voltage transformer.
15. Poor transformer primary connection.
16. Leaky HV diode.
17. Defective triac.
18. Shorted HV capacitor.
19. Defective magnetron tube.

Intermittent operation

1. Bad ac plug in house.
2. Defective extension cord.

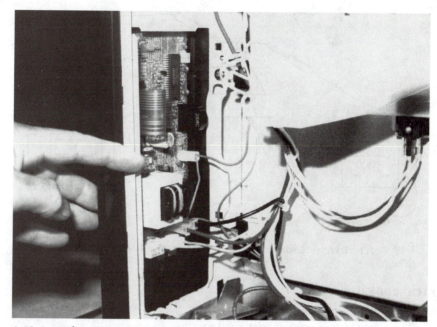

4-18 In a Sharp R5A50 model, you might find a defective control board with no cooking.

3. Intermittent latch or interlock switches.
4. Poorly crimped connection on power transformer leads.
5. Intermittent control board (replace entire board assembly).
6. Defective magnetron tube.
7. Improper setting of door interlock switches.
8. Door realignment.
9. Defective thermal cutout.

No heat/no cooking

1. Leaky magnetron.
2. Leaky diode.
3. Defective thermal cutout.
4. Defective triac.
5. Replace upper switch strike assembly.
6. Shorted HV capacitor.

Slow cooking

1. Defective magnetron tube.
2. Leaky diode (just runs warm).
3. Defective triac assembly (figure 4-19).
4. Change electronic control board.
5. Shorted power transformer.
6. Overheated thermal cutout.
7. Improper or low ac line voltage.

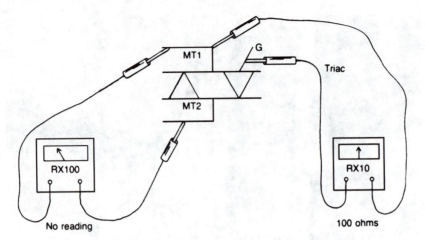

4-19 A defective triac can cause the oven to run all the time or cook too slow.

Erratic cooking

1. Defective magnetron.
2. Poor transformer crimped connection.
3. Poor connection to control board.
4. Defective control board.
5. HV diode.
6. Overheated thermal cutout on magnetron.
7. Erratic interlock or latch switch.
8. Dirty oven or on/off switch.
9. Broken door spring to engage interlock switch.

Lights up/no fan/no cooking

1. Defective interlock and latch switches.
2. Defective control board assembly.
3. Door alignment.

No cooking

1. Defective magnetron tube.
2. High voltage present /no current/tube defective.
3. No HV/defective capacitor or leaky diode.
4. Shorted or open power transformer.

Can't shut oven off

1. Leaky triac (figure 4-20).
2. Defective oven relay.

No cooking /only a hum noise

1. Replace leaky triac assembly.

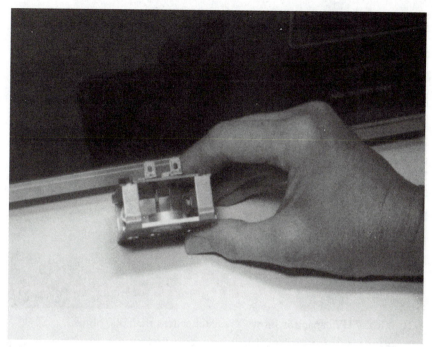

4-20 A leaky triac will cause the oven to operate all the time until power cord is pulled.

Loud pop and oven quit

1. Leaky or shorted capacitor (figure 4-21).
2. Leaky or shorted HV diode.

Keeps blowing line fuse

1. Hung-up interlock switch.
2. Loose door and latch assembly.
3. Requires door alignment (too much door play).
4. Leaky HV diode.
5. Shorted magnetron.
6. Arcing ac line varistor.

Blows fuse when open door

1. Replace interlock switch and monitor or safety switch.
2. Realignment of door/Open door.

Went dead

1. Replace defective monitor and interlock switch.
2. Replace control board assembly.
3. Replace oven relay.
4. Replace defective triac.

4-21 A leaky HV capacitor shows a resistance less than 500 ohms.

Oven stops operating when door is closed

1. Blown fuse.
2. Defective interlock switch (figure 4-22).
3. Defective oven relay.
4. Defective triac.
5. Open low-voltage transformer.
6. Defective touch-panel control board.
7. Defective control board.
8. Realignment of door.

Fuse blows when oven turned on

1. Check interlock and safety switch. Replace monitor switch if it will not open up with the door opened.
2. Inspect ac power line varistor. These parts are designed to arc over during power outage or when lightning strikes the power line to the oven. (Outage is when excessive voltage is found on the power line, usually in a storm or sometimes when the ground wire is broken on the power line.)
3. Replace safety switch. This switch might hang up or stay closed when the oven door is closed and might not open when door is opened.

Oven runs too hot

1. Defective magnetron.
2. Too-high ac line voltage (above 130 Vac).
3. Burns food/defective magnetron.

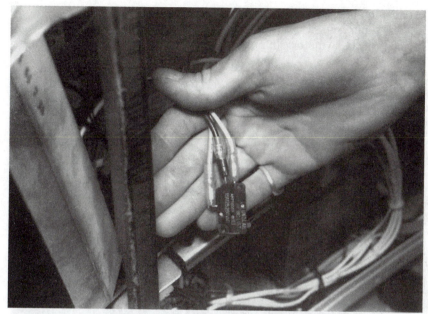

4-22 When the monitor or safety switch blows a fuse, replace both the monitor and interlock switch.

Something burning in oven

1. Never use regular paper bags for cooking.
2. Grease behind shelf guide and metal screws.
3. Grease behind waveguide cover.
4. Replace waveguide cover.

Hot spots

1. Improper cooking.
2. Defective magnetron tube.

Arcing in oven

1. Exposed or loose oven bolt.
2. Loose front door screw.
3. Grease behind shelf metal screws.
4. Loose nut or screw on browning element inside oven.
5. Waveguide cover (replace).
6. Sparking from defective magnetron tube.

Very noisy

1. Fan assembly loose (figure 4-23).
2. Noisy pulley or stirrer assembly.
3. Constant buzzing of the power transformer (replace).
4. Noisy fan (needs lubrication).

5. Loose fan bracket.
6. Noisy timer motor assembly (lubricate or replace).
7. Noisy turntable motor (lubricate or replace).

4-23 A loose fan assembly can cause a vibrating noise when operating.

No fan operation

1. Fan blade jammed.
2. Poor socket from fan to control assembly.
3. Bad fan motor lead (poorly soldered connection).
4. Won't turn, frozen fan, needs lubrication.
5. Intermittent fan operation, bad cable plug.
6. Slow coming on, needs lubrication.

Door doesn't close properly

1. Door alignment and adjustment (figure 4-24).
2. Replace damaged door.
3. Door sticks, realign door guide assembly (figure 4-25).
4. Replace loose screw in switch housing assembly.

Loose door latch

1. Broken spring in door.
2. Loose screw.
3. Loose pop-rivet.

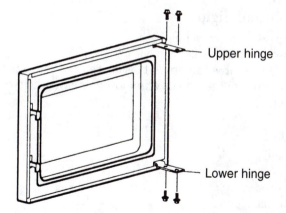

4-24 Check the front door for proper alignment.

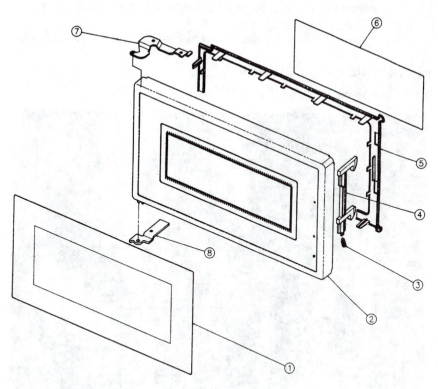

4-25 The various door components in the Samsung MW2500W oven. Courtesy
Samsung Electronics America, Inc.

Oven cooks/no cook light

1. Check for a defective oven bulb.
2. Inspect the oven light socket and wiring.
3. If lamp socket has arced and pitted inside, replace with a new lamp socket.
4. Check connections or terminals where oven or cavity light connects to the oven circuit.

No oven light

1. Defective bulb.
2. Defective socket.
3. Defective oven switch.
4. Wire off/on cable plug.
5. Oven light doesn't go off, replace latch or oven switch.

Oven goes into cook cycle but timer does not move

1. Check the voltage going to the timer (120 Vac). Replace timer if voltage is found at the terminals.
2. Test for open timer motor. Inspect winding where it connects to terminals for broken connections.
3. Replace timer if voltage is present and mechanism seems to be jammed.

Timer runs slow or erratically

1. Replace timer assembly (figure 4-26).

4-26 Suspect a defective timer with erratic or no operation in the small Goldstar oven.

Timer cord will not operate oven

1. Replace cord assembly.
2. Check cord inlet assembly.

Turntable doesn't operate

1. Replace broken turntable plastic coupling.
2. Replace turntable motor.
3. Check wiring on motor (measure ac voltage).
4. Check turntable bearings.

No defrost cycle

1. Replace defective defrost switch.
2. Replace defrost timer motor assembly.

Browner doesn't function

1. Defective switch.
2. Open heating element.
3. Bad flexible wire connection at end of heating element.

Oven shock

1. Check for three-prong ac grounding plug (figure 4-27).
2. Check grounding of oven.
3. Static elements between kitchen carpet and oven.
4. Discharge static electricity in oven metal before touching the control panel (might damage IC components in touch-control board).

Oven operates/no temperature probe control

1. Check switch/must be turned to probe operation.
2. Defective probe. Test for correct resistance.
3. Inspect jack for poor connections or wires broken off of jack terminals.
4. Do not overlook a defective control board assembly. Replace it.

No variable-cooking control

1. Determine if failure is found in controller variable relay circuits or switch contacts.
2. Discharge HV capacitor.
3. Clip a 120-volt pigtail light or bulb across the ac switching terminals of variable power supply (figure 4-28).
4. If the light comes on and does not go off (cooking begins), suspect poor terminal contacts or defective variable controller.
5. Check to see if relay is energizing, as oven comes on and off. If not, suspect a defective variable control circuit in control board.
6. Suspect poor or dirty switch contacts if the relay energizes.

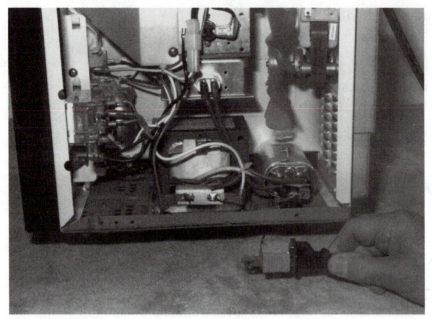

4-27 Check the center terminal ground plug to oven chassis. Plug in a ground 3-prong plug and clip wire to help ground the oven.

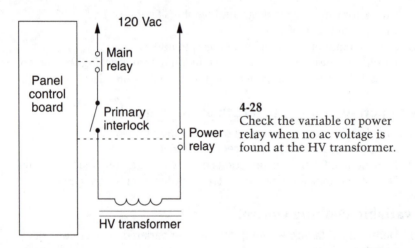

4-28
Check the variable or power relay when no ac voltage is found at the HV transformer.

Proper polarization and proper grounding

For the safety of oven operators and service technicians, the microwave oven with a three-prong plug must be grounded and polarized at the wall outlet. The round hole should be grounded at the receptacle through three-wire cable back to the switch box. Polarization means the larger slot of the receptacle must be neutral and the small slot must be the hot wire. Check the color of the wire screws. The small slot (hot) should have copper-colored screws and the large slot (neutral) should have white-colored screws.

Do not cut off the ground round terminal of the ac oven plug if the outlet will not accept it. Use a grounding adapter temporarily until the electrician can run a new three-wire cable to the outlet. All microwave ovens should be operated from a separate outlet to prevent overloading and ensure proper oven operation.

Defective triac

The defective triac might become leaky, letting the oven run all the time, even when the oven is turned off. Leakage occurs between MT1 and MT2. An open triac will not let ac voltage pass through the triac with no oven operation. Improper or no gate voltage applied to the triac makes the triac act as an open component (figure 4-29). Check the control board for missing voltage and triac control signal. Remove the triac from the oven circuits. But first, discharge the HV capacitor. Mark down each wire terminal. Check the color code of each wire. This is very important in replacing the new triac. Check the suspected triac with resistance measurements or a triac tester.

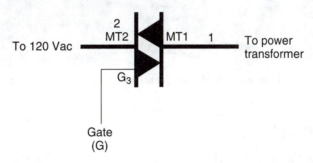

4-29 The correct terminals of the triac assembly in the microwave oven.

Tri-Check II

The Tri-Check II tester was designed to test the suspected triac assembly. This tester will test the triac for open, shorted, leaky or normal conditions (figure 4-30). Before removing the wires from the triac, mark down all color-coded wires on a piece of paper or schematic. Now remove all wires from the triac assembly.

Both lights will light when the test button is pressed and turn off when button is released if a normal triac. With an open triac, no lights will light up. Both lights will light with a shorted triac. Turn off the triac tester when finished testing, as two batteries are needed to operate the Tri-Check II.

Triac checker

A triac tester will quickly test the condition of the triac assembly found in the microwave oven. The checker consists of only 10 small parts that can be picked up at the local Radio Shack store. No on and off switch is necessary, as the tester can be plugged into the ac outlet when needed. Build the checker into a small plastic box so it can fit into the tool caddy.

4-30 Remove all the wires to the suspected triac and connect to the Tri-Check II tester.

The circuit

T1 is a 12.6 Vac step-down power transformer that provides ac voltage to a half-wave rectifier D1 (figure 4-31). After the voltage is rectified to dc, C1 smooths out the ripples for the triac circuits. R1 provides dc voltage to the MT2 terminal. R2 limits the dc voltage applied to the red LED (light emitting diode) indicator. SW1 is a momentary open push-button switch, while R3 (50 k) resistors vary the voltage applied to the gate terminal. MT1 connects to the negative side of the power supply.

Testing

Remove the leads from the triac to be tested. Clip the black alligator clip to the MT1, red clip to MT2, and green lead to the gate terminal of the triac. The normal triac will not light up with these connections. If it does, the triac is leaky (figure 4-32). Push down on the momentary switch and rotate R2 until the red light remains on no matter where the control is adjusted. Let up on SW1.

With the normal triac, the red light will remain bright when the gate is triggered on. Now, remove the green gate clip and the red (LED) remains on with a normal triac (figure 4-33). If the light goes out, suspect a defective triac.

Use only the red and black terminal leads to check for a leaky triac. The red light is on and normal when the leads are attached to the MT1 and gate (3) terminals. Does the LED light with reverse leads between black and green terminals? If the LED lights when the red and black leads are connected to the gate and MT2, the triac is

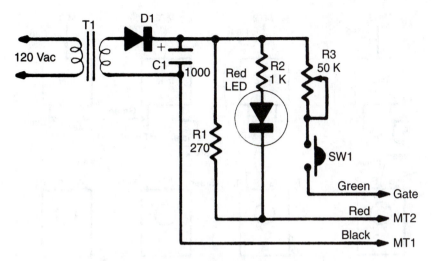

4-31 Circuit schematic of triac checker with only a few parts.

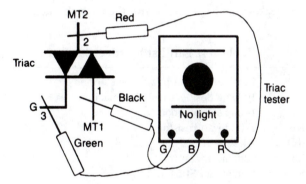

4-32 With a normal triac, the light stays on after
SW1 is pressed and released.

shorted or leaky. The defective triac will show leakage between MT1 and MT2. This
little tester can also be used as a continuity tester by using the red and black clip
leads (Table 4-4).

Conclusion

The results of basic troubleshooting procedures are to take the trouble symptom, diagnose the problem, and then isolate the suspected component, or locate the
suspected component from the manufacturer's diagnosis service procedures or typical trouble procedures. Isolate or break down the circuits where the trouble might
occur. Test the suspected component with those given or found in the service literature. Replace the defective part or component with the original manufacturer's replacement parts.

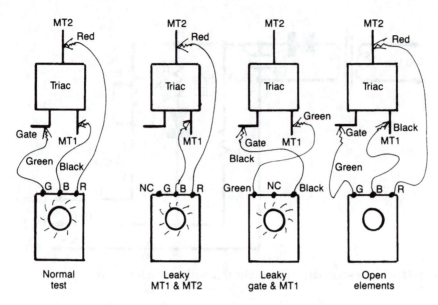

4-33 Normal, leakage, and open tests with the triac checker.

Table 4-4. Part list for triac checker

T1	12.6 Vac 450 mA power transformer
C1	1000 µF 35-V electrolytic capacitor
D1	1-amp silicon diode
R1	270 ½-watt resistor
R2	1-k ½-watt resistor
R3	50-k linear control
SW1	Momentary SPST switch
LED	Red
Misc.	Hookup wire, three different-colored alligator clips, plastic box, etc.

While trying to isolate and service the various microwave oven circuits, you might find more than one trouble within the oven. For instance, you might have a blown fuse, defective interlock switch, and electronic control board. When lightning or power line outages occur, several different components might be damaged. Excessive lightning charges can cause the fuse to keep blowing because of an arcing power line varistor. Lightning charges can also damage the electronic control board, power switches, and small power transformers. (See figure 4-34.)

4-34 Excessive lightning damage can destroy the varistor and parts in the control board.

5
CHAPTER

Troubleshooting microwave oven switches

More problems are caused by a defective interlock switch than any other component in a microwave oven. Perhaps it's because several interlock, defect, monitor, and safety switches are activated when the door is constantly opened and closed (figure 5-1). These switches are located at the top and bottom of the door area, and some are directly behind the door handle. The many interlock switches are installed primarily to protect the operator from injury and exposure to RF radiation. The interlock switches are used to shut off the oven should the door be opened while the oven is in operation.

5-1 The interlock and monitor switches were built as one unit in early oven circuits.

The small interlock switch has a microswitching action that is triggered with only a light touch of the small plastic plunger. These switches come in normally open and normally closed positions. Look closely; they are all marked on the side of the switch area. The switch might be an SPST or SPDT type. A three-prong or terminal interlock switch is an SPDT type. Usually, the common connection is located at the bottom terminal of the switch.

These interlock switches have different voltage and current ratings. The voltage reading might be the same, but different current ratings are marked on the body of the microswitch. Always replace the interlock switch with one having the very same current rating or one with a larger current rating. These switches are all about the same size. They can be interchanged from one oven to the next as long as the current rating is correct with a normally closed or normally open interlock switch. One or two models of ovens have larger interlock switches that are not interchangeable because they will not mount in the same area.

A microswitch can develop poor or open internal contact points and become defective. Dirty or pitted contacts might let the oven operate intermittently in an erratic manner. The switch contact spring might become broken, causing the small plunger to rest even with the surface switch area. Suspect a defective microswitch when an SPDT switch can make contact in one position and be open in the other.

A defective interlock switch might become locked in one position and not release when it should. Sometimes the defective interlock can be located by inspecting the appearance of the switch terminals. Check the interlock switch terminal for burn marks. If the plastic insulation is curled back or appears black, suspect a poor terminal connection or poor internal switch points (figure 5-2). Suspect food or liquid alongside the small plunger locking it in the downward position. Replace both the switch and connection when this condition occurs. You can solder the old connection to the new switch terminal for good contact. Otherwise, the new switch could be destroyed by a poor terminal connection and erratic oven operation.

All interlock switches can be easily checked with the low-ohmmeter range of the VOM. Make sure the power cord is pulled out and the high-voltage (HV) capacitor is discharged before attempting any ohmmeter measurements. Set the ohmmeter to R×1 and connect both leads across the switch terminals (figure 5-3). It's best to have alligator clips on both VOM terminal leads so they can be clipped to the switch terminals. In some cases, you might have to remove one terminal to make a good connection. Always write down the color codes of the wire terminals when three wires are connected to the switch. It is not necessary to record the connection of a two-terminal interlock switch.

Open the door and notice the meter movement of the VOM. A NO switch will not show a reading, while the NC switch will have a shorted measurement. If any type of resistance is noticed between the two terminals with an NC switch, suspect dirty contacts. Replace the switch at once. A good switch will open and close when the door is opened and closed in their respective switching modes.

Several names are used for the micro-interlock switches you find in ovens. Some oven manufacturers list the interlock switches as primary and secondary interlock switches, while others might call them latch or defeat switches. Besides being called the monitor switch, some ovens might call it a safety switch. Although the names are

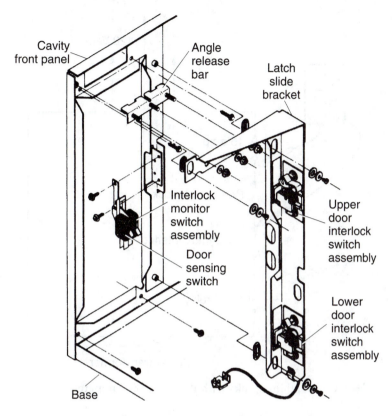

Cavity
front panel

Angle
release
bar

Latch
slide
bracket

Interlock
monitor
switch
assembly

Door
sensing
switch

Upper
door
interlock
switch
assembly

Lower
door
interlock
switch
assembly

Base

5-2 The upper and lower interlock switch assemblies are mounted behind the oven door. Courtesy Norelco Corp.

different, the small microswitches perform the same task in microwave ovens. Upper and lower interlock switches are used in some ovens, and you could find one or two interlock switches ganged with an oven lamp switch.

Before leaving the microwave oven, be sure to check for interlock action. Simply open the door and listen for the switches to make or break contact. The door should be adjusted for a very tight fit with interlock action. Never bypass or remove an interlock switch. Always replace the defective switch before attempting to operate the oven. Often the oven will not operate with a defective interlock switch.

GE pull-down door interlocks

Although the GE JEBC200 oven has a pull-down door, the interlock switches are mounted to the left and right of the oven cavity (figure 5-4). The primary interlock switch is located in right body latch, inside switch on the bottom. The secondary interlock switch is located in the left body latch, outside switch on the bottom. A monitor switch is located in right body latch top assembly. The door sense interlock is located in the right body latch, outside on bottom and located in the left body latch, inside the switch in the bottom.

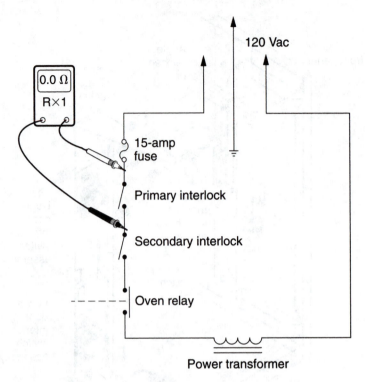

5-3 Check all interlocks with R×1 scale of ohmmeter.

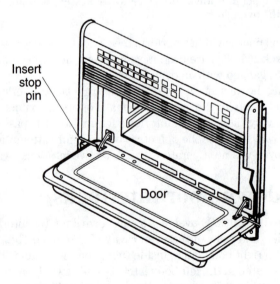

5-4 The interlocks are found behind the pull-down door in the GE oven. Courtesy General Electric Co.

These door interlock switches are activated by the latch heads on the door. When the door is opened, the switches interrupt the circuit to all components, except the oven lamp. A cook cycle cannot take place until the door is firmly closed thereby activating both interlock switches. The primary interlock system consists of the door sensing switch and primary interlock relay located on the control board circuit.

Interlock action

The interlock switches are often located with the primary and secondary circuits. You might locate them in both sides of the power line or in only one side going to the power transformer (figure 5-5). In early Sharp models, the interlock monitor and fuse are mounted on one replaceable assembly (figure 5-6). The interlock switches are so placed within the circuit that no matter what happens, the oven will not operate with the door open.

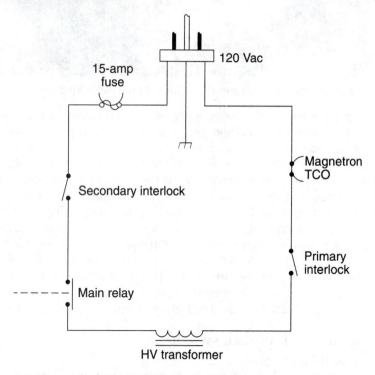

5-5 You might find interlocks on both sides of ac power line.

Even in a simple microwave oven, all the door interlock switches are activated when the door is closed. Either the switches are located right behind the door, or they can be tripped by a metal lever at the bottom of the door. In that case, the interlock switch assembly is located on one side of the oven cabinet. The primary and secondary switches are activated when the door is closed by the contacts of the monitor or safety switch opening up.

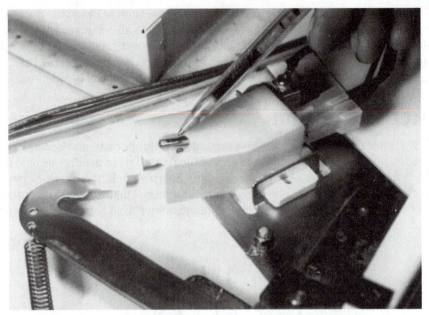

5-6 The front door arm activates this enclosed interlock switch.

A latch or interlock switch-hook assembly is mounted on the door—when closed, it trips the small microswitches. Some of these switchhook or trip assemblies are movable or flexible, while others are mounted stationary on the door. A warped or out-of-line door might not let the hook assembly strike the monitor switch assembly, resulting in no interlock switch action. So don't overlook the most obvious problems, such as a broken hook spring, misaligned hook assembly, and extended switch assembly.

You might be able to locate a defective oven operation by taking interlock switch testing procedures or checking the door alignment. Most ovens have an interlock switch testing procedure using an ohmmeter. If the service literature is not handy for a given oven, take ohmmeter tests across the suspected interlock switch. Remember that the interlock switch controls are closed when the door is closed, except for the monitor switch. Suspect a defective interlock switch when you have no reading across the switch terminals. The following is a typical oven interlock switch testing procedure.

Checking the left interlock switch

1. Remove the power plug.
2. Connect one lead of the ohmmeter to the blue lead at the bottom timer terminal.
3. Connect the other ohmmeter lead to the orange wire of the timer switch.
4. Firmly close the door. The meter hand should show a direct short. Replace the interlock switch if the reading is erratic or if there is no measurement.
5. Open the door slowly, and when the door is about 1/4 inch open, the meter hand should show an open circuit. If the meter hand does not show an open circuit, check for correct interlock switch adjustment.

Checking the right interlock switch

1. Remove the power cord and close the door.
2. Attach one lead of the ohmmeter to the terminal board on the bulkhead.
3. Attach the other meter lead to the white wire on the surge relay or the fuse.
4. Open and close the door. The meter should show continuity when the door is closed. The switch should open the circuit before the door exceeds $7/16$ of an inch.
5. If the switch doesn't close the circuit, or if it is erratic in operation, suspect an open or dirty interlock switch contact. Check the operation of all interlock switches after completing the service of the microwave oven.

Meter clip-on tests

It is much easier to clip the alligator meter probe to each interlock terminal for accurate switch continuity tests. Remove the slip-on connector for a good connection. If the wires are soldered directly into the interlock switch terminals, clip over the connections (figure 5-7). With this method, you can work the closing of the door and watch the ohmmeter. Leave slack in the probe wires so they will not pull off while opening and closing the door.

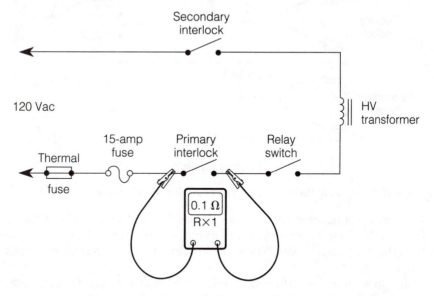

5-7 Remove the wires from the interlock and check continuity across the switch.

Getting down to where the interlock switch is located might be difficult. If several interlocks are on one whole assembly, remove the assembly. Then clip the meter leads to each interlock terminal and trip it with a small screwdriver. The interlock assembly might have to be readjusted when replaced.

GE interlock test circuit operation

Make sure the power cord is pulled before testing or checking the interlock switches. To test the interlock switches, connect a temporary jumper across the re-lay contacts, primary and secondary interlock switches to simulate shorted switch contacts. Locate the convenient connections in circuit to be certain COM and NO or NC terminals are used (figure 5-8).

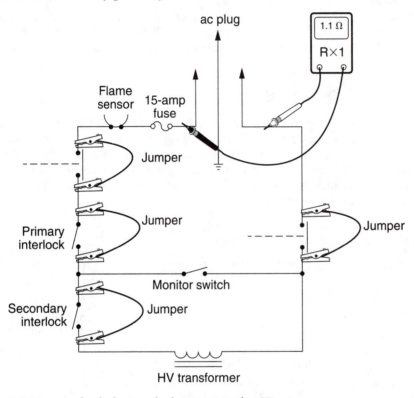

5-8 How to check the interlock system in the GE microwave oven.

Connect the ohmmeter low ohm scale across the two line terminals of the power cord. When the door is closed, you will measure some ohms (less than 2 ohms), but with the door open, 0 ohm. The zero ohm represents the resistance across the monitor switch.

Now remove the 15-amp fuse and the circuit should test open (infinite ohms). If not, check wiring of monitor and interlock circuits. In this oven, the main relay switch contacts are closed without any power to the oven.

Norelco model S7500 interlock switches

The latch (primary) protective switch is activated by the latch when it is closed. The switch is in an open position (when the door latch is open) and interrupts the current to the magnetron and probe circuit. Closing the door latch will close the latch (primary) switch, and normal operations can be continued.

The door (secondary) switch is located behind the front frame and is activated when the door is closed. When the door is open, it will act as a safeguard to interrupt the circuit if, for some reason, the latch (primary) switch should cease to function properly and remains closed when the door is open (figure 5-9).

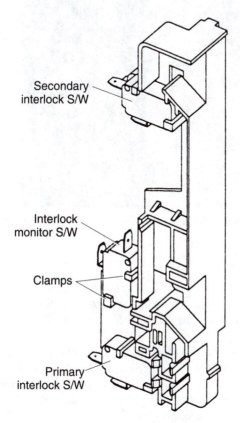

Secondary interlock S/W

Interlock monitor S/W

Clamps

Primary interlock S/W

5-9
The interlock assembly can include the primary and secondary interlocks and the monitor interlock switch.

The sense interlock switch disables the control panel if the door is open or if the switch is out of adjustment. This switch is ½ of the secondary switch. It is considered a low-level interlock, meaning that a full 115 V does not flow through it as it does in the regular secondary interlock.

The interlock sense switch is normally closed. If the lead is put on the normally open terminal, the control panel can be operated with the door open, but the magnetron will not power up under this condition.

The safety switch (which is normally closed) and the latch switch and door switch (both normally open) are activated by closing the door. As the door is closed, the plunger or safety switch is pushed and the switch is opened (figure 5-10). As the door continues to close, the pin on the door closes the door switch. When the latch is closed, the safety switch is closed and the circuit is complete.

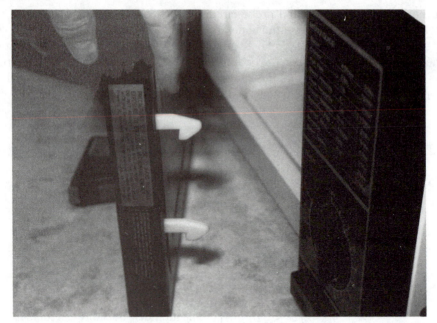

5-10 The door interlock switches can be activated with the plastic latch heads on the door.

If the latch switch and secondary switch were stuck in a closed position and you tried to open the door, the circuit would be shorted through the normally closed safety switch. This would blow the fuse. In some ovens, the safety switch is called the monitor or short switch.

If a switch assembly is replaced, it must be adjusted. If the oven fuse blows due to an oven fault or a circuit breaker trips (often from being reset one time with the house circuit unloaded except for the oven) the three switches must be replaced as an assembly. This is because the short circuit current will have damaged these important switches. Throw the defective switch assembly away.

Small oven interlock system

The small oven interlock system might have only a primary and secondary interlock switch. The primary interlock is at the top and the secondary is in series with timer switch and fuse (figure 5-11). When the timer is rotated to turn to the correct cooking time, the timer switch closes and provides ac voltage to the primary winding of power transformer.

The timer assembly also might control the blower motor directly from the timer circuits. A monitor switch will blow the fuse if either primary or secondary switches hang up or become defective. This prevents oven radiation when the door is opened and the oven continues to cook. The 15-amp fuse is blown by the monitor switch when the door opens and shuts down the oven circuits.

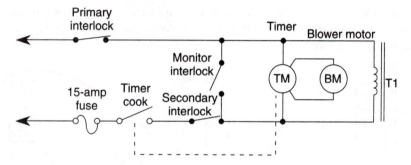

5-11 You might find only primary and interlock switches on a small microwave oven.

Replacing a defective interlock switch

After locating a defective interlock switch with the ohmmeter, remove and replace it. Although some of the latch or interlock switches are out in the open, you might encounter some that are difficult to remove. Besides removing components around the switch area, you might have to loosen up or pull out the front control assembly. Most manufacturers list the components that must be removed to get at the interlock switches. With no available service literature, check the area over carefully before removing any other components.

The defective interlock switch should be replaced with the original manufacturer's part number. When an interlock or latch switch is not readily available, you might be able to use another oven microswitch instead. Make sure the switch current and voltage is the same with proper normally closed or normally open operation. You might find the defective switch riveted to a metal switch assembly. Simply grind off the metal rivet. Mount the new switch with small bolts and nuts.

You might find that some interlock switches just clip into a plastic switch assembly or are held in place with one metal screw. These interlock switches are easily exchanged with other oven switches. Do not replace a smaller interlock switch with a large one. Remember, the switch must mount so the door latch will trip the small microswitch plunger. When available, always replace with original parts for safety and easy replacement. Realign the switch and door assembly.

Samsung MG5920T interlock switch measurements

When the interlock system is repaired or adjusted, a continuity check for each interlock and interlock monitor switches should be performed. Before checking the continuity, disconnect the two wire leads from the main relay mounted on the control circuit board, and short these two wires with a jumper wire. Make sure ac cord is pulled.

With an ohmmeter (low scale), check the continuity of primary interlock, between one of power cord blades (A) and high voltage transformer (B). With the door closed, a normal reading should be 0 ohm. Now open the door and the measurement should be infinite resistance (figure 5-12). Likewise, check the secondary interlock switch in the same manner.

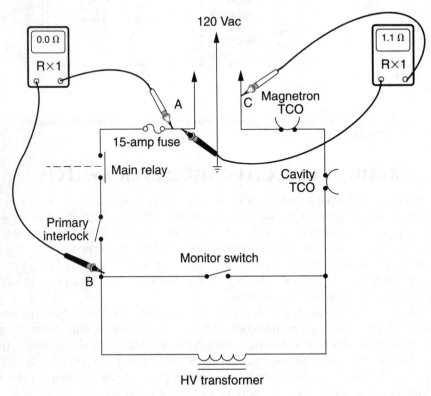

5-12 Check the continuity of a Samsung oven between A and B—and A and C with an ohmmeter.

Short out the primary and secondary interlock switches with a jumper wire and measure the continuity between the two power cord terminals A and C. With the door open the resistance should be zero ohm. This indicates the monitor switch has shorted contacts and is normal. After making the above continuity tests, remove the jumper wires from interlock and across main relay. Replace the two terminal leads to the main relay.

Samsung MW5820T interlock adjustments

To make adjustments of the primary interlock switch, door-sensing, and interlock monitor switch, loosen the screws securing the body latch switch and adjust the body position so that the interlock monitor switch opens before the primary interlock switch closes. The door-sensing switch will open when the door is closed tightly to the front cavity. Fasten the screws tightly.

Make sure the energy leakage is within the limit of the regulation (5 mW/cm^2) when measured by a detector. All service adjustments should be made for minimum RF emission readings.

When replacing faulty switches, make sure the clamps are not bent, broken, or deficient to secure the switches in place. Disconnect all wire leads from the primary interlock switch, interlock monitor switch, and secondary interlock switch.

Remove screws holding the body latch (figure 5-13). Release the clamps securing those switch bodies. After replacing the switches, make sure that the clamps hold the switches properly.

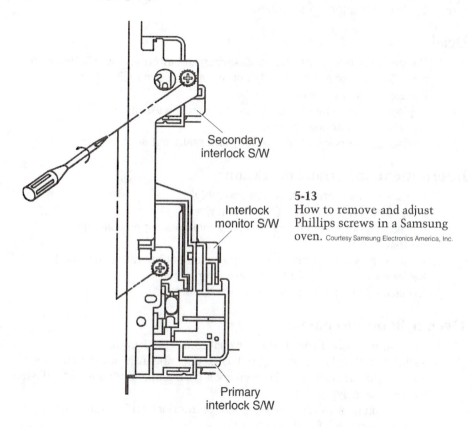

Secondary interlock S/W

Interlock monitor S/W

Primary interlock S/W

5-13
How to remove and adjust Phillips screws in a Samsung oven. Courtesy Samsung Electronics America, Inc.

Interlock switch replacement

Proper adjustment of the interlock switch assembly might also help to adjust the door for any play between the door and oven. After replacing the interlock switch, check the switch assembly for proper latch head alignment. The latch or interlock switch head must trip the microswitch with no play of the door. Usually the interlock switch assembly can be moved back and forth with metal screws. Pull the assembly back away from the door with hand pressure against the door. Tighten the screws of the interlock switch assembly. Check for play in the door by pulling on the door.

Especially check the bottom interlock door switch assembly. Do not press the door release button while making any adjustments. Make sure the latch or interlock key or hook assembly moves smoothly after each adjustment is completed, and make sure the mounting screws are tight. Sometimes a dab of cement or glue on the screw heads or nuts will keep the interlock switch assembly from loosening up. Double-check for smooth operation of the upper latch and bottom latch head when activated by one long bar assembly. Improper latch head action might be caused by a binding lever or missing spring of the latch head assembly. Follow the manufacturer's adjustment procedure. Here is a list of various microwave oven problems that were caused by defective interlock switches.

Dead

1. Replaced broken door latch head, spring, and interlock switch. Installed new 15-amp fuse and adjusted door and switch assembly.
2. Installed new latch switch assembly.
3. Replaced 15-amp fuse with new monitor switch assembly.
4. Defective top interlock switch.
5. Replaced interlock switch assembly and realigned door.

Intermittent and erratic operation

1. Replaced upper interlock switch assembly.
2. Door latch assembly loose (check pop rivets).
3. Realigned door, as latch head would sometimes not engage interlock microswitch.
4. Installed new door spring at the top (would not operate door switch).
5. Replaced latch switch and excessive door play.
6. Replaced defective interlock switch assembly.

Oven light on/no cooking

1. Dead when pushed cook button (replaced interlock switch).
2. Lights up but no fan or cooking (adjusted door and interlock switch assembly).
3. Have to pull up door to make oven cook (realignment of door and adjusted switch assembly).
4. Press in on door to cook (realignment of interlock and monitor switch assembly on side of oven cavity).

Panasonic safety switch system

While there might be a slight difference in construction, most microwave ovens have three safety switches, namely the primary, secondary, and safety switches. These switches are activated by the movement of the door. The primary and secondary interlock switches shut off the power supply to the appliance when the door starts to open. The safety switch monitors both switches and blows the line fuse to de-energize the appliance in case both interlock switches malfunction and do not turn the power off (figure 5-14).

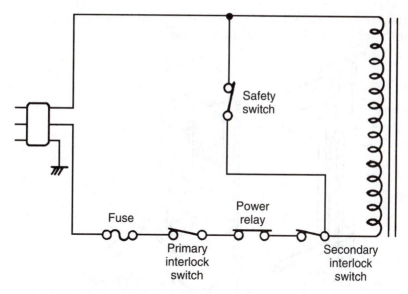

5-14 The safety switch is called a monitor switch in some ovens.

The basic function of the primary and secondary interlock switches is to interrupt the power supply to the magnetron, thereby stopping the microwave oscillation when the door starts to open during operation. The safety switch, which is wired to close the circuit when the door is opened, creates a short circuit when both or either interlock switches (depending upon the applicable regulation) fail to operate and cause the line fuse to blow and shut down the operation of the appliance. There is a time lag between the operation of the primary and secondary interlocks and the safety switch so that the line fuse remains intact during normal operations of both interlock switches.

Samsung MW2500U interlock switch replacement

To replace the primary interlock switch, door-sensing switch, and interlock monitor switch, disconnect all wire leads from each switch. Remove two (2) screws securing the body latch. Release the clamps holding the switch bodies (figure 5-15). Make sure that the clamps hold the switches properly after replacing switches. Make adjustments and check microwave emission tests.

The monitor switch

The monitor or safety switch is intended to render the oven inoperative by means of blowing the 15-amp fuse when the contacts of the interlock switch fail to open when the door is opened. In case the oven keeps on operating with the door open, the monitor switch blows the fuse, shutting the oven down to prevent radiation burns or injury to the operator. The monitor switch contacts are always open when the door is closed. Because the 15-amp fuse is in one side of the power line, the monitor switch automatically blows the fuse (figure 5-16).

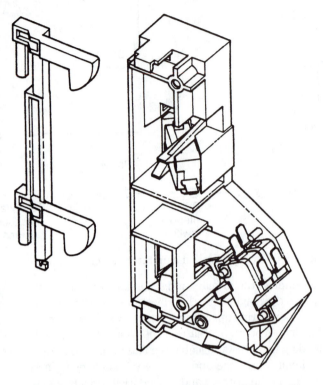

5-15 To replace the interlock plastic assembly in Samsung MW2500U, release the clamp holding the body of the switch.

Look at figure 5-17. If the contacts of interlock switch 1 and the contacts of latch switch 2 malfunction (or only interlock switch 1 remains closed) the 15-amp fuse will open due to the large current surge covered by the short or monitor switch. The magnetron tube stops oscillating (figure 5-18).

Always replace both latch switch 1 and the safety or monitor switch when the fuse keeps opening. Full power line voltage is placed across the contacts of each switch before the fuse opens. In some models, the whole monitor-interlock fuse assembly is replaced when the shorting condition occurs.

If the fuse blows after you open the front door, you can assume that the interlock switch contacts are closed or arced over and will not open up. Sometimes the fuse will open if there is too much play in the front door. This might let the interlock switch stay closed with the monitor or safety switch closing—thus knocking out the fuse. The interlock switch assembly should be replaced and properly adjusted so the door has no play between the door and cabinet.

The interlock switch arm assembly might be hanging up, keeping the interlock switch contacts closed and causing the fuse to blow. Sometimes the interlock switch plunger contacts are held down by liquid or food spilled down inside the switch area. Always replace both the interlock and monitor switch. Clip the test light and socket across the fuse holder so that you don't keep blowing fuses while you are trying to locate the defective safety switch. The light will be bright with a shorted interlock switch.

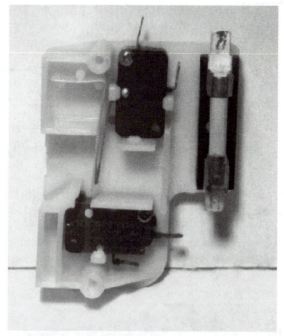

5-16 The monitor switch can be found with the primary interlock and fuse in the early ovens.

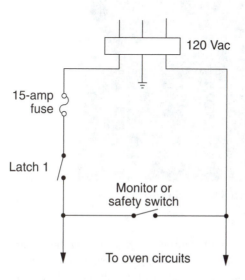

5-17
If the latch or primary interlock switch hangs up, the monitor switch blows the 15-amp fuse.

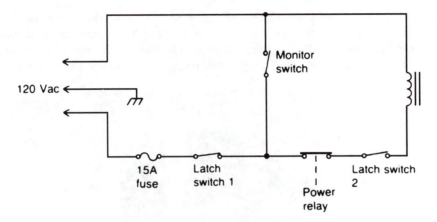

5-18 Replace both the monitor and safety interlocks if one is found defective.

The monitor or safety switches can be checked with the R×1 range of the ohmmeter. Simply place the meter leads or clips across the suspected safety switch (figure 5-19). When the door is closed, the switch contacts are normally open. If not, the safety switch is defective. Now, open the oven door. The switch contacts will close with the door open. Check the service literature for the location of the monitor or safety switch.

5-19 Check the interlock switches with R×1 scale of a DMM.

In early Sharp microwave ovens, a monitor interlock switch assembly is located in the left side of the oven (figure 5-20). A long door-lever bar triggers or operates the interlock assembly. The 15-amp fuse, interlock switch, and monitor switch are located on the same metal switch assembly. Simply adjust the whole monitor switch assembly by loosening two nuts. Usually, the oven cannot be turned on because the door lever will not trigger the interlock switch. Snug up the mounting bolts very tightly. Place a dab of glue over each nut so the monitor switch assembly will not loosen up.

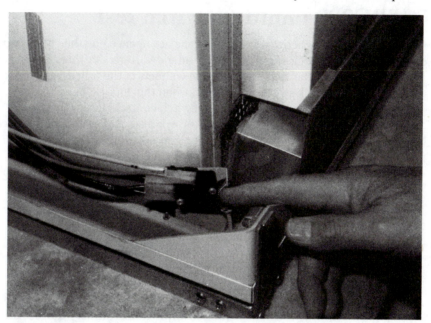

5-20 The door sensing switch is closed when the door is closed.

Some monitor or safety switches are in a fixed position, so no adjustment is needed; the secondary interlock monitor switch assembly of other ovens must be adjusted for correct operation by inserting this spacer material (0.004 inch or 0.008 inch) between switch plunger and stopper; this energizes the secondary interlock switch before the monitor switch is activated. If it is not activated, when the door is opened the secondary switch will not open before the monitor switch closes, thus opening up the 15-amp fuse.

Most safety switch adjustments are made by loosening the two mounting screws or metal screws holding the whole switch assembly. Slide the switch assembly until the lever trips the microswitch and then tighten down the mounting nuts. Follow the manufacturer's monitor switch adjustment procedure for correct oven operation.

Tappan primary interlock and monitor/door sense switch removal

Disconnect the power cord, remove the wrapper, and discharge the HV capacitor before checking and removing interlock switches. Remove the control panel to gain access to the primary interlock switch. Gently spread the mounting bracket tabs and left mounting bracket from the bezel. Remove the switch by spreading switch hold tabs and lift out switch.

To remove the monitor/door sense switch, remove the wires from the switch terminals. Just squeeze the tabs on both sides of the switch. Push the switch through the front frame area.

GE JEBC200 monitor switch test

The monitor switch is activated (the contacts are opened) by the latch head on the door while the door is closed. The latch is intended to render the oven inoperative by means of blowing the monitor fuse if the contacts of the primary interlock relay and secondary interlock switch fail to open when the door is opened.

1. When the door is opened, the monitor switch contact closes (the "on" condition). At this time the primary interlock relay and switching interlock switch are in the "off" condition (contacts open).

2. As the door goes to a closed position, the monitor switch contacts are first opened and then the door-sensing switch and the secondary interlock switch contacts close. (On opening the door, lack of these switches operates inversely.)

3. If the door is opened and the primary interlock relay and secondary interlock switch contacts fail to open, the monitor fuse blows simultaneously with the closing of the monitor switch contacts. Always check the primary interlock relay, door-sensing switch, monitor switch, and secondary interlock switch for proper operation when a monitor fuse is blown.

Disconnect the oven from the power supply. Before performing the monitor switch test, make sure the secondary interlock switch and primary interlock relay are operating properly. Disconnect the wire lead from the monitor switch (NC) terminal. Check the monitor switch operation by using an ohmmeter as follows:

When the door is open, the meter should indicate a closed circuit (figure 5-21). When the monitor switch actuator is pushed by a screwdriver through the lower latch hole on the front plate of the oven cavity, with the door opened (in this case the plunger of the monitor switch is pushed in), the meter should indicate an open circuit. If not, the switch might be defective. After testing the monitor switch, reconnect the wire lead to the monitor switch (NC) terminal.

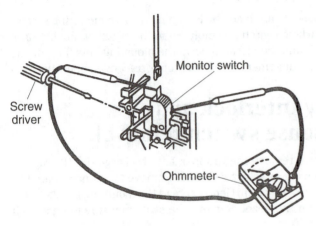

Monitor switch

Screw driver

Ohmmeter

5-21
Check the monitor switch with screwdriver and ohmmeter to determine if switch is defective.
Courtesy General Electric Co.

Samsung MW2500W door sensing adjustment

Make sure the power cord is pulled before trying to remove or adjust interlock switches. To adjust the door sensing switch, loosen lower screw securing the body latch switch and adjust switch body position so that the interlock monitor switch opens before the primary interlock switch closes. Close the door tightly (figure 5-22).

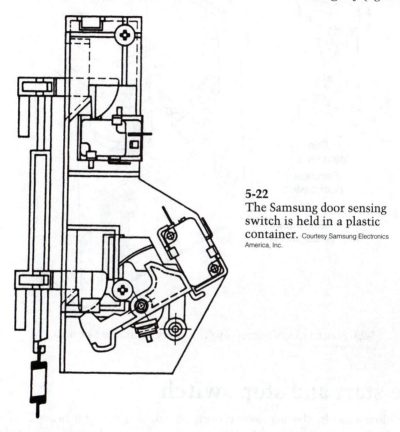

5-22
The Samsung door sensing switch is held in a plastic container. Courtesy Samsung Electronics America, Inc.

Now secure or fasten screws tightly. Take an energy leakage test and so the reading does not exceed the limit of 5 mW/cm^2, when measured by a calibrated and certified leakage tester. All service adjustments should be made with a minimum RF emission reading.

Norelco S7500 safety switch adjustment

Loosen the two screws that secure the switch on top until the assembly can be moved freely back and forth (figure 5-23). Close the door and secure the latch. Remove the wire from one side of the switch and place the leads of the ohmmeter across the terminals of the switch. Move the switch toward the door until it is lightly stopped (switch is normally closed). Check by opening and closing the door. As the door is opened, the meter should show an open circuit. Tighten and secure the two switch screws and replace the wire.

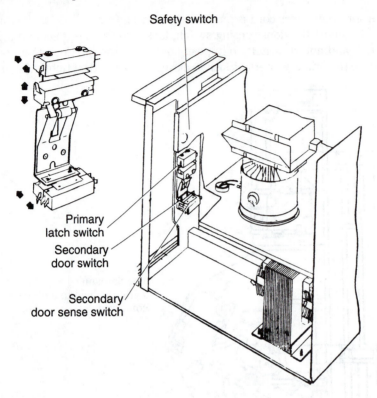

5-23 Loosen the two screws so the safety switch can be adjusted.
Courtesy Norelco Corp.

The start and stop switch

In older models, the microwave oven was turned on by a heavy-duty off/on switch. Because the switch supplied the entire oven current load, these switches had a tendency to arc internally, and after several years they had to be replaced. They would arc and sputter sometimes, making poor contact, and then again the oven was inoperative. These switches must be replaced with the original part number.

You might find a separate start and stop switch in the controller assembly (figure 5-24). The start switch is pushed to start up the control assembly, while the stop switch is a closed switch. The start switch is normally open until punched. When the stop button is pushed, the switch disengages the control assembly from the circuit.

Both switches can be checked with the R×1 scale of the ohmmeter. Clip the meter leads across the switch terminals. Always pull the power plug and discharge the HV capacitor when taking ohmmeter readings. Push in on the start switch. The meter should read a direct short or less than 1 ohm with the start button pushed in. Do likewise with the stop switch. A directly shorted reading should be obtained with the meter leads clipped across the switch terminals. Now, push in on the stop switch. The meter should show an open circuit.

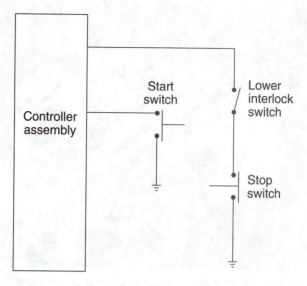

5-24 A separate start and stop switch can be found in early ovens.

The oven or cook switch is normally open and when pushed will make contact energizing the oven relay or triac assembly, applying line voltage to the power transformer (figure 5-25). In some lower-priced or commercial ovens, the cook switch might be in series with one side of the power line, connecting voltage directly to the power transformer (figure 5-26). You might find another rotary type switch that provides voltage to the defrost circuits or cooking modes. Actually, the defrost switch contacts are in parallel with the cook switch—except when the oven is set to the defrost position and the switch is controlled off and on by the defrost timer assembly.

Like all switches, the oven or cook switch can be checked with the ohmmeter. Remember: the cook switch that supplies voltage to the oven relay only makes contact while pushed in. When you let up on the cook switch, the oven or cook relay contacts take over and are in parallel with the cook switch contacts. Suspect defective oven relay contacts if the oven will only operate by holding in on the cook button (figure 5-27). Either a wire is broken off or the relay contacts are defective.

Norelco S7500 on/off switch tests

Remove the on/off switch and wires and tag them for identification. Set the multimeter on R×1 ohm scale and the meter on 0. Place one test lead on terminal 7 and the other on terminal 8. Depress the ON push button. The meter should defect to the 0 mark. With the leads on the same terminals, depress the OFF button. The meter should indicate an infinite amount of resistance. Remove the leads and place one on terminal 3 and the other on terminal 16 (figure 5-28).

5-25 The oven or cook switch is normally open—it closes when pushed.

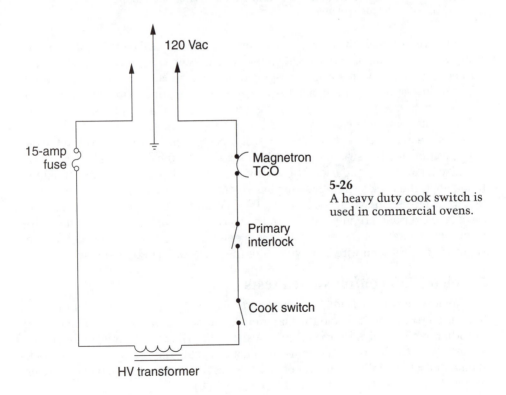

5-26
A heavy duty cook switch is
used in commercial ovens.

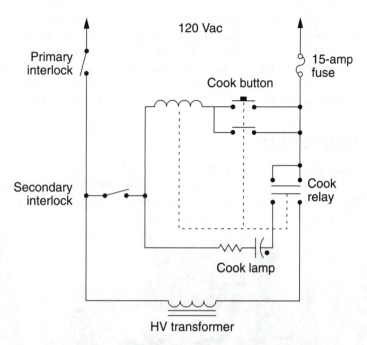

5-27 Suspect defective oven relay contacts if the oven will stay on with cook switch held in the on position.

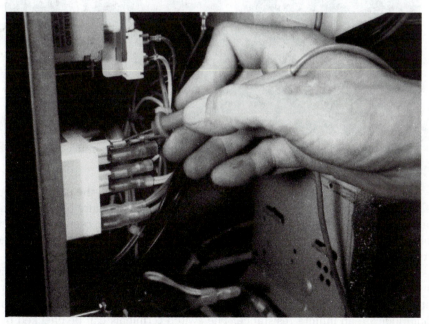

5-28 Check the continuity of the oven switch with the low ohm scale of DMM.

Depress the ON push button and watch the multimeter. When the button is pushed, the meter will go to 0. Push the OFF button and the meter should indicate an open circuit. Place the leads on terminal 1 and 17. Depress the ON push button and watch the multimeter or DMM. When the button is pushed, the meter will go to 0. Push the OFF button. The meter should read an open circuit. If the meter hand changes and appears erratic, the switch terminals might be dirty, burned, or pitted.

The oven relay

Line voltage is applied to the primary winding of the HV transformer by the cook switch, timer controlling switch, oven relay, or triac assembly (figure 5-29). Some ovens might call the relay a power relay, as power is applied to the power transformer. Often, you can hear the oven relay energize.

5-29 All interlocks, cook switch, relays, timer, and sensors must operate before power is applied to the HV transformer.

The cook or power relay can be checked with a VOM. With the power off, locate the solenoid coil winding terminal and check for continuity (figure 5-30). Most oven relays have a resistance between 150 and 300 ohms. If in doubt, remove the external cable leads from the solenoid terminals. Then take another resistance measurement. Of course, if the coil is open, the relay must be replaced.

A continuity check between switch contact terminals should indicate an infinite resistance. Only when the coil is energized should you measure a short across the switch terminals. A continuity check between each switch terminal and ground should indicate an infinite resistance. The switch terminals can be checked by removing the top cover and using an insulating tool to push the solenoid together.

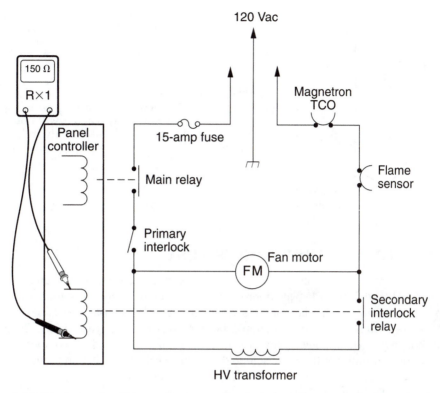

5-30 Discharge the HV capacitor. Measure the solenoid resistance of suspected relays with an ohmmeter (150-650 ohms).

In case voltage is not reaching the solenoid terminals, the oven relay will not operate. Some oven relays operate from the ac power line, while those with a control board assembly operate from 10 to 24 Vdc. The voltage can be checked at the solenoid terminals when the oven will not turn on (figure 5-31). Always clip the meter leads to these terminals. First, determine if the voltage is ac or dc. Low or no voltage at the relay indicates a defective control board circuit. Correct voltage at the solenoid terminal indicates an open or defective relay solenoid.

When everything else has been checked and the relay will not energize, remove the solenoid terminal wires. Mark where each wire goes before removing them. Now apply 14 to 20 Vdc from a bench power supply to the low-voltage relay. A normal relay will operate the cooking cycle. With a suspected open solenoid winding, the contacts can be checked by pushing down on the contact point assembly with an insulated tool with the oven in operation. Use this method as a last resort, being careful not to touch any other components.

The relay contact voltage can be measured at the primary winding of the HV power transformer (120 Vac). Clip a VOM across the primary leads of the transformer to monitor the applied voltage. When the oven relay operates normally, the line voltage is measured with the voltmeter. Besides a defective oven relay, you might encounter a leaky triac assembly.

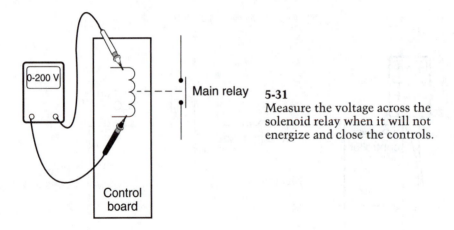

5-31
Measure the voltage across the solenoid relay when it will not energize and close the controls.

Samsung power relays test

The relays found within the Samsung MW5820R oven are mounted on the control panel board. When the oven door is open, the main relay contacts are closed and the secondary interlock relay contacts are open. Check the resistance of relay points when the switch is closed and should be zero resistance (figure 5-32). Suspect burned points when a few ohms are measured across the switch contacts. Now check the continuity of the relay solenoid winding with the R×10 ohm range.

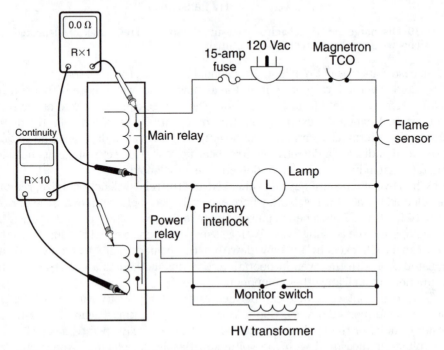

5-32 Check the power relays with ohmmeter test on the switching contacts and the continuity of relay solenoid winding.

Remove the relay wire leads when making resistance measurements. Make sure the power cord is pulled when taking resistance or continuity measurements. The relays can be tested by operating the oven with a water load in the oven and power level at high selection.

Relay contact layout

The oven relay might have two or three different circuits, switched in or out, and a solenoid winding. The oven relay winding might be controlled by the ac power line voltage or the dc voltage from the control circuits (figure 5-33). The contacts and winding spade terminals might come out at the top of the relay. Locate the relay winding by visual means or with the ohmmeter. The resistance across the oven relay winding might be from 150 to 650 ohms. Check each set of switching contacts with the ohmmeter across them by manually depressing the oven switch to locate poor or open contacts.

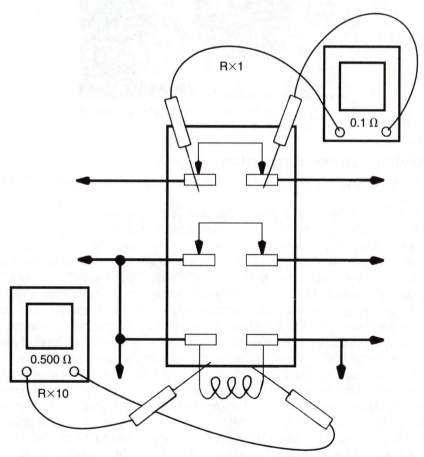

5-33 The approximate measurements of the relay contacts and solenoid. Check the manufacturer's literature for solenoid resistance.

In some ovens you might encounter another auxiliary relay (figure 5-34). Here, the auxiliary relay turns on the stirrer and blower motor. The relay is controlled from the electronic controller assembly. Notice that the resistance of this particular relay has a total of 1.3 kilohms. All relays should be replaced with the original part number.

5-34 You might find an auxiliary relay to control stirrer, fan blower or heater element.

Panasonic's power relay control

There are basically two types of magnetron duty control systems available in the field:

1. To control the on/off time of magnetron in the primary circuits.
2. To control the on/off time of magnetron in the secondary circuits.

The primary duty cycle circuit controls the duty cycle of the magnetron in the primary circuit (figure 5-35). The primary power supply to the HV transformer is interrupted intermittently by means of a control circuitry and a switching device. The switching device might be either a power relay or triac. The control circuitry controls the interrupting periods of the magnetron within a specified duty cycle and varies the average power output of the microwave oven. Because this system controls operations in the primary circuit, the switching device requires a lower level of insulation. However, a higher stress to the components in the secondary circuit would result due to transient currents, which could possibly shorten the component's life.

The secondary duty control circuit controls the duty cycle of the magnetron in the secondary circuit (figure 5-36). The high voltage to the magnetron is interrupted intermittently by means of an associated control circuitry and a switching device. The control system of the magnetron duty cycle is equivalent to the primary duty cycle control circuit, except that the switching device is in the secondary circuit.

The advantage of this system is that it results in less stress to the components in the HV circuit. It switches the circuit when no high voltage is being applied, so no transient currents are generated. The switching device, however, requires a higher grade of insulation because the device is located in the HV circuit.

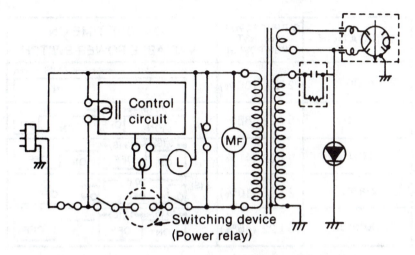

5-35 The control unit powers the switching relay of ac power line.
Courtesy Matsushita Electric Corp. of America

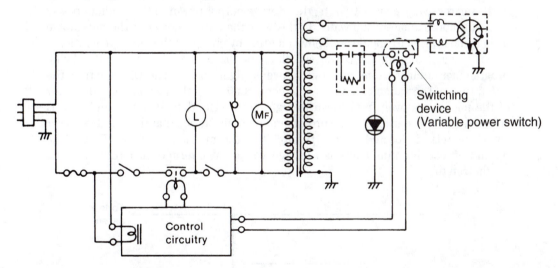

5-36 The secondary control circuit powers the variable power switch applying HV to the magnetron. Courtesy Matsushita Electric Corp. of America

When the power selection is set at any power setting other than high power, the variable power switch (switching device) is energized intermittently by signals from the variable power control circuit. The variable power control circuit controls the on/off time of the variable power switch contacts within a specified duty cycle. With most Panasonic ovens, one complete duty cycle is approximately 22 seconds (Table 5-1).

**Table 5-1. The medium indication of the Panasonic power operation
is on 15 seconds and off 5 seconds with a 68 percent output power**

Courtesy Matsushita Electric Corp. of America

INDICATION	OUTPUT POWER AGAINST HIGH POWER	ON. OFF TIME ON VARIABLE POWER SWITCH			
High	22 / 22 (100%)	⊢— 22 (S) —⊣⊢— 22 (S) —⊣ ON / ON			
Med	15 / 22 (68%)	⊢—15 (S)—⊣7 (S)⊢ ON / OFF / ON / OFF			
Low	9 / 22 (41%)	⊢9 (S)⊣⊢13 (S)⊣ ON / OFF / ON / OFF			
Warm	2 / 22 (10%)	2 (S) ⊢—— 20 (S) ——⊣ ON / OFF / ON / OFF			
Defrost	10 / 22 (45%)	⊢10 (S)⊣⊢12 (S)⊣ ON / OFF / ON / OFF			

Assuming that the full power of the microwave oven is 700 watts, the average output power at medium power setting is approximately 477 watts. If the full power of the microwave oven is 600 watts, the average output power of the medium power setting is approximately 409 watts. In this way, the output power of the microwave oven can be varied to any desired power setting by changing the duty cycle ratio.

The coil of the variable power switch is energized intermittently by the digital programmer circuit in some microwave ovens (figure 5-37). The DPC controls the on/off time of the variable power switch contacts to vary the average output power of the microwave oven from warm to high power. The defrost power and time is selected with the start pad of the programmer circuit. Notice that the variable power switch or relay is located in the HV circuit of the magnetron (figure 5-38). The on and off cooking time controls the high voltage at intermittent intervals to the magnetron tube.

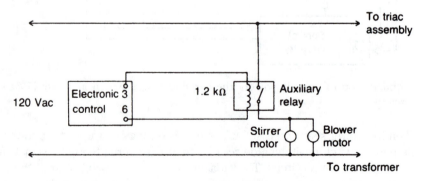

5-37 The auxiliary relay is controlled by the panel control circuit to turn on stirrer and blower motors.

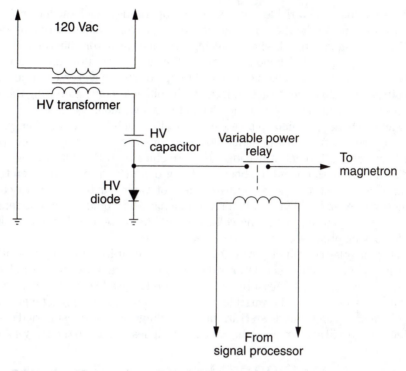

5-38 A variable power relay is found in the HV circuits.

No variable power switch/Panasonic oven

In the Panasonic variable control circuits, the variable power switching is done within the high-voltage side, between HV diode and magnetron tube. The variable power relay is controlled by the controller unit. The variable relay switches off and on the high voltage to the magnetron tube (figure 5-39).

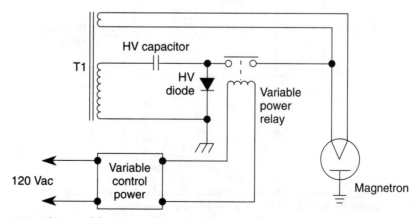

5-39 The variable power relay switches HV off and on in a Panasonic oven.
Courtesy Matsushita Electric Corp. of America

To determine if the variable power system is operating, check the high voltage at the anode terminal of the HV diode and at the magnetron filament terminal. Before attaching the Magnameter, discharge the high-voltage capacitor. Clip the red lead of meter to the top side of HV diode where the high-voltage capacitor connects and clip the black lead (gnd) to the chassis. You might clip both test leads on each side of the high-voltage diode, if easier to get at. If the high voltage is constant during the cooking process, proceed to the test at the filament or heater side.

Again discharge the high-voltage capacitor after each test. Remember, you are making hot connections in the high-voltage circuits. As usual, the variable power contacts are difficult to get at because they are shielded with a plastic cover. With the red lead into position and the black lead at ground (G), fire up the oven to provide variable cooking. Make sure a cup or beaker of water is in the oven cavity before making these tests. No high voltage on the magnameter could indicate a defective control unit or open relay coil. The voltage should be there and then cut off in the variable-cooking process.

Check the relay coil winding with the ohmmeter set at R×1. Be very careful here. Make sure the high-voltage capacitor is discharged and the power plug is pulled before taking any resistance measurements. If the relay has low continuity, the controller must be defective. The variable relay should energize (applying high voltage to the magnetron) and shut down (turning high voltage off in the rest period). Again, the high voltage will appear when the controller applies voltage to the relay solenoid.

Samsung MG5920T browner relay tests

The browner relay provides continuity between the 1000-watt heating unit and heater sensor. The heater operates from the 120-volt power line and should have a resistance under 50 ohms. The heater sensor cuts out the ac voltage to the heater terminal when too much heat is found in the oven (figure 5-40).

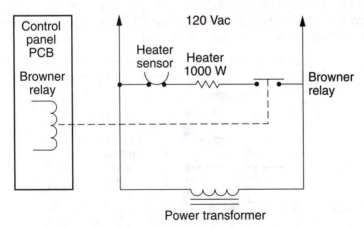

5-40 The panel control circuit activates the browner relay in a Samsung MG5920T oven. Courtesy Samsung Electronics America, Inc.

Check the browner or heater relay points when convection unit does not heat up. No resistance should be found here. Erratic heater operation can be caused by burned or pitted heater relay contacts. Check the continuity of the browner relay solenoid when relay will not energize.

Samsung MG5920T relay tests

All three relays (power, main, and browner) are located on the control circuit board. Isolate the relays from the main circuit by disconnecting the leads. Operate the microwave oven with a water load in the oven and power level at high selection. Check continuity between terminals of relays with the digital multimeter (DMM). If there is no continuity, replace the PCB assembly (figure 5-41).

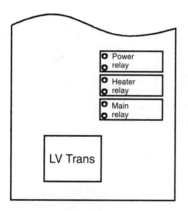

5-41
The power, heater and main relays of the Samsung MG5920T oven are mounted on the PCB.
Courtesy Samsung Electronics America, Inc.

Pattern side of relays

Tappan 56-4851 browner and turntable relays

The browner relay switches the browner element off and on at the preprogrammed times by applying and removing power to the browner relay, which is mounted to the plastic control panel housing. The turntable relay switches the turntable off and on each time the turntable pad is touched (during any kind of cooking operation) by sending a dc signal to the turntable relay coil through terminals L4 and L5.

Using a straight-blade screwdriver or similar tool, gently pry against the plastic snap that secures the relay, while pulling outward on the relay, to remove relay assembly. Do not flex the snap more than necessary to release the relay. When the relay is free of the snap, slide the relay from under the fixed tab (figure 5-42). Reverse the procedure to install a new relay.

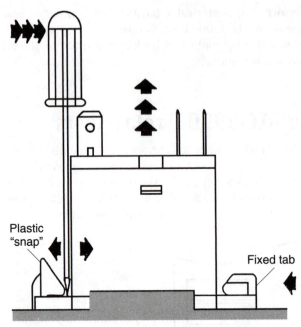

5-42 Release the plastic snap with a screwdriver to remove suspected relay. Courtesy Tappan, brand of Frigidaire Co.

Timer controls

You will find one or two timer control assemblies in the manual or commercial microwave ovens. The top timer might have a scale from 0 to 5 minutes, while the second timer has a longer cooking time. Set the timer dial for the desired cooking time,
and the blower motor will rotate. The oven light will come on at once (figure 5-43).
Actually, when the correct cooking time is set, the timer switch contacts close in the
timer assembly. In this particular circuit, notice that the timer will not start to operate until the cook switch is pushed, as the timer motor connects into the ac circuit after the oven or cook switch.

The timer switch contacts can be checked with the ohmmeter. Pull the power
plug and discharge the HV capacitor. Clip the meter leads across the timer switch
terminals. Now rotate or set the timer for three or four minutes. The meter hand
should indicate short or continuity if the switch is normal. An erratic ohmmeter
reading might indicate a dirty switch contact. Clean or spray the switch with cleaning fluid. If the switch remains intermittent, replace the timer assembly.

A defective timer might become intermittent in operation or not shut off. If the
timer motor fails to run, check for line voltage (120 Vac) across the motor terminals.
The timer motor should rotate when the cook button is pushed. In case the motor
does not operate with the correct ac voltage applied to the timer motor terminals,
suspect jammed gears or an open motor coil. Remove the one terminal from the motor coil and check for continuity with the ohmmeter. Replace the timer assembly if
the motor is open or is erratic in operation.

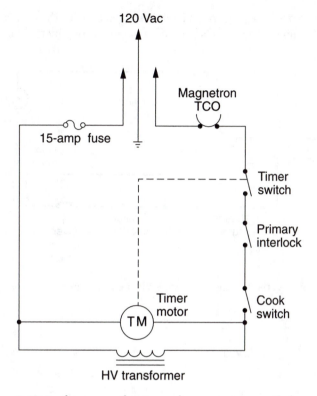

5-43 In this oven, the timer does not start until the cook switch is pushed.

Oven lamp switch

Although most oven bulbs are turned on when the door is opened, you might find a separate lamp switch (figure 5-44). In this particular circuit, the lamp is turned on when the door is opened and when the power relay is energized. All of this occurs with a small microswitch ganged with the latch and short switch.

You can find one or two oven lamps in some ovens. Try replacing the lamp bulb before tearing into the oven. If the oven lamp does not come on, check for a defective lamp switch. Suspect a defective lamp socket or connecting wire after replacing the bulb and checking the lamp switch. You should measure the entire line voltage across the lamp socket terminals (120 Vac).

The same oven lamp might serve as a cook lamp in figure 5-44, or a separate neon indicator might be found (figure 5-45). Here the neon bulb and voltage dropping resistor are included in one component mounted on the front panel. When the cook lamp comes on, you know line voltage is applied to the primary winding of the HV transformer.

The oven lamp is located in the light switch circuits or after the oven relay is energized. In some ovens, when the light switch is turned on, the oven lamp is on.

When the oven is operating, the oven lamp is turned on with the control relay. In the Norelco RR7000 model, the oven lamp is controlled entirely by the control panel circuit (figure 5-46).

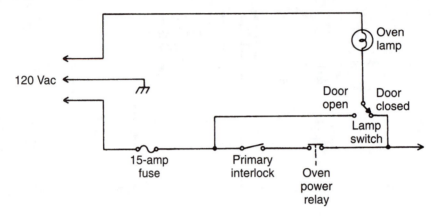

5-44 In some ovens, the light can be switched on when the door is opened or closed.

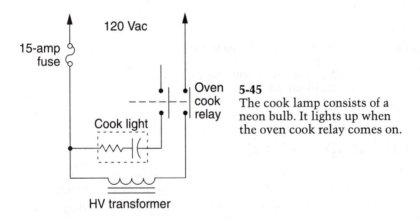

5-45
The cook lamp consists of a neon bulb. It lights up when the oven cook relay comes on.

Tappan oven light replacement

Disconnect power and remove the wrapper and lamp access cover screw (figure 5-47). Turn the lamp counterclockwise and remove the lamp from the socket. Replace with a 40-watt candelabra-base 125-volt bulb.

Tappan interlock switch removal

If the oven is rendered inoperative due to failure of one of the interlock switches, replace the interlock switches and monitor switch. After the interlock switches have been replaced, check the monitor switch for continuity.

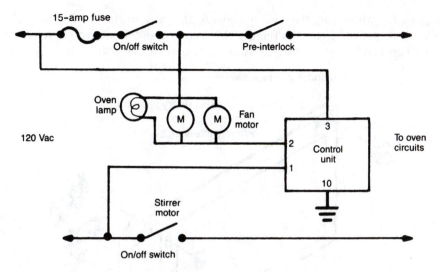

5-46 The oven light is controlled by control panel in a Norelco oven. Courtesy Norelco Corp.

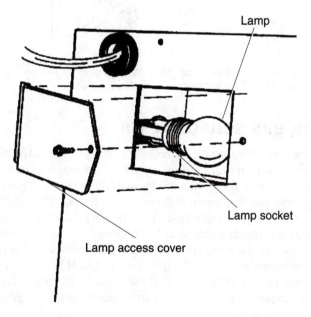

5-47 Replace the oven lamp after removing lamp access cover. Courtesy Tappan, brand of Frigidaire Co.

To remove the primary interlock switch, disconnect power, remove the wrapper, then discharge the capacitor. Remove the control panel. Gently spread mounting bracket tabs and lift mounting bracket from the bezel. Spread the switch-holding tabs and lift out the switch (figure 5-48). Remove the monitor/door sense switch by removing the wires from the switch; squeeze the tabs on the switch and pull the switch through front frame. To remove the secondary interlock switch, remove the control panel and bezel. Spread the switch tabs and lift the switch out.

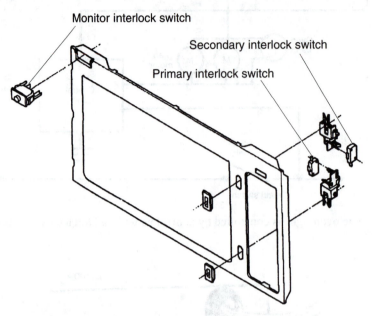

Monitor interlock switch

Secondary interlock switch

Primary interlock switch

5-48 Remove the mounting tabs to check or replace interlock switches in the Tappan oven. Courtesy Tappan, brand of Frigidaire Co.

Samsung gas sensor tests

The gas sensor in the Samsung MW5820T oven is located at the rear of the oven at rear vent. The gas sensor plugs into an assembly harness that connects to the control panel circuits. To give the oven a gas sensor test, touch and hold the number 1 and 2 pads. Observe the diagnostic number in the display. The number 5 to 213 is normal. If the sensor is open, unplugged, the wiring is poor, or the smart board is defective, the diagnostic number will be 214 or higher.

Disconnect the sensor wires from the gas sensor holder and connector to test the gas sensor. Measure the continuity between H1 and H2 of the sensor. Now measure the continuity between the A and B of the sensor. Both normal readings should be infinite ohms. Replace gas sensor with an ohmmeter measurement between the above terminals.

Thermal cutout switches

The microwave oven, convection oven, and magnetron thermal units are actually heat-related switches. When the oven reaches a certain temperature, the thermal switch cuts out and shuts down the oven voltage to the primary winding of the power transformer. If the blower motor cannot produce enough cold air to protect the magnetron, the thermal unit cuts off the power until the magnetron cools down. These are heated-related control devices to protect components from overheating.

The oven thermal unit can be found in the oven relay circuit; the unit opens up when the oven becomes overheated. The oven cavity thermal unit is found in the low-voltage circuits in series with the 15-amp fuse. The convection oven thermal switch is usually found in series with the heating element. A separate thermal fuse heating element can be controlled by the digital programmer (figure 5-49). Often, these thermal units vary in temperature from 150 to 300°F, depending on the manufacturer. The open and close temperature might be within 30°F.

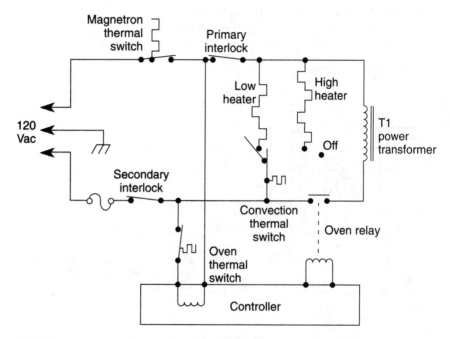

5-49 The convection heater is controlled by the panel control unit.

The magnetron thermal switch is located directly on the magnetron tube. When the magnetron becomes too hot, the thermal switch opens and must cool down before it switches on. The magnetron thermal switch might be located in the low-voltage circuits. These magnetron thermal units' cutoff temperatures vary from 80 to 300°F, depending on the wattage of the microwave oven.

GE JEBC200 interlock switch adjustment

The switches are attached to two plastic latch bodies that are mounted in each side of the cavity. The primary switch is located in the right body latch, inside switch on the bottom. A secondary interlock is located in the left body latch system, outside switch on the bottom. The monitor switch is found in the right body latch at the top. A door sense switch is located in the right body latch, outside switch on the bottom and in the left body latch, inside switch on the bottom.

Disconnect power and pull the oven out to remove the top front panel. Remove the side access cover to the switches. Loosen the latch switch bracket through the holes in the side unit. Adjust each housing for proper switch operation and door fit, and retighten screws. After switch adjustments, check for leakage.

Left bracket switch replacement

1. Disconnect power and pull oven out to remove top front panel.
2. Remove front top cover.
3. Discharge high-voltage capacitor.
4. Remove two bracket screws (do not drop).
5. Open door and work bracket up through the top.
6. Switches are held in by plastic mounting tabs. (See figure 5-50.)

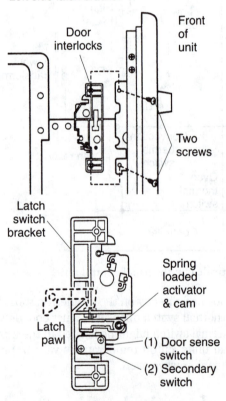

5-50
How to adjust and replace the door interlock assembly in a GE oven. Courtesy General Electric Co.

Right bracket switch removal

1. Follow steps 1 and 2 for the left bracket.
2. Remove from the wall.
3. Remove outer panel.
4. Remove two bracket screws and bracket.
5. Switches are held in by plastic tabs; make sure they are not bent or broken. (See figure 5-51.)

Right side latch bracket

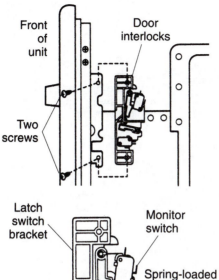

5-51
How to remove the right
bracket switch in a GE oven.
Courtesy General Electric Co.

GE JEBC200 interlock switch tests

Disconnect power and pull the oven out from the wall to remove the top front access panel and interlock access panels; discharge the high-voltage capacitor. The conditions of a small interlock switch are given in Table 5-2. (Also see figures 5-52 and 5-53.)

Table 5-2. The conditions of a small interlock switch

Primary interlock	Check continuity of COM and NO Door closed—0 ohms Door open—infinite ohms
Secondary interlock	Check continuity of COM and NO Door closed—0 ohms Door open—infinite ohms
Monitor switch	Check continuity of COM and NC Door closed—infinite ohms Door open— 0 ohms
Door sense switch	Check continuity of COM and NC Door closed—infinite ohms Door open—0 ohms

Right latch switch bracket

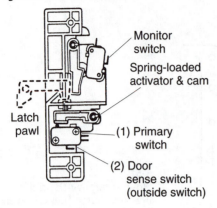

Monitor switch

Spring-loaded activator & cam

Latch pawl

(1) Primary switch

(2) Door sense switch (outside switch)

5-52
Testing the right primary interlock switch in a GE oven.
Courtesy General Electric Co.

Left latch switch bracket

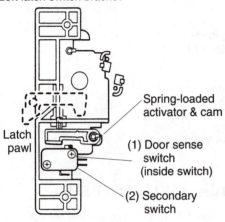

Spring-loaded activator & cam

Latch pawl

(1) Door sense switch (inside switch)

(2) Secondary switch

5-53
Locating and replacing the door sense switch in a GE oven.
Courtesy General Electric Co.

Test circuit operation

Connect the temporary jumper across the relay contacts and primary and secondary interlock switches to simulate shorted switch contacts. Locate a convenient connection in the circuit to be certain COM and NO, or NC terminals are used.

GE JEBC200 popcorn key test

The popcorn feature is a programmed function that uses a humidity sensor to automatically select the correct cooking time. Do not use the metal shelf with the popcorn pad. This feature works best when the popcorn contents are in the 3.0–3.7-ounce range (figure 5-54).

5-54 The popcorn key test in a GE JEBC200 oven. Courtesy General Electric Co.

1. Remove the outer wrapper from the microwave popcorn.
2. Open the oven door and place the package of popcorn in the center of the oven floor. Close the oven door.
3. Touch the popcorn pad. *POP* flashes.
4. Touch start. The popcorn sensor automatically calculates the cooking time. After the popcorn sensor detects steam, the oven sounds and displays the remaining cooking time needed.
5. When cooking is complete, the oven signals and flashes *END*. Open the door and remove the popcorn.
6. When popcorn is undercooked, you can make an adjustment by adding time. Touch the popcorn pad and then touch number 9. The word *POP* will appear on the display with a plus (+) sign of 20 seconds more cooking time.
7. When popcorn is overcooked, make an adjustment by subtracting time. Touch the popcorn pad and then touch number pad 1. The word *POP* will appear with a minus sign (-), indicating 20 seconds less.
8. Connect the ohmmeter (low-ohm scale) across the terminals of the power cord. Continuity must show: Door closed-Some ohms, Door open-No ohms.
9. Remove the 15-amp fuse; the circuit must be open (infinite ohms). If not, check the wiring of the monitor and interlock circuits. Remember to remove the temporary jumpers before starting up the oven.

6
CHAPTER

Servicing low-voltage circuits

All low-voltage problems with microwave ovens are related to the power line voltage. When low voltage or no voltage is found at the primary winding of the large power transformer, check for low voltage problems (figure 6-1).

6-1 The low-voltage components consists of a fuse, triac, relay, sensor, transformer, and interlock switches.

In other words, a defective component is preventing ac voltage from being applied to the primary winding of the power transformer. The defective component could be located on either side of the power line. Any part from the power cord to the transformer winding could be the defective component (figure 6-2).

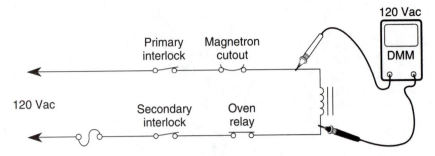

6-2 Check the power line voltage at power transformer to determine if the low-voltage circuit is normal.

Pictorial ac circuits

The pictorial schematic is somewhat like the wiring diagram, except that it is a more detailed drawing. This type of schematic makes it easy to see how the parts are tied together in the oven and examine what they look like (figure 6-3). Even though some of the new oven parts have changed in appearance, they work in the same type of oven circuit. Except for the power transformer and magnetron, you will find that the latest oven components are smaller in size.

Low-voltage procedures

To determine what component might be defective, use existing lights and the fan motor as indicators. You will find many lamps and motors throughout the low-voltage circuits. No light or fan motor rotation might indicate an open fuse, a defective power cord, or a defective switch. If the oven lamp is on, the fuse is okay.

When the oven lamp is on, you can tell that the power line voltage is normal to the oven relay. This can also indicate that all the interlock switches, thermal cutout, and timer circuits are normal. If the oven relay does not come on, it could be due to a defective cook switch, cook relay, or poor contacts on the oven relay.

In ovens with digital programmer control (DPC) circuits, a defective control board, thermal cavity fuse, thermal protection, triac, or power relay can prevent line voltage from being applied to the primary winding of the power transformer. First, you should check all fuses and triacs with the ohmmeter tests. Next, check the power relay components and circuits before replacing or checking out the control board. Information on servicing the control board and replacement data is given in Chapter 10.

A normally lighted cook lamp indicates that power line voltage is being applied past the oven relay assembly (figure 6-4). Remember that in some ovens, the control programmer turns on the oven cavity light. Check the defrost lamp and defrost motor for a low-voltage indicator in manually operated ovens. The turntable motor rotation

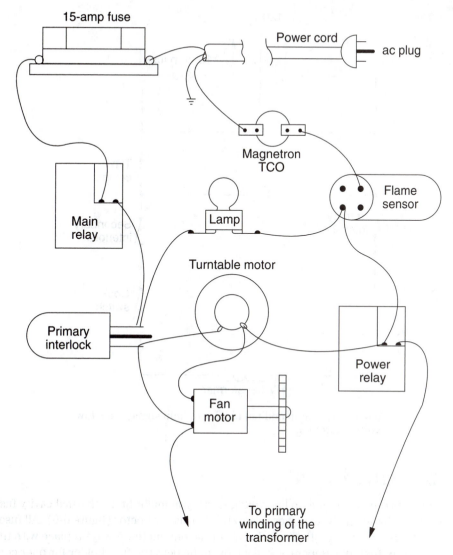

6-3 A pictorial diagram of the microwave oven.

of a Sharp oven is another component that indicates whether the power voltage is being applied up to the primary winding of the power transformer.

Usually the oven's fan blower will come on, indicating that low-voltage is being applied to the transformer. In a Quasar oven, the rotation of the fan blower indicates that voltage is applied to the power transformer circuits (figure 6-5). By simply checking the light bulbs and various motor operations, you could quickly isolate the defective component in the low-voltage circuits.

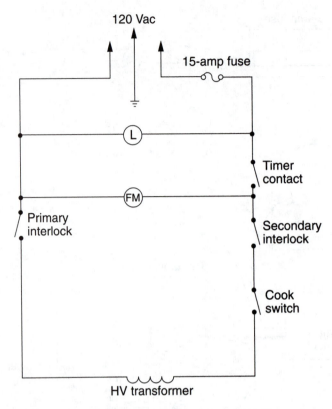

6-4 Use the lamp and fan blower to help signal trace low-voltage circuits.

Checking fuses

Besides the 15- or 20-amp line voltage fuse, you might find a thermal cavity fuse in series with the line fuse, power relay, and thermal protector (figure 6-6). All fuses can be quickly checked with the R×1 scale of the ohmmeter. Always replace with the exact type of fuse and amperage in each oven. Inspect the fuse holder for poor connections. Sometimes the fuse clips are spread and might not fit tightly around the end of the fuse, causing intermittent oven operation. The thermal fuse is enclosed in a metal case.

Samsung MG5920T low-voltage circuit

The low-voltage ac circuit can consist of a 15-amp fuse, main relay, primary interlock switch, low-voltage transformer (LVT), magnetron thermal cutout (TCO), thermal cavity cutout, lamp, drive or turntable motor, fan motor, heater control sensor, 1000-watt heater, browner relay, power relay, monitor interlock switch, and primary winding of HV transformer (figure 6-7).

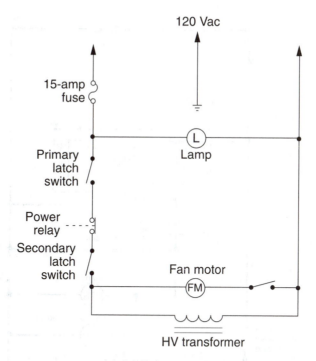

6-5 The rotation of the fan blower indicates that voltage is applied in a Quasar oven.

The low-voltage transformer, main relay, browner relay, and power relay are found mounted on the controller board. The fuse holder, drive or turntable motor, heater, and power transformer are found on the metal chassis. A fan motor, control board, lamp, cavity cutout, and interlock switches are found on different areas of the cabinet. The primary, secondary, and monitor interlocks are found on a plastic container behind the door. The magnetron TCO is mounted on the metal shell of magnetron tube. The 1000-watt heater coil is found at the top of the oven cavity.

Most all oven low-voltage parts are wired in series of the power line. The low-voltage transformer, lamp, turntable and fan motor, and heater components are in parallel or across the 120-volt ac line. When the low ac voltage components are normal and in closed or on position, a low continuity measurement at the power plug indicates components are normal. This resistance should be below 5 ohms.

Tappan 56-9431 and 56-9991 low-voltage circuits

In the low-voltage input circuits of the Tappan ovens, a regular fuse and fire protection thermal fuse provide power line protection. A primary and secondary interlock switch is found in each leg and a monitor switch is located across the low-voltage circuits. A separate power and browner relay provides line voltage to the power transformer and browning heater unit (refer to figure 6-5).

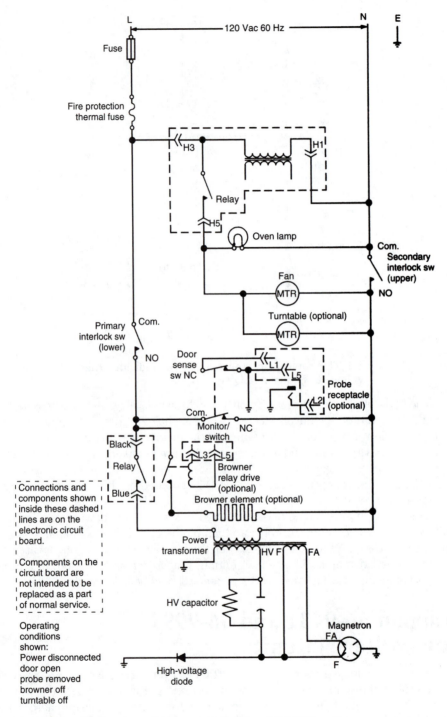

6-6 Schematic diagram of Tappan's 56-9431 and 56-9991 microwave oven.
Courtesy Tappan, brand of Frigidaire Co.

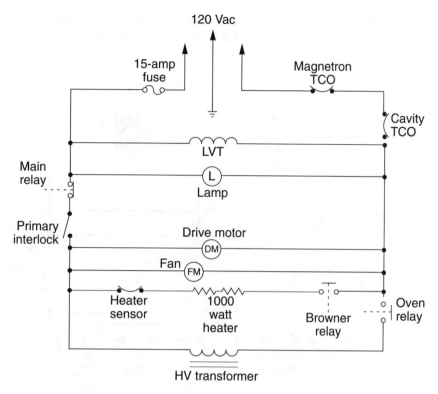

120 Vac

15-amp fuse

Magnetron TCO

Cavity TCO

LVT

Main relay

L

Lamp

Primary interlock

Drive motor

DM

Fan

FM

Heater sensor

1000 watt heater

Browner relay

Oven relay

HV transformer

6-7 Low-voltage circuit of a Samsung MG5920T oven.

Keeps blowing fuses

Besides the interlock switch, the 15-amp fuse is replaced more often than any other component in the microwave oven. Most ovens use a 15-amp chemical cartridge fuse (figure 6-8). The fuse could be located at the top or bottom of the oven area. You can't go wrong, as only one fuse is found in the primary circuits.

A leaky magnetron or high-voltage diode might keep blowing the fuse in the high-voltage circuits. Excessive arcing inside the magnetron will blow the main fuse (figure 6-9). Remove the old fuse and clip the Circuit Saver across the fuse clip-terminals. Monitor the line voltage across the primary winding of the high-voltage transformer with a pigtail light or ac voltmeter.

Pull the ac cord and discharge the high-voltage capacitor. When discharging the capacitor, notice if a loud arcing noise is heard. If so, you know that high voltage is present. No arcing of the high-voltage capacitor might indicate a leaky magnetron or HV diode. Remove the heater terminals and take the three magnetron resistance tests. Suspect a leaky high-voltage diode when other magnetron resistance tests are normal.

Ask several important questions when the owner complains that the fuse keeps blowing in the oven. Does the fuse blow when the door is opened or closed? Does the fuse blow after the oven has operated for several hours? Does the fuse blow when the power cord is plugged in or the oven is turned on? Does the fuse open when the door is moved? Has the fuse been replaced only once or several times? These questions might help you to locate the defective component.

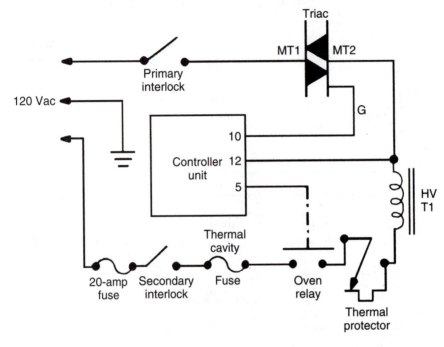

6-8 A thermal cavity fuse and thermal protector is found in this oven.

If the fuse blows when the door is opened, look for a defective interlock and monitor switch. In case the fuse blows when pulling up on the door, suspect a defective interlock switch. Check for a defective cord, plug, or component on the power line when the fuse opens as the power cord is plugged in. If the fuse blows after operating several hours, suspect a high-voltage (HV) component that is operating quite hot or warm and breaks down under heat.

Check the fuse for open conditions with the Rx1 ohm scale of the ohmmeter. You cannot see inside these chemical type fuses as you can with regular glass fuses. An open fuse will show no continuity reading. If the ends of the fuse appear loose, replace it. After replacing the open fuse, start the oven. It might operate perfectly without any further service. If the fuse blows at once, inspect the oven for possible overloaded conditions.

Inspect the interlock and monitor switches for erratic operation (figure 6-10). Always replace both the interlock and monitor or safety switch if the fuse blows when opening the oven door. Replace the interlock switch if it is frozen or has poor switching contacts. You should also replace the monitor switch, which is placed directly across the power line with the 15-amp fuse (figure 6-11).

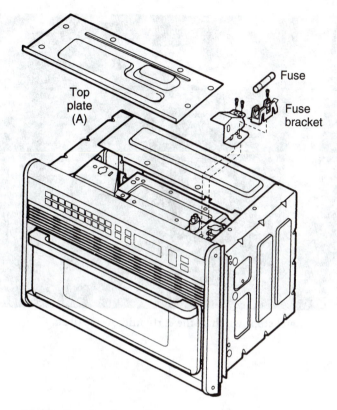

6-9 The fuse bracket is located on top of magnetron in a GE oven. Courtesy General Electric Co.

Tappan fire protector thermal fuse

The fire protector thermal fuse is found in one leg of the power line. This oven is equipped with a protective device that monitors the exhaust temperature inside the plastic sleeve; open exhaust duct. Should the temperature rise above 140°C, this device will open, disabling the oven. This device opens only when a fire occurs in the oven cavity. The fuse is located inside a plastic exhaust duct and has lead wires permanently attached (figure 6-12).

To replace the thermal fuse, a special Torx tool with a hollow center (#TX20H) is required. Remove the outer wrapper and remove the plastic duct by removing two special screws. Disconnect the two lead wires from the 20-amp fuse and magnetron thermal fuse terminals. Pull the two wire leads out through the hole in the back of the oven. (First remove the rubber grommet in the hole.)

Test the fire protector with the R×1 scale of the ohmmeter. The fuse is normal if the resistance measurement is less than 2 ohms. If the circuit is open, replace the thermal fuse, and reverse the procedure for replacement.

6-10 The start and stop switch in the early microwave oven.

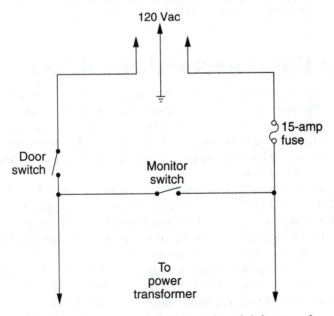

6-11 When the primary interlock is found defective, also replace monitor switch.

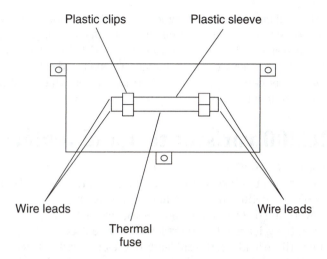

6-12 A thermal fuse component found in a Tappan microwave oven. Courtesy Tappan, brand of Frigidaire Co.

Erratic thermal tests

Clip a pigtail light bulb across any thermal unit in the low ac voltage circuits to test for erratic operation (figure 6-13). When the oven keeps kicking off and on or if oven gets too warm, the thermal units might shut down. You can clip the analog ac voltmeter across the thermal unit and watch the hand go up to 120 volts ac or line voltage and clear down. Never use the ac scale of a DMM for this test, as the numbers on the voltage reading will rapidly change, making it hard to read.

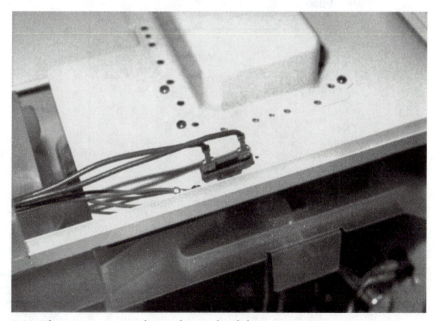

6-13 The oven sensor is located outside of the oven cavity.

Often the thermal cutouts of the magnetron, heater sensor, and cavity +CO will shut down at a certain temperature. For instance, the magnetron thermal cutout will open from 140 degrees to 302 degrees Fahrenheit in GE JEBC200W oven. The flame sensor operates 120 degrees Celsius/0 degrees Celsius and 248 degrees Fahrenheit/32 degrees Fahrenheit. Replace any thermal cutout when it shuts the oven down before cooking time is ended.

GE JEBC200 varistor test and replacement

The varistor is located on the smart board. However, it can be replaced without having to replace the board. Take a visual check to see if it has been burned. Replace the varistor by disconnecting the power and removing the oven from the wall. Remove the top covers and discharge the high-voltage capacitor. Clip off the old varistor wires at the green body, leaving two short leads. The replacement varistor has push-on terminals. Push these leads on the cut leads of the old varistor. Check for tightness.

Suspect a defective varistor or surge absorber in ovens that keep blowing the fuse when the oven is turned on or plugged in. These small varistors are located within the plastic sleeve across the power line. They are included across the power line to prevent excessive damage to the oven components (figure 6-14). This same type of varistor is found in the power line circuits of TV receivers. So don't overlook one of these small varistors when the fuse continues to open. When a power outage occurs or lightning strikes the power line, the varistor can be damaged and cause the fuse to open.

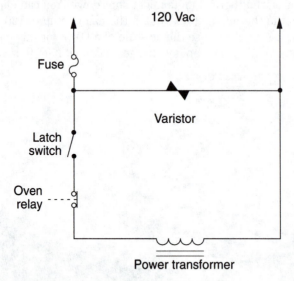

6-14 The varistor protects oven circuits from lightning and power outage conditions.

Varistor test

If the varistor is suspected and it has a few black marks or missing areas, test it with the R×1000-ohm scale. It's best to remove the varistor from the primary terminals of the controller transformer or ac power line. Inspect the varistor for damaged areas. Replace it if it's damaged. Connect the ohmmeter leads across the varistor leads—a normal varistor should read infinite ohms. If the varistor has any type of resistance measurement, replace it.

If the fuse keeps blowing, insert a 100-watt bulb across the fuse terminals. Remove the defective fuse and just clip the bulb leads to each fuse clip. You could save several dollars this way because chemical fuses are not cheap. Just leave the bulb in place until you locate the defective part. At times, however, the light bulb will not indicate when a shorted component in the HV section is causing the fuse to open.

You might find an open fuse, replace it, and the oven begins to cook. Then in a few minutes, the cooking quits and you have a blown fuse. The magnetron could be leaky; open up the thermal cutout assembly. Although the magnetron is often the prime suspect, don't overlook a jammed blade in the cooking fan. If the fan blade doesn't rotate, the magnetron tube overheats. Likewise, the thermal protector switch shuts down the circuit if the fan motor is overheating, causing the fuse to blow. Don't overlook a defective fan motor when the fuse opens after several minutes of oven operation.

A shorted magnetron or leaky HV diode could also cause the fuse to open. The fuse might blow after a few minutes or when the oven is shut off— this is sometimes caused by a leaky triac assembly. Check each component for possible damage if the oven stops working after an electrical storm. Usually, those components that tie into the power line and the control-board assembly receive the most damage.

Intermittent fuse saver tests

The Circuit Saver is the ideal tester to use when servicing microwave ovens with an intermittent blown-fuse problem. Instead of replacing those expensive fuses each time, this tester can save a few dollars in extra fuses. Connect the Circuit Saver across the fuse clip terminals (figure 6-15). Then each time the oven shuts down, you can reset the fuse saver and try again.

This tester comes in handy when trying to locate parts that intermittently blow the fuse. Sometimes the oven will operate for days without blowing a fuse. Just leave the Circuit Saver connected and reset when needed. Operate the oven at different times to try to find the intermittent problem, especially after the oven is quite warm.

Although the cost of the tester is less than $20, within a few months it will pay for itself because you will not have to replace the fuse each time. Do not forget to discharge the HV capacitor each time the oven shuts down.

6-15 If the fuse keeps blowing, clip in the Circuit Saver to determine the defective component.

Intermittent oven lamp

The oven lamp can be used as an indicator to show that power is being applied to the oven and the fuse is good. The lamp might come on when the door is opened or closed. In some ovens, an SPDT switch is used for this purpose (figure 6-16). Here, the oven lamp is on with the door open so the operator can load the oven cavity. With the door closed, the oven lamp goes out until the power relay is in operation. When the lamp stays on with the door closed, replace the lamp switch, or suspect that a switch lever is hung up. Again, with this circuit, the oven lamp can be used as an indicator showing that power is applied up to this point.

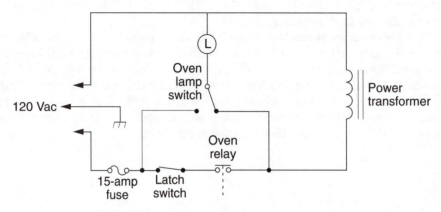

6-16 The oven lamp switch operates with the door opened or closed.

You will find two separate light bulbs in some microwave ovens. Often, the cavity lamp is 30 watts or smaller and can be replaced with a regular appliance-type bulb (figure 6-17). These bulbs are available from hardware, department, and electronics stores. All microwave oven distributors and manufacturers supply oven lamps for their particular ovens.

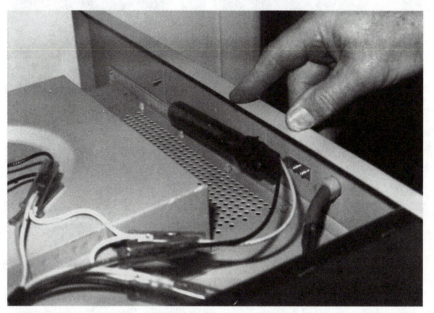

6-17 A 25- to 40-watt bulb lights up the oven cavity.

An erratic or intermittent oven lamp could be caused by a defective bulb, lamp, or interlock switch. You might find that the bulb is loose in its socket. Always replace the oven lamp while servicing the oven if the bulb appears black—even if it comes on. Usually these bulbs do not last very long, and this replacement might prevent a callback to replace just a lamp bulb.

Replace a suspected lamp switch when the bulb is intermittent or does not stay on. In some instances, the bulb comes on with the door open, but will not come on when the oven is operating. The oven lamp switch might have burned or dirty switch contacts. The cavity light switch can be found separately or ganged with other interlock switches (figure 6-18).

If the oven lamp flickers with the oven operating, suspect a loose bulb or a defective interlock switch. The light might flicker with the door open, but operate normally when the door is closed. Replace the secondary interlock switch. If the lamp does not go out with the door open or closed, replace the lamp or latch switch. A low, flickering lamp with the oven in operation might indicate a heavy current being pulled in the magnetron tube circuit.

Don't overlook a defective lamp socket or loose wiring harness when the cavity lamp will not come on. Use the VOM and check for 120 Vac across the lamp terminals. The 100-watt pigtail lamp can be clipped across the lamp socket as an indicator. Suspect a defective harness or wiring cable if the light appears intermittent

6-18 A defective oven switch can cause intermittent operation of the oven.

when the wiring is moved with a long plastic tool. Lift the cable with a wooden or plastic tool. Do not use a metal tool or screwdriver inside the oven area.

Next, open and close the oven door with the power connected to the oven. If the light socket is intermittent or arcs internally, replace it. Unplug the microwave oven and discharge the HV capacitor. Unsolder the leads from the light socket. Check the resistance across the light socket with the R×1 scale of the ohmmeter. The meter should read 0 ohms. Replace the socket with actual oven part numbers. To reinstall, reverse the preceding procedures.

Intermittent oven cooking (cutout)

Suspect a defective thermal protector switch if the oven stops after it has been cooking for several minutes. The fan and turntable motor might be operating, except there is no cooking. Although the thermal cutout might be bolted to the magnetron tube, the cutout assembly controls the low-voltage circuits. In some ovens, you might find two separate thermal protector assemblies. One is located on the magnetron and the other on the metal oven cavity (figure 6-19).

The magnetron thermal protector switch prevents the magnetron from overheating and shuts down the cooking process. Inside the thermal unit, a switch contact connected to a bi-thermal strip opens and closes with heat. If the magnetron tube has an internal leakage or short, the metal shield becomes hot and in turn opens the thermal cutout switch. The oven might start to cook once again after the magnetron has cooled down. Suspect a defective thermal cutout assembly, magnetron, or triac when intermittent cooking is noted.

Magnetron thermal
cutout

6-19 The TCO on the magnetron shuts down the oven if the magnetron runs too hot.

Check the thermal protector with the VOM on the low R×1-ohm scale. Make sure the power plug is pulled and the HV capacitor is discharged. Connect the ohmmeter lead directly across the thermal switch terminals. You should read a direct short/no resistance. If 2 to 5 ohms of resistance is noted, replace the thermal switch. Although the oven might operate, the thermal protector will soon produce erratic heating or cooking (figure 6-20).

If the cooking cycle starts cutting on and off within a few seconds, the thermal switch should be checked with a 120 Vac measurement across the terminals. An open thermal unit indicates the entire power line voltage (120 Vac). The thermal switch is usually found in series with one side of the low-voltage circuit. No light or voltage measurement indicates the thermal switch is normal. Be careful with these voltage measurements. Clip the VOM leads across the thermal cutout and then fire the oven up (figure 6-21). The 100-watt pigtail bulb can be used as a monitor light across the thermal unit.

The thermal protector assembly can be located in a low-voltage control board circuit. Here, the thermal switch is in series with the power oven relay. The voltage furnished to the relay by the low-voltage transformer is controlled by the DPC (figure 6-22). An open thermal cutout assembly in this type can measure from 20 to 30 volts supplied by the control board.

The thermal protector should be replaced when it is defective or when installing the magnetron tube. Simply remove the two mounting screws and mount the thermal unit on the newly installed magnetron. Replace the connecting wire clips to the thermal component. No harm is done if these two wires are interchanged. The magnetron thermal switch is interchangeable, but not so with the oven thermal assembly.

6-20 The thermal cutout is located on the metal part of a magnetron tube.

When the oven has been operating for 15 minutes, the thermal switch will open up in some ovens, indicating the magnetron is overheating; this would normally indicate that the cutout or magnetron is defective. First, replace the thermal protector switch (figure 6-23). If the new unit opens up after several minutes, check the magnetron for leakage. Both the magnetron and thermal cutout might be normal. Slip a piece of insulation between the thermal switch and body of the magnetron. Some manufacturers provide thermal insulation kits for this purpose. Replace the magnetron tube if it is running extremely hot.

Replacing thermal protection

Before testing or replacing thermal protectors, unplug the oven and discharge the HV capacitor. With the ohmmeter, check the thermal protector for open or erratic conditions; the meter should read 0 ohms. Remove the wires from the thermal protector. Note how they are connected. Remove the screws holding the defective thermal unit on the magnetron tube. Install a new thermal protector of the same type. To install it, reverse the previously discussed procedures.

Intermittent thermal cutout

An intermittent thermal cutout can be difficult to locate. The best method is to clip the ac voltmeter, pigtail light, or neon circuit tester across the thermal switch terminals. When the oven is operating correctly, the light or meter will be out—or the meter will not register. If the thermal unit kicks out, the light will come on. Clip meters or testers with long test leads so the meter will not be inside the oven for tests.

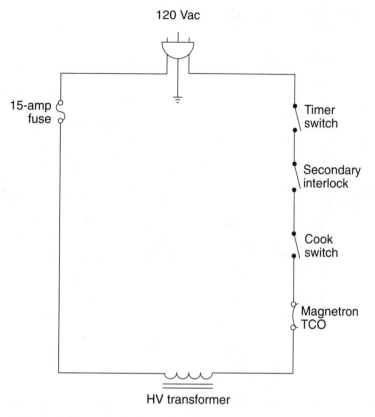

6-21 Switches, interlocks, and cook relay controls can be checked with ac voltage tests.

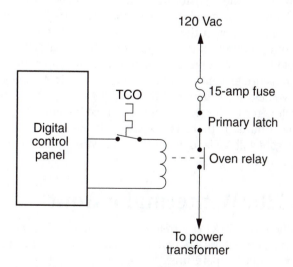

6-22 The thermal cutout is located in series with the oven relay solenoid in some ovens.

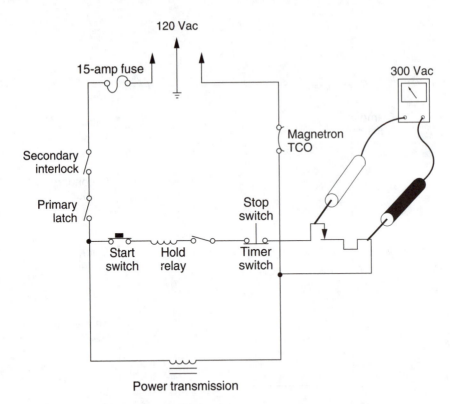

6-23 Monitor the suspected cutout with an ac voltage meter or pigtail light.

In the intermittent oven, you might want to monitor the oven relay or triac terminals, across the primary winding of the power transformer, and on the thermal cutout (figure 6-24). If the transformer light goes out and the thermal unit light comes on, you know the thermal cutout is open or defective. Check the oven for overheating with an open oven thermal cutout. Suspect an overheated magnetron with an open thermal cutout on the magnetron tube. Usually the defective thermal unit will only hold for a few minutes or seconds and then open up.

When the transformer light goes out and the triac light comes on, suspect the triac or no gate voltage at the gate (G) terminal. Discharge the high-voltage capacitor and clip a voltmeter to the gate terminal. Notice if voltage is present at the meter when the oven quits. If there is no gate voltage, suspect a defective control panel. When the gate voltage is found at the triac without oven operation, check for a defective triac.

GE JEBC200W thermal cutout

The convection oven TCO (thermal cutout) is located at the back wall next to the convection fan motor. It is a safety device that is responsible for not allowing runaway convection heater temperatures.

1. To remove the heater TCO, disconnect power and remove oven from wall.

2. Remove top panel.
3. Discharge high-voltage capacitor.
4. Remove screws that hold sides to back panel.
5. Remove outer back plate.
6. Rotate back.
7. Disconnect leads to thermal cutout and 2 screws that hold it in place.

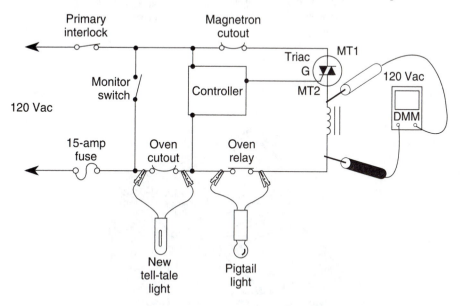

6-24 Check oven intermittent operation with a pigtail light and ac DMM.

Samsung heater sensor

The Samsung MG5920R convection oven contains a heating element that is located in the top area of the oven cavity. The 1000-watt heating element is switched into the 120-volt ac circuit with a browner relay, controlled from the panel control board. A heater control sensor acts like a thermal cutout to shut off the heating element when the oven reaches a certain temperature (figure 6-25).

To remove heating element, disconnect all lead wires from the heater terminals. Remove hex bolts securing the heater. Take out the heater and stopper heater from the heater cover (figure 6-26). Remove screws securing the heater cover.

Samsung cavity cutout

Besides the thermal cutout upon the magnetron, you may find a thermal cutout in some of the Samsung microwave ovens (figure 6-27). The thermal cavity cutout is found in series with the magnetron TCO and power relay. Most thermal cutouts are located in the outside of the oven cavity. When the oven cavity becomes excessively hot, the cavity cutout opens up shutting down HV to magnetron.

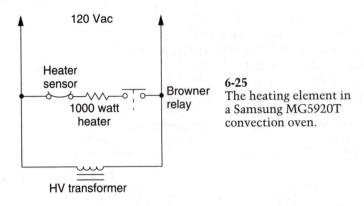

6-25
The heating element in
a Samsung MG5920T
convection oven.

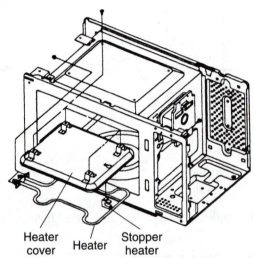

6-26 The location of the heating element in a
Samsung MG5920T oven. Courtesy Samsung Electronics Corp.
of America, Inc.

Oven does not go into cook cycle

In the low-voltage circuits, a defective timer switch, thermal protector cook
switch, component wiring, or cook relay might cause a no-cook cycle. Suspect an
open cook-relay assembly after checking all other components with the ohmmeter.
This relay is sometimes called a cook, hold, oven power, or latch relay. The cook relay provides a switch in one side of the power line to the primary winding of the
power transformer, completing the circuit. A power relay can operate from the
power line or from the low voltage of the controlled assembly. The cook or power relay is energized during the entire cooking cycle.

Often, the cook relay can be heard when energized, or you can see the switch
point contacts close after the cook button is pushed (figure 6-28). No action from the
cook relay could indicate that voltage is not applied to the relay solenoid winding.

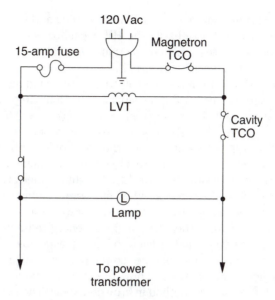

6-27 The Samsung thermal cavity and magnetron TCO in the ac circuits.

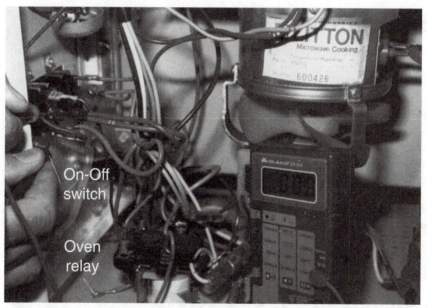

6-28 Often, you can hear the oven relay energize.

Pull the power cord and discharge the HV capacitor. Clip leads to the solenoid terminals and connect to the 300-Vac VOM scale. Prepare the oven and push the cook button. No voltage measurement (120 Vac) could indicate a defective interlock micro or cook switch.

When ac voltage is measured at the solenoid winding and the relay does not energize, suspect an open coil. Discharge the HV capacitor with the power plug removed. Remove one side of the cable connection of the solenoid winding so you will not measure any other resistance in the circuit. No resistance measurement indicates that the coil is open. Check for a broken coil wire at the winding. Sometimes these small wires are broken off right at the terminal connections. Replace the oven relay if it tests open or if a very high winding resistance is measured. Take an ohmmeter reading across the coil terminals. This resistance can vary between 150 to 300 ohms.

Oven or auxiliary relays operating from the electronic control board are often low-voltage types. You should have a 10- to 24-volt measurement across the solenoid winding. The coil resistance could be from 1 to 2 kilohms. Check the manufacturer's service literature for correct resistance. When little or no voltage is measured at the solenoid winding, suspect a defective controller or power transformer.

In very difficult situations, the relay might be energized manually by pushing down on the top section of the relay with a wooden or plastic dowel. Actually, you are making the switch terminals come together so the oven will begin the cook cycle. If the magnetron begins to operate, you know the relay or the applied voltage is at fault. Do not attempt to use this method in crowded areas. The power relay must be out where you can get to it. If ac voltage is applied to the primary winding of the transformer, you know either the relay is defective or it has no applied voltage.

Intermittent oven relay

You can usually hear the oven relay click in when voltage is applied to the solenoid. The intermittent oven relay can be checked by clipping a pigtail light, neon circuit tester, or ac voltmeter across the oven relay switching terminals. Discharge the high-voltage capacitor before attaching any instruments or reaching into the microwave oven.

When the relay light comes on or the ac meter reads 120 volts, the oven shuts down. Poor or pitted oven relay contacts might cause poor oven operation. An intermittent solenoid winding or poor connecting wires might produce intermittent cooking. Determine if the points are defective, solenoid, or controlled voltage.

Check the solenoid for poor or open winding. Check the voltage applied to the solenoid. Some oven relays work directly from an oven switch and power line voltage, while others are controlled from the panel control board. If the oven relay light comes on, the oven shuts down, the solenoid continuity test is normal, with improper solenoid voltage, check the control board.

Samsung MG5920T relays

This microwave oven contains the power, main, and browner relay. All three relays are mounted on the control panel and are operated by the control circuits. The main power relay points are in a closed position while the power and browner relay contacts are open. The main power relay operates in series with the fuse and primary winding of transformer. A power relay contact is connected in the other power leg with the magnetron TCO and cavity TCO.

The power relay is in several transistor circuits of the relay driving circuits of A, B, C, and D (figure 6-29). The main relay is turned on by transistor Q10 from relay driving circuit E. Driving circuit F operates Q12 that operates the browner relay. Notice a –12 volt dc source is connected at leg of browner relay and power relay. This –12 volts comes from the low-voltage transformer supply of control panel.

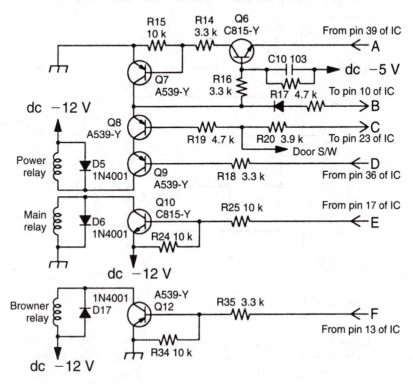

6-29 Schematic configuration of relay driving circuits. Courtesy Samsung Electronics America, Inc.

The driving circuit waveforms are found at A, B, C, D, E, and F. When the start pad is pressed, waveform A applies a dc –5 volts until a cancel pad is pressed from pin 39 of microcomputer IC1 (figure 6-30). Waveform D shows how the power level relay is working approximately 50% of the time: it is on 16 sec and off 14 sec. The waveform of E and F is on all the time from the start of browner relay or when the cancel pad is pressed.

Oven keeps operating (defective triac)

A defective triac might let the oven operate after the program controller is shut off and might cause many different problems in a microwave oven. The gate voltage of the triac assembly is controlled by the controller or DPC. When a voltage is fed from the control board to the gate terminal of the triac, the triac is turned on. In some ovens, this voltage might vary from 4 to 6 Vdc. The triac acts as a switch and connects one side of the power line to the primary winding of the transformer.

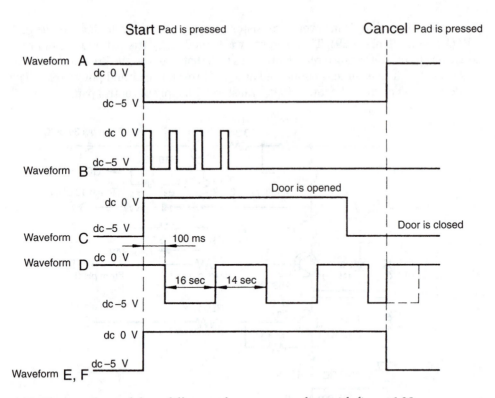

6-30 The waveform of three different relays corresponding with figure 6-29. Courtesy Samsung Electronics Corp. of America, Inc.

The triac can be checked with applied voltage or ohmmeter tests. To determine what component is defective, monitor the ac voltage at the primary winding of the power transformer (figure 6-31). No voltage at the primary winding with the oven in the cooking mode indicates a defective triac, electronic controller, thermal protector, or faulty cable connections. You can assume that the electronic control board is working if the unit counts down properly. Quickly check the contacts of the thermal switch with the R×1 ohmmeter scale.

Either the electronic controller is not feeding voltage to the gate terminal or the triac is defective. To check the triac in the circuit, measure ac voltage across one side going to the power transformer and the other to the power line (figure 6-32). When 120 Vac is measured across these two terminals, the voltage is present but the triac is not switching. A 100-watt pigtail lamp can be used as an indicator across these triac terminals.

To determine if the HV circuits are operating, simply clip a lead across these triac terminals (1 and 2). Always pull the power plug and discharge the HV capacitor before reaching into the oven. Use clip leads when connecting leads or test instruments to the oven circuits. With the clip lead in place, start the oven up. The oven should produce heat and cook without any time control (figure 6-33).

Remove the suspected triac assembly. You might find several different triac units or modules in different microwave ovens. Usually the triac assembly is bolted to the

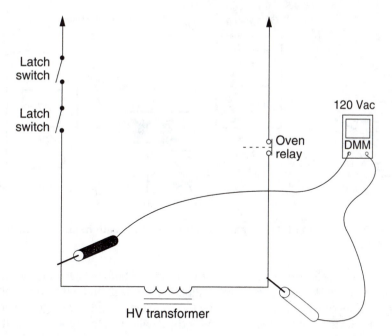

6-31 Monitor the voltage at the primary winding of the transformer to test for trouble in the primary circuits.

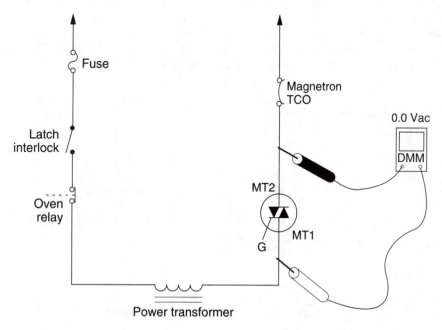

6-32 Measure the power line voltage across a suspected triac to test for an open or leak.

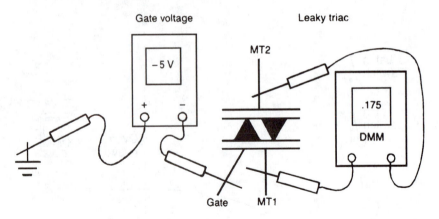

6-33 MT2 and MT1 has a shorted ohmmeter measurement and must be replaced.

oven cavity wall. All terminal leads must be removed before checking the triac with the ohmmeter. Mark down each wire connection. You will find a different color code for each connecting cable. Interchanging the triac leads might damage the new triac or control board. Often the gate terminal is the smallest space-type lug and it cannot be interchanged with the others.

The triac module or assembly connections are easy to trace out. The MT2 terminal connects to one side of the power transformer (figure 6-34). Terminal MT1 connects to one side of the power line. The gate terminal (G) connects to the connecting wire from the electronic controller or control board. The gate voltage supplied from the controller in the Norelco R7700 oven is –5 volts.

Check the continuity between any triac terminals with a VOM, VTVM, digital VOM, or the diode test of a digital meter. Switch the meter to the diode test of the

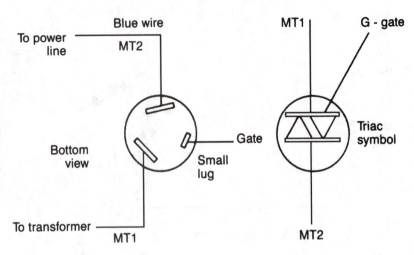

6-34 MT2 always connects to the power line voltage and MT1 to transformer.

digital meter and read the resistance between G and MT1. This reading will be somewhere around 0.554 + in either direction (figure 6-35). Reverse the test leads and the reading should be the same for a normal triac. You will not read any resistance between other combinations of two terminals for a normal triac. If a reading other than the gate to MTI is noted, the triac is leaky.

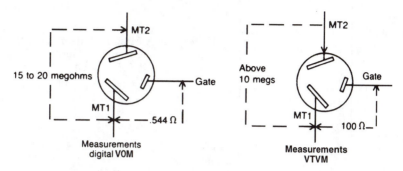

6-35 Resistance measurements and tests on the suspected triac assembly.

The low-ohm scale of a VTVM should measure around 100 ohms between gate and MT1. In a Hardwick oven, the same terminal resistance might be 22 ohms. A reading above 10 megohms from MT1 and MT2 is normal. Replace the triac assembly when a 5 megohm or lower measurement is noted between any other terminals than MTI and G. A digital ohmmeter test between MTI and G should read 100 to 110 ohms. It's best to check each triac reading with the manufacturer's data, if handy. Usually, the defective triac will appear shorted and leaky. Very seldom does the triac go open.

When the oven turns itself off, the fuse is blown and will not start up again until the fuse is replaced. Sometimes the oven will run for 10 to 15 minutes before this condition occurs. You might waste a lot of service time on this one. Check the oven and circuit for a defective triac assembly and replace it.

A defective triac might cause the oven to hum under load when the oven is turned on. You just hear a humming noise and have no cooking. You might also notice that the electronic control board is counting down rather fast. The triac does not turn on the voltage to the primary winding of the power transformer. If this is the problem, simply replace the leaky triac with the original type.

You might find another type of triac assembly subbed or modified from the original one. No doubt, the manufacturer has found the new triacs are reliable and should be installed when the old one is found defective. The main thing to remember is to connect each wire as shown in the new schematic. Improper connections of the triac might damage the control circuit board. With triacs marked MT2, MT1, and G1, MT2 goes to the power transformer, MTI to one side of the power line, and G goes to the gate connection from the control board.

Goldstar triac connections

In a Goldstar ER-505M microwave oven, the three triac terminals tie in the oven circuit and directly to the Micom Controller circuits (figure 6-36). Usually in most

6-36
A Micom controller controls
the triac in a Goldstar
ER-505M oven.

triac circuits, the gate connection is only connected to the control panel, while MT1
and MT2 tie into the ac input and one leg of high voltage transformer. Instead of MT1
and MT2 of triac terminals, they are listed at T1 and T2, leaving off the letter M.

Norelco triac measurements

Set the ohmmeter to the highest range (10 k). Remove leads from the triac assembly for accurate tests. Mark down where each wire lead connects to the triac.
Place one lead of meter on each terminal, and then reverse leads for another measurement. Check Table 6-1 for correct resistance measurement of the triac assembly.

**Table 6-1. The correct resistance and test
methods of checking a Norelco triac assembly**

Terminals	Forward	Reverse
MT2 to Gate (G)	∞	∞
MT2 to MT1	∞	∞
Gate (G) to MT1	Small	Small

∞ = Infinite resistance.
MT1 = Cathode or mounting terminal 1.
MT2 = Anode or mounting terminal 2.
G = Gate-silver color-coded lead.
Gate voltage = 5 volts dc pins 11 and 12.

Reversed triac connections

Before replacing a suspected triac, mark down the color-lead-wire connections.
In the Norelco ovens, the MT2 connection has a blue wire, the MT1 has a white wire,
and the gate is silver. With the Hardwick ovens, the triac MT1 lead is a white wire,

the MT2 is a black wire, and the gate terminal is a red wire. Remember the color code of the various triac lead wires might be different in various ovens. If MT2 and MT1 wires are reversed, the microwave oven will operate 180° out of phase and that can damage the control microprocessor assembly circuits.

Always keep the gate wire connected to the triac. If the gate wire is removed or accidentally knocked off, the microprocessor might be damaged in the control circuits.

Lamp triac tests

You might check the oven triac assembly with a lamp-type tester (figure 6-37). Build the tester with rubber-coated alligator clips so you do not get shocked. Clip the leads to the triac with the power plug pulled on the tester. The triac leads must be removed for these tests. It's best to remove the triac and test on the bench.

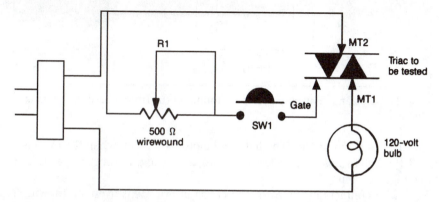

6-37 This triac tester is made up of a few ac parts and is shockproof.

After the triac is connected, check it for a short between MT1 and MT2. Plug the tester in—if the lamp does not light up, the triac is not shorted. Now press down on the momentary SW1 and rotate R1 until the bulb lights. Let up on the SW1; if the light stays on, the triac is shorted. If the light does not light up at all, suspect an open triac assembly.

Amana triac location

In a number of Amana microwave ovens, it might be difficult to locate the triac assembly or oven relay mounted on the metal chassis. Some models have the triac, relay, and low-voltage transformer mounted on the digital timer control assembly (figure 6-38). Just remove the wires from the suspected component to make accurate tests. These parts should be replaced with original part numbers.

Intermittent power transformer

Very seldom does the power transformer become intermittent, but when it does, unusual problems exist. Monitor the primary winding of the transformer with

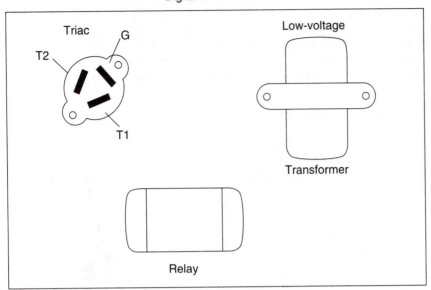

6-38 The triac and relay assemblies are found on the timer board in the Amana oven.

a pigtail light or ac voltmeter. The light and meter will be on when the oven is operating. If the light stays on and the oven stops cooking, suspect a transformer or high-voltage problem.

Poor wire contacts in the oven can cause intermittent oven shutdown. Sometimes when the primary leads and coil wire are crimped together, the enameled coil wire makes a poor contact with intermittent connections. Quickly discharge the HV capacitor and make a resistance test across the primary winding. If there's an open or erratic reading, wiggle the black wires and pull on them to determine if connections are bad.

You might be able to repair the intermittent connection by cutting into the paper cover of the winding. Be careful not to cut too deep. If the primary winding is wound on the outside, you are in luck. Remove layers of fabric paper to get at the terminal connections. Remove crimped wires, scrape off the enamel and resolder the connection. Repair both leads in the same manner.

The secondary winding of the transformer might cause intermittent operation, but this can be checked with the ac VTVM or Magnameter.

Caution: Be careful; discharge the high-voltage capacitor each time.

Leave the meter connected until the meter reading lowers, with the pigtail light on the primary winding. Because both the secondary and filament or heater windings have large wire, the only problem is caused where the filament wires connect to the magnetron. These filament connections run quite warm and cause a poor heater connection. This results in poor, intermittent, and no oven cooking.

You might find a secondary high-voltage winding that shorts between turns, overheats, smokes, and doesn't cook. This can happen when the magnetron tube shorts out between filament and anode terminal. Of course, this turns into a very expensive oven repair. Both the magnetron and the transformer must be replaced.

Samsung touch control power supply

The low-voltage power supply within the control circuits of a Samsung MG5920T oven connects directly across the power line, after the fuse and cavity TCO. The negative dc voltage sources are –5 volts, –12 volts, and –31 volts (figure 6-39). The low voltage converts ac to dc voltages with transistor regulation in the –5 and –12 volt sources. The power supply generates a half wave and transfers it to timer pulse shaping circuit. Also, the ac transformer provides low filament voltage to the fluorescent tube display. The –31 volt source is applied to the fluorescent tube circuits.

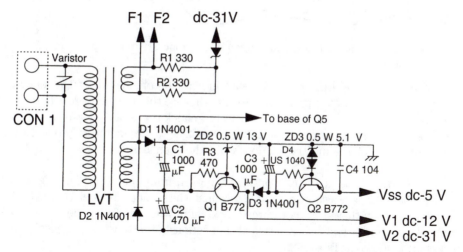

6-39 The negative low voltage supply of control circuits in a Samsung MG5920T oven.
Courtesy Samsung Electronics America, Inc.

Everything lights up but no cooking (meat probe)

In some ovens, the temperature probe must be inserted when the cooking switch is turned to the meat probe operation. If the meat probe is not in use and the switch remains in that position, the next time the oven is turned on, it will not cook. Undoubtedly, this situation is caused when the probe is used but isn't switched to the probe position (this might occur in a few ovens). The oven might come in for repair with nothing wrong—you'll only have to turn off the probe switch to get it to operate normally.

Improper meat probe temperature cooking might be caused by a defective temperature probe, probe jack, or faulty wiring. The temperature probe can be tested with a 100-kilohm scale. Connect the ohmmeter leads to the temperature probe male jack. This jack looks like any standard earphone jack with the point and common terminal connected to the meter (figure 6-40). Normal resistance can be measured from 30 to 75 kilohms. If infinite or extremely low resistance is measured, replace the temperature probe.

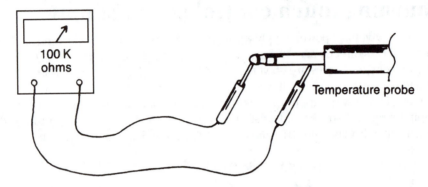

6-40 Check the resistance of temperature probe at probe tip (30 to 75 kilohms).

Check the temperature probe once again with the ohmmeter leads connected and immerse the tip of the probe in hot water. In case the ohmmeter resistance decreases with the probe immersed in hot water, check the probe jack for problems. However, if the resistance does not decrease, replace the temperature probe. Never let the probe lay on the metal oven floor while the oven is in operation. Remove the probe when not in use.

A defective probe jack might prevent the oven from operating. You might find some probe jacks have a short-circuiting terminal while others are open (figure 6-41). In a Norelco oven there is a short-circuit jack. Simply use the R×1 scale of the VOM and make continuity tests of the terminal connection and cable wires. A typical probe test chart is shown in Table 6-2.

Different temperature probe tests

Different microwave ovens might have probes that have different elements with different resistances. The defective probe might appear leaky or open. Measure the resistance across the probe jack. If the resistance is lower than normal, the probe element is leaky. Check the wiring and plug for open measurement. The various oven probe resistances at room temperature are shown in Table 6-3.

GE JEBC200 temperature probe

The temperature probe for the GE oven is located on the left-hand side of the oven cavity. The probe consists of a sensor that has a thermistor swagged in the tip

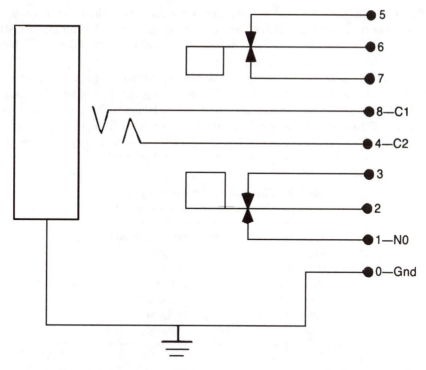

6-41 A special probe socket with the short circuitry terminal connections.

of a stainless steel tube. Check the probe at room temperature with a resistance measurement of 60,000 ohms. If the probe is inserted into a container of hot water, the resistance will visibly drop.

Disconnect the plug from the ac outlet. Pull oven from the wall to remove the top front panel. Unscrew the nut holding receptacle in the oven cavity. Disconnect wiring from connector 4 on the smart board and remove receptacle.

**Table 6-2. A typical probe test chart
is shown for various connections**

Condition of probe	Terminal test points	Normal indication
Probe out	G to C1	Open
	C2 to N0	Open
Probe in	G to C1	30 k to 75 k ohms
	C2 to N0	Continuity

Panasonic's probe thermistor tests

To test the sensor probe's thermistor, fill a container partially with lukewarm water and place a thermometer in it. Then take the sensor probe out of the oven and place it in the container next to the thermometer so they both will sense the same temperature.

Now take another container with boiling water and gradually pour this into the first container to achieve graduated temperature steps. At each temperature step, you should get the corresponding resistance with the ohmmeter (Table 6-4). If not, the probe is defective and should be replaced.

Check to see if the probe thermistor is working under cooking conditions. Set the oven cooking temperature at 150°F. Fill the testing cup with about 9 ounces of tap water. Place the container and probe in the oven. Time the oven to where it shuts down at 150°F. Now check the temperature of the water with the thermometer; the water temperature should be close to 150°F. Check the resistance of the probe to see if resistance has changed. If not, replace the temperature probe.

Table 6-3. The different oven temperature probes' resistance at room temperature

Manufacturer	Resistance in ohms
Amana	50,000
General Electric	60,000
Hardwick	50,000
Norelco	55,000
Samsung	46,600
Sharp	49,500

Table 6-4. Panasonic sensor thermistor probe test with the different temperatures and resistance

Temp.		Thermistor resistance nominal	Temp.		Thermistor resistance nominal
(°C)	(°F)	(kΩ)	(°C)	(°F)	(kΩ)
30	86	44.4	60	140	13.7
35	95	35.9	65	149	11.5
40	104	29.3	70	158	9.7
45	113	24.0	75	167	8.2
50	122	19.8	80	176	7.0
55	131	16.4	85	185	6.0
			90	194	5.1

Defective browning unit

A convection oven has two types of cooking methods enclosed in one oven. Often, the ac coil heating element is located at the top of the oven, and in some cases it might be lowered over the food to be cooked. When the element is enclosed, a rotating fan blows the hot air on the food. You might find a browning or cooking element at the very top of the oven cavity (figure 6-42).

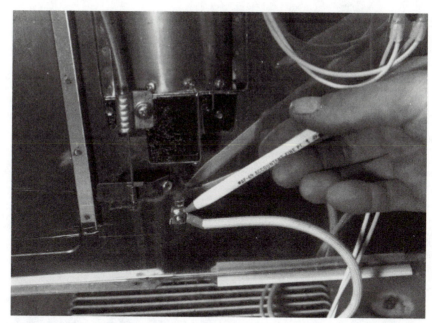

6-42 Check the bolted heater connections for poor contacts causing intermittent convection cooking.

When the heating element does not appear to become hot, suspect a burned cable, bad switch, poor wire-to-element connection, or an open heating element. The heating element can be checked with the Rx1 scale of the VOM. This element is always insulated from the metal oven cavity. No ohmmeter reading should be obtained from the heating element to the cavity walls.

Check the screw terminals where the insulated cable connects to the end of the heating element. Poor terminal connections or a burned-off wire can prevent the element from heating up. Often, the screw connections will become burned and brown-like, indicating poor terminal connections. If the cable is burned off and is too short to make adequate connections, replace with the original cable harness from the manufacturer. These are a special type of asbestos-insulated cables. Loose or sharp points on element mounting screws might cause arcing inside the oven when the microwave oven is cooking. Also, check for sharp edges or loose screw terminals in the oven cavity (figure 6-43).

Panasonic heating element replacement

In models NN-9807 and NN-9507, only one heating element is located in the top of the oven. While in the NN-8507/NN-8807/NN-8907 CPH oven, a lower and upper heating element is powered by relays controlled by the DPCs. After finding either element open or shorted, the unit must be replaced with the correct part number.

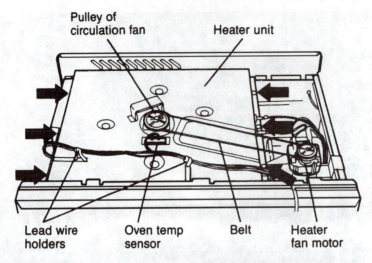

Pulley of
circulation fan

Heater unit

Lead wire
holders

Oven temp
sensor

Belt

Heater
fan motor

6-43 To remove heating element, remove the hardware, holder, and belt pulley in a Panasonic oven. Courtesy Matsushita Electric Corp. of America

In the NN-9807 and NN-9507 oven, replace the heater as follows:

1. Remove the lead wires from the lead wire holders (figure 6-44).
2. Remove the belt from the pulleys of the circulation fan.
3. Remove the two screws holding the temperature sensor.
4. Remove the two lead wires from the heater terminals by removing the screws (figure 6-45).
5. Remove the six screws holding both sides of the heater unit and lift the heater unit carefully.
6. Remove the two spring nuts from the heater terminals.
7. Remove the two screws holding the heater supports and detach the heater.

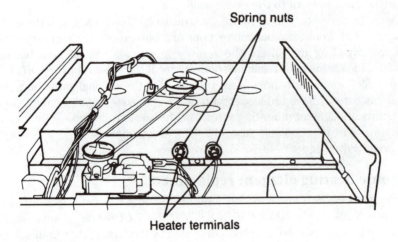

Spring nuts

Heater terminals

6-44 Remove the heater terminal nuts and wires to remove the heater in a Panasonic oven. Courtesy Matsushita Electric Corp. of America

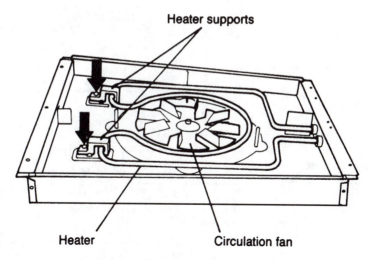

Heater supports

Heater   **Circulation fan**

6-45 Remove heater supports in order to detach heater in Panasonic NN-9807 oven. Courtesy Matsushita Electric Corp. of America

GE JEBC200 convection oven

The 1400-watt heater element operates from the ac power line (120 Vac). The
element is turned on with a heater relay located on the PCB assembly. A heater thermal cutout (TCO) is found in series with the heating element (figure 6-46).

A damper door motor closes the damper door during convection and combination cooking. Of course, the regular fan blower operates all the time the oven is on.
The convection fan motor has two separate fan blades moving across the heater and
the outside fan to cool the fan blower motor. The motors are controlled with a separate motor relay within the PCB assembly. The heater thermal cutout (TCO) is a
safety device to prevent runaway convection heater temperatures.

Samsung MG5920T convection oven

Besides a regular microwave oven, this oven has a browning element that operates across the low-voltage power line. A browner relay applies ac voltage (120 Vac)
to a 1000-watt heating element and is controlled by the PCB assembly. The heater
control sensor is found in the other leg of the power line, which senses heat and fire
conditions (figure 6-47). Remember, the browning element remains dark in color
when real hot. Do not touch the browning element.

Samsung MG5920T heater replacement

To replace the heater cover and heating element, disconnect all lead wires from
the heater terminals. Remove hex bolts securing the heater. Take out the heater and
stopper heater from the heater cover (figure 6-48). Remove the screws holding the
heater cover.

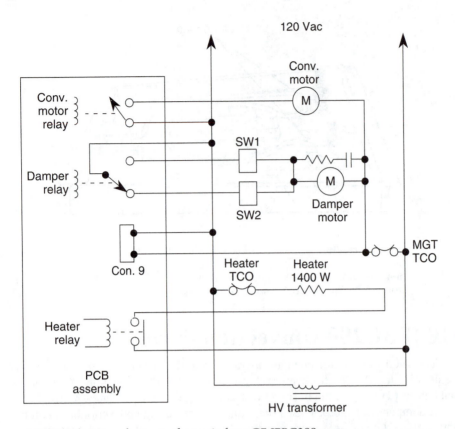

6-46 The heating element schematic for a GE JEBC200 oven. Courtesy General Electric Co.

Tappan convection oven tests

Disconnect the power, remove the wrapper, and discharge the high-voltage capacitor. Remove the wires from the browning element. Connect the ohmmeter across the terminals; if the element is open, it will have an infinite reading. The element is good with a 45- to 60-ohm measurement.

To remove the browning element, remove the wires from the ends and remove the nuts from the threaded studs holding the element in place. Remove the element by pulling out one side at a time. Be careful not to scratch the paint when removing the element (figure 6-49).

Summary

Monitor the power line voltage across the primary winding of the transformer with a pigtail light or ac meter. Measure the ac voltage (120 Vac) across the primary winding of the power transformer to determine if problems are within the low-voltage or high-voltage circuits. No ac voltage or pigtail light indicates low-voltage problems. Leave the ac meter or pigtail light on the primary winding as a low-voltage

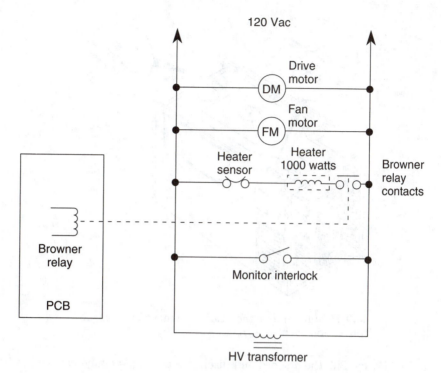

120 Vac

Drive motor

DM

Fan motor

FM

Heater sensor

Heater 1000 watts

Browner relay contacts

Browner relay

PCB

Monitor interlock

HV transformer

6-47 The convection oven circuits in Samsung's microwave oven.
Courtesy Samsung Electronic America, Inc.

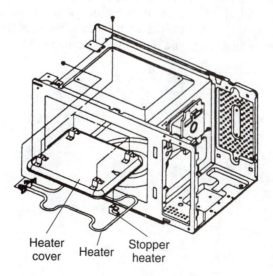

Heater cover   Heater   Stopper heater

6-48 How to remove and replace heating element in a Samsung oven. Courtesy Samsung Electronics America, Inc.

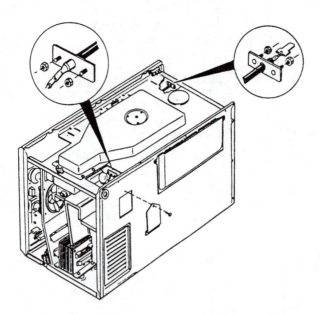

6-49 How to remove the heating element in a Tappan convection oven. Courtesy Tappan, brand of Frigidaire Co.

monitoring device. Pull the ac cord and discharge the high-voltage capacitor when doing tests of the oven circuits.

Use lights and various motors to determine what part of the low-voltage circuits are functioning. The defective component might be between the last working part and the ac meter monitor. A pigtail light or neon circuit tester across the suspected component will light up with an open switch, thermal cutout, and fan motor. When the power line voltage is measured at the primary winding of the power transformer, with no heat or cooking, suspect a defective component in the high-voltage circuits.

7
CHAPTER

Troubleshooting high-voltage problems

The high-voltage (HV) basic circuits consist of a magnetron, HV capacitor, diode, and power transformer. Power line voltage is applied to the primary winding of the power transformer and is then boosted with a voltage-doubler circuit. When a high dc voltage is fed to the magnetron, the tube oscillates, sending RF energy from the antenna to the waveguide assembly. The RF energy enters the oven cavity and cooks the food in the oven (figure 7-1).

7-1 In early microwave ovens, the fan blower is mounted on the magnetron.

Low- and high-powered ovens

The high voltage and current is much lower in the low-powered microwave ovens. It can vary from 1.5 to 2.2 kV, with a current below 200 mA. The high-voltage transformer might be wound with smaller-diameter copper wire, and the transformer can also be smaller in physical size. The magnetron tube is also smaller and operates a lot cooler than those in high-wattage ovens.

The domestic, high-powered ovens operate in the 700- to 900-W range. They use between 2 kV and 3 kV, and their current is 250 to 350 mA. You will find larger power transformers and magnetron tubes in the high-powered ovens. Notice that the high-voltage capacitors have an increase in capacity and a higher working voltage.

In the commercial microwave ovens, the wattage can vary from 1000 to 1300 watts (Table 7-1). The high voltage is usually between 2.5 kV and 3.75 kV, with a plate current of 350 to 750 mA. All high-voltage components are larger and made of higher breakdown material. You might find two different magnetron tubes in some high-powered models. These microwave ovens operate a lot hotter and have very warm exhaust features. Always replace the magnetron and transformer with the same part number. If a universal magnetron is used for replacement, make sure it has the high-voltage and wattage capabilities the oven requires.

Table 7-1. Commercial ovens with plate current and voltages with oven wattage

Model	Magnetron current	Magnetron voltage	Oven wattage
Amana RC-10	400–500 mA	3.0–4.5 kV	1000
Litton 70-40	600–650	3.0–3.25 kV	1000
Litton 80-50	600	2.0–2.75 kV	1300
Litton 550	500–600	3.2–3.75 kV	1300
Sharp R-22	350–450	2.5–3.75	1000

Samsung HV circuit operation

When the START pad is touched, the main relay and power control relay are energized by the touch control circuit. The oven lamp comes on. The fan motor rotates and cools the magnetron by blowing the air coming from the intake on the back panel over the magnetron fins. After cooling the fins, this air is directed into the oven to blow out the vapor and odor in the oven through the vents.

The drive motor rotates for even cooking. The rotation speed of the turntable is 3 rpm. The ac voltage is applied to the primary winding of the power transformer.

3.5 volts ac is generated from the filament winding of the high-voltage transformer. This 3.5 volts is supplied to the magnetron to heat the magnetron filament through two noise-preventing choke coils.

High voltage of 2120 volts ac is supplied from the high voltage secondary. This voltage is then increased by the HV diode and the charging of the high-voltage capacitor. This resultant dc voltage is then applied to the anode of the magnetron (figure 7-2). The first half cycle of the high voltage produced in the high-voltage transformer secondary charges the high-voltage capacitor.

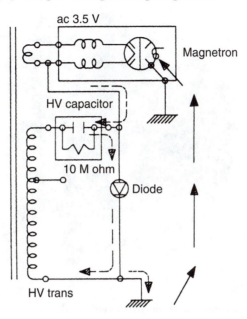

7-2 The dotted lines indicate current flow in the HV and magnetron circuit.

The dotted lines indicate the current flow. During operation of the second half cycle, the voltage produced by the secondary transformer and the charge of the high-voltage capacitor are combined and applied to the magnetron—as shown by the solid line—so that the magnetron starts to oscillate. The interference wave generated from the magnetron is prevented by the choke coils of 1.6 µH, filter capacitor of 500 pF, and the magnetron shielded case so not to interfere with radio or TV set.

The magnetron

A magnetron tube consists of a heater, cathode, metal cylinder block, and antenna (figure 7-3). The inner cylinder represents the heater and cathode. The heater is called the filament in some microwave ovens. An outer cylinder represents the anode block, strap ring, and vanes. When the heater voltage is applied, the cathode emits electrons. With a higher voltage on the anode block area, electrons start to travel from the cathode to the outer cylinder.

The magnetic field produced by an external magnet makes the electrons spin faster about the inner cylinder. These spinning electrons moving between the cathode and anode generate microwave power at 2450 MHz. A small antenna picks off the RF energy, which is transmitted to the oven cavity via the waveguide assembly.

In figure 7-4, the various parts of the magnetron tube are shown. The large heater terminals feed into the bottom of the magnetron. The metal fins are spot-welded to

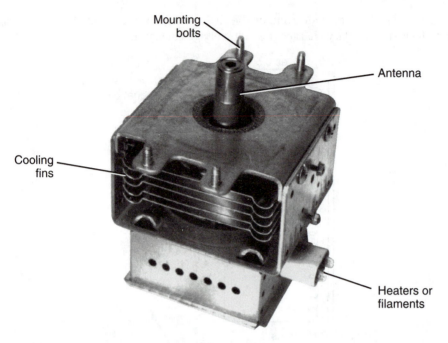

7-3 The magnetron antenna is on the top side of the tube.

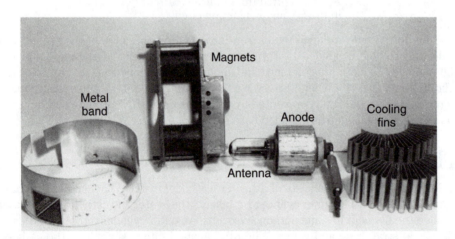

7-4 The many parts of a magnetron tube.

the outer anode cylinder, which operates at ground potential. Besides heater voltage, a high negative dc voltage is fed to the heater or cathode terminals. Notice the small metal antenna enclosed within the glassed area. The air within the magnetron is pumped out, like any vacuum tube.

A heavy magnet is placed around the anode cylinder to provide a highly concentrated magnetic field. Here two smaller, fixed magnets are used on each end, while powdered iron core magnets are found in the new magnetron tubes. Be careful when replacing or working around the magnetron tube. These strong magnets might pull a metal tool from your hand and in the process, break the glass area. Magnetron replacement is one of the most costly microwave oven repairs.

To operate, the magnetron must have a very high dc voltage (up to 4500 Vdc) and a heater or filament voltage of 3.1 volts ac (Vac). When two separate power transformers are found in the older ovens, the smaller one might furnish only ac voltage to the heater circuit, while the larger power transformer completes the voltage-doubler circuit. Usually the newer ovens have only one power transformer that furnishes ac to the voltage-doubler circuit with a separate heater winding (figure 7-5).

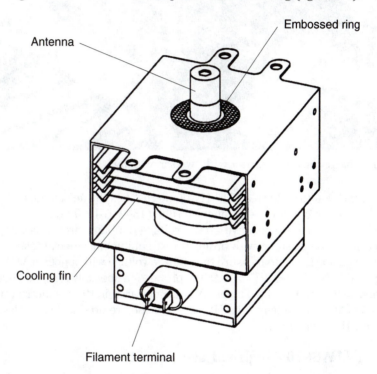

7-5 The embossed metal ring around the antenna element of the magnetron.

The voltage-doubler circuit

The ac voltage is supplied to the voltage-doubler circuit by the secondary winding of the large power transformer. The power transformer can supply 1800 to 3000 Vac to the voltage-doubler circuit. Of course, the output dc voltage is not double the ac input voltage, due to the loss in the circuit (figure 7-6).

When ac voltage is supplied to the voltage-doubler circuit, the HV capacitor is charged. As the ac current reverses the direction of current flow, the HV diode prevents the capacitor from being discharged back through the diode. The peak voltage

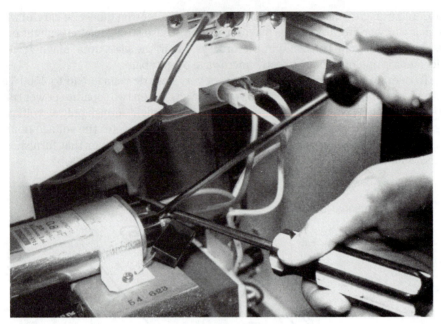

7-6 Discharge the HV capacitor with two screwdrivers.

could be from 2000 to 6000 volts. The HV diode rectifies the ac voltage and provides a high-dc source to operate the magnetron tube (1800 to 4500 Vdc).

The high voltage developed within the voltage-doubling circuits is very dangerous and cannot be measured with ordinary test equipment. Special test equipment is needed. Never attempt to measure the high-dc voltage with a pocket VOM. Use either a special HV dc meter, VTVM with HV probe, or a specially designed microwave oven Magnameter. Notice the danger HV warnings around the magnetron tube area (figure 7-7). Always discharge the HV capacitor before attempting to clip the test leads to any HV component.

Samsung MW5820T high-voltage circuits

The high-voltage circuits consist of a power transformer, capacitor, diode, and magnetron (figure 7-8). The high-voltage winding of the power transformer generates 2120 Vac into a voltage-doubler circuit. The high-voltage capacitor (0.91 μF) has a 10-megohm parallel bleeder resistor to bleed off the high voltage when the oven is shut down. An HV diode rectifies the ac voltage with the cathode terminal tied to the common ground. A separate beater or filament winding is fed through RF chokes and small bypass capacitors to the magnetron. Always discharge the high-voltage capacitor before servicing the oven.

Panasonic HV circuits

In most microwave ovens, the high voltage is generated by the action of the diode and the charging of the HV capacitor. This circuit is called a half-cycle double circuit. This circuit is commonly used because it is as economical as a smaller transformer, while still producing sufficient voltage.

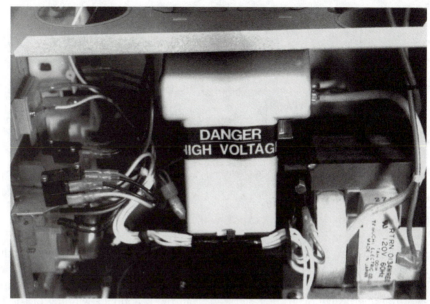

7-7 The DANGER—HIGH VOLTAGE sign reminds the technician to discharge the HV capacitor.

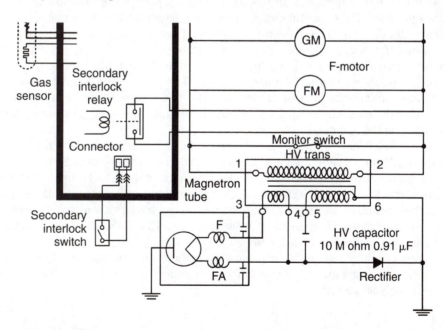

7-8 The high-voltage circuits in a Samsung MW5820T oven. Courtesy Samsung Electronics America, Inc.

The typical half-cycle doubler circuit with the capacitor and diode are connected in the HV transformer secondary (figure 7-9). Generated from the filament winding on the HV transformer, 3.3 Vac is applied to the magnetron filament through noise suppression chokers and capacitors. Two chokes and capacitors are enclosed

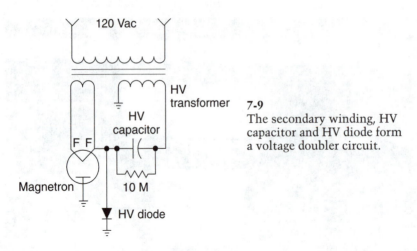

7-9
The secondary winding, HV capacitor and HV diode form a voltage doubler circuit.

within the magnetron shield case to help prevent microwaves from affecting radios and television sets.

The ac voltage of approximately 2000 volts or more (depending on the output power of the microwave oven) is generated from the secondary winding of the HV transformer. The capacitor charges through the diode during the first positive cycle of the ac from the transformer (figure 7-10). The charge path of the capacitor is shown by the dash lines. The capacitor charges to approximately 2000 volts or more. During the capacitor charging period, the magnetron is off because the diode shunts it.

During the negative half cycle, the voltage on the capacitor and the voltage across the transformer secondary winding are combined and applied across the magnetron anode, shown by the solid lines (figure 7-11). This resultant potential of approximately 4000 volts is used to oscillate the microwave from the magnetron. Notice that the magnetron is pulsed on and off at a rate of 6050 cycles per second, depending on the line rate used.

Some models have a varistor across the diode. This varistor protects the circuit in case of HV surges generated at the magnetron. The varistor normally has a very high resistance. However, when a high pulse (approximately 10 kV) is applied across it, the internal resistance decreases. This low-resistance characteristic is used to shunt the magnetron and diode. Some models do not have this varistor externally. These models have a diode called an avalanche type, which is similar to the varistor but built into the molded diode assembly.

In order to discharge the HV capacitor when the oven is turned off, a bleeder resistor (9 or 10 megohms) is provided. The time required for complete discharge is approximately 30 seconds.

Caution: Be careful: Do not assume that the bleeder has discharged the capacitor. Always discharge the capacitor before working on the HV circuit or magnetron. In some ovens, two HV capacitors are used for high-voltage output. The power selector switch must be set in the high power position to discharge both capacitors.

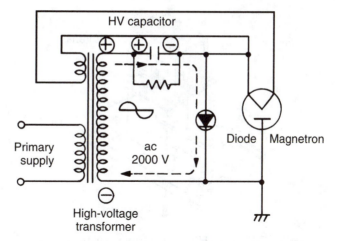

7-10 The capacitor charges through the diode during the first half cycle.

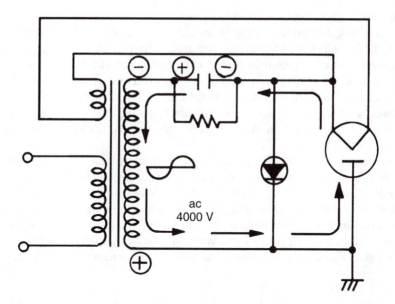

7-11 The voltage from the capacitor and high-voltage winding are applied across the magnetron.

The high-power transformer

The high-power transformer might have a primary and two secondary windings. One of these secondary windings supplies ac voltage to the heater or filament circuit of the magnetron. The large HV winding supplies high ac voltage to the voltage-doubler circuit (figure 7-12).

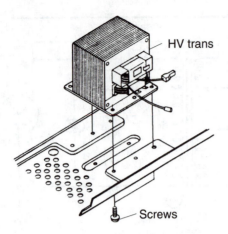

7-12 The HV transformer supplies high voltage to the voltage-doubler circuit and filament voltage to the heater of the magnetron.

You will find a separate heater transformer in some ovens. The defective power transformer might go open, have shorted turns within the HV winding, or arc through the HV winding to the metal core area. When no high-dc voltage is present at the magnetron, the power transformer can be checked with the ohmmeter. Before taking resistance measurements, pull the power cord and discharge the HV capacitor. The primary winding can have a resistance of 0 to 2 ohms, depending on the oven transformer and what ohmmeter you are using. Most primary windings have a resistance of less than 1 ohm. A resistance measurement with a digital VOM might read a fraction of 1 ohm.

The secondary HV winding might measure from 50 to 100 ohms. Some manufacturers give a reading of 20, 50, and 70 ohms for the secondary winding resistance. Use the Rx1 scale of the VOM (figure 7-13). The filament or heater resistance might show a short or 0 ohms. Always pull off one of the heater leads to the magnetron for correct resistance measurement. If not, the heater of the magnetron and power transformer winding is in parallel and you might have an open transformer winding or connection, resulting in erroneous filament or heater measurement (Table 7-2).

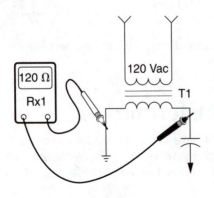

7-13
The secondary winding of T1 should have a resistance of 50 to 150 ohms.

**Table 7-2. Resistance measurement of
the HV transformer in several ovens**

Manufacturer	Primary	Secondary	Filament or heater
Amana	0–1	109–130	0–1
Goldstar	0.4–0.6	130–150	0 ohms
Hardwick	0.16	60	0–1
Panasonic	0–1	80–120	0
Samsung	0	50–70	0
Sharp	0–1	53	0–1
Whirlpool	0.2	55–61	0

In some older ovens if you look down inside the top of the magnetron, you might be able to see the filaments light up—but most are covered with a metal guide cover and will not show the heater lighting up. To determine if the heater of the magnetron or the heater winding of the transformer is open and has a higher than normal resistance reading, take an accurate resistance measurement. Remove the filament transformer lead from the magnetron to make an accurate resistance reading. The normal heater or filament should have a resistance of less than 1 ohm. Suspect a high resistance heater measurement when the oven symptom is slow cooking.

If the heater winding from the power transformer remains in doubt, disconnect both heater leads from the magnetron and take an ac voltage reading. Make sure the filament leads are disconnected from the high-voltage diode and capacitor circuits. When checking the filament voltage, keep meter leads away from the oven. Clip the meter into the circuit with extra test leads. The heater voltage of most ovens range from 2.8 to 3.6 Vac (figure 7-14).

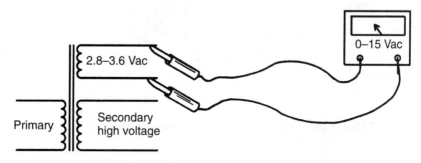

7-14 The filament voltage range from 2.8 to 3.5 volts in the HV transformer.

Damaged high-voltage transformer

The primary winding of the high-voltage transformer might overheat and damage the winding. Lightning or an outage on the power line might damage the primary winding, or a shorted magnetron. The HV capacitor also might destroy the secondary winding before the fuse opens. Remove the secondary leads from the HV transformer. Fire up the oven and take notice if the transformer runs hot and produces an overloaded noise or groaning sound. Often the lights in the oven and on the service bench will dim when the oven starts up.

Replace the transformer if it runs hot and begins to smoke. Usually, a breakdown in the high-voltage winding will produce a crackling noise and a pungent odor will come from the windings. Excess tar or transformer drippings indicate an overheated transformer. Check other high-voltage components to determine which part damaged the high-voltage transformer.

Panasonic HV transformer tests

Remove the connections from the transformer and check the winding continuity. The spade lugs pull off the primary and secondary winding terminals. The secondary HV winding should have a cold resistance measurement of approximately 80 to 120 ohms. The filament winding should have a measurement of approximately 0 to 1 ohm (figure 7-15). Before making any resistance or voltage measurements within the HV section, discharge the HV capacitor.

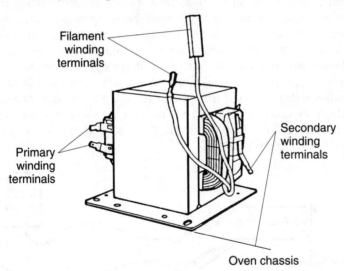

7-15 Check the resistance of the power transformer with a continuity test.

Samsung HV transformer tests

Remove the connectors from the transformer terminals for a continuity measurement. Mark down the color-coded leads. Measure the resistance of the primary winding (0.33 ohms). Now take a low resistance measurement of the high-voltage leads. You will find only one HV spade terminal, as one side of winding is grounded to the framework of transformer (65 to 85 ohms). The ac voltage of secondary winding should not be measured with the small VOM or DMM. Use the Magnameter for this test.

Filament transformer tests

Some microwave ovens have a separate filament transformer in the HV circuits. In other words, there are two separate transformers providing voltage for the magnetron (figure 7-16). The primary winding should have a resistance from 0 to 20 ohms, and the filament terminals should have a resistance of less than 1 ohm. This filament transformer has HV insulation, just like the large power transformer, as it is connected to the HV dc source. Of course, it is physically much smaller than the HV transformer.

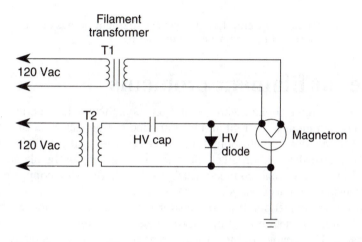

7-16 A separate filament transformer can be found in early ovens.

Microwave ovens with digital controls have a low-voltage transformer to supply ac voltage to the control circuits. You might find several secondary windings furnishing 2.5 to 3.3 Vac and 20 to 28.5 Vac. The low-voltage power transformer can be checked with ac voltage and resistance measurements. Check the manufacturer's oven schematic for correct voltage and resistance readings.

Besides open conditions, the power transformer might cause intermittent and erratic oven operations. Check the transformer filament lug terminals on the magnetron heater connections for burned crimped connections and wiring. When found, replace the crimped connections if possible. Sometimes these connections can be cleaned up and soldered to the component terminals.

Poorly crimped solder lugs on the power transformer's enameled wiring can cause intermittent or dead oven conditions. Often, these lugs are crimped right through the enameled wires. In time, this can cause a poor connection. Simply scrape the wire with a pocketknife where the crimped connection ties to the wire winding (figure 7-17). Apply rosin core soldering paste to the scraped area. Now, solder the crimped lug to the scraped wire. Always check these transformer connections when you find that an oven operates erratically or intermittently.

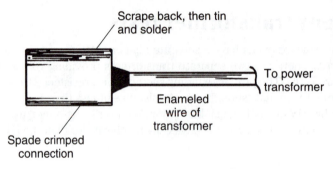

7-17 Check the clip that fits over the filament prongs of the magnetron for poor-burned connections.

Heater or filament problems

Slow or improper cooking can result from poor heater connections. Because there is a great amount of current and heat applied to the heaters of the magnetron tube, poor connections result. Sometimes excessive heat on the filament connections build up tarnished or burned connections. Overheating of the filament connections result in poor voltage connection (figure 7-18). Brass or copper connectors become tarnished and overheated.

Remove the connections from the heater terminals on the tube. Scrape or file down the heater contact connections of the magnetron. Replace with new connectors if possible. If the cable wire is burned too badly, replace the whole cable assembly. This must be obtained from the manufacturer. You might end up, after cleanup, soldering the connections to the heater pins. Always replace burned connectors when a new magnetron is installed.

GE low-voltage transformer tests

The low-voltage transformer is mounted to a service tray on top of the oven. Check the continuity of low-voltage transformer leads at the smart board. There are four different windings on this transformer. The primary winding (W-W) has a resistance of 43.3 ohms. There is a 13-volt secondary winding (R-R) of 1.3 ohms, a 3-volt winding (Y-Y) of 1.1 ohms, and an 8.6-volt winding (BL-BL) of 3.1 ohms.

You will find the resistance of the low-voltage transformer winding will have a different resistance in various models of microwave ovens. In Samsung's MG5920T oven, the primary has a resistance between 270–285 ohms. While in the Samsung MW2500W oven, the primary resistance is 15–30 ohms. But in the Goldstar oven, the low-voltage transformer resistance is approximately 160 ohms.

You can check the low-voltage transformer by disconnecting all lead connections. Clip a test lead with alligator clips to the primary winding (figure 7-19). This winding has the most resistance in ohms. Plug the ac cord into the receptacle and measure the ac voltage at each secondary winding. If the transformer begins to heat up, suspect a leaky primary winding. Replace it with the original.

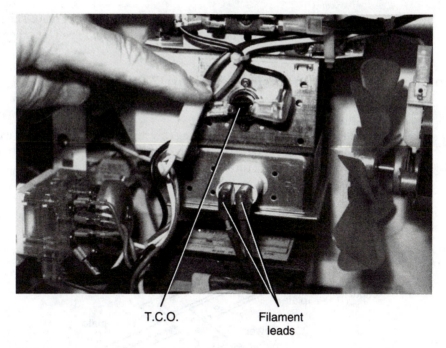

T.C.O.   Filament
leads

7-18 Remove the filament leads and check for low resistance of the heater in magnetron tube.

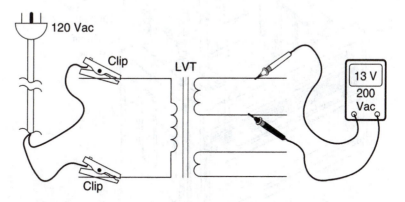

7-19 Remove leads of the low-voltage transformer, clip the ac lead to primary, and measure the ac voltage at the secondary windings.

GE JEBC200 power transformer test

Always discharge the high-voltage capacitor before touching any oven component or wiring. Disconnect the primary input terminals and measure the resistance of the transformer with an ohmmeter. Check for continuity of the coils with an ohmmeter.

On the R×1 scale, the resistance of the primary coil should be less than 1 ohm. The resistance of high-voltage winding should be approximately 60 ohms. The resistance of the filament coils should be less than 1 ohm.

To remove the power transformer (figure 7-20):

1. Disconnect power, remove top covers, and remove from the wall.
2. Discharge the high-voltage capacitor.
3. Remove the four screws that secure serving tray and ground wire next to cap.
4. Disconnect and mark the wire leads from power transformer and magnetron.
5. Lift the servicing tray up and turn sideways towards low-voltage transformer side. Be careful not to damage the smart board.
6. Remove the four screws that hold transformer.
7. Remove the four screws from the bottom holding transformer to base plate.

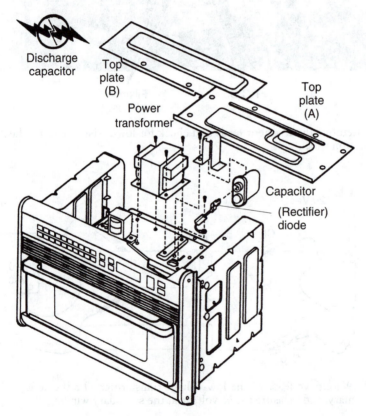

7-20 Remove top plates A and B to get at the HV transformer in a GE oven. Courtesy General Electric Co.

Panasonic's positive lock connector

The positive lock connector is a specially designed locking connector. You will find this connector in many lead wire connections. To remove the connector, pull the lead wire by pressing an extruded lever in the center of the receptacle terminal (figure 7-21).

In the NN-9807 oven, when connecting two filament leads to the magnetron, be sure to connect the lead wires in the correct position. The lead wire of the HV transformer should be connected to the F terminal and the lead wire from the HV capacitor should be connected to the FA terminal (figure 7-22).

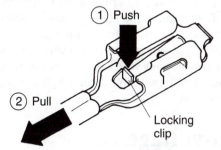

7-21 Panasonic's positive lock connection is found on the magnetron and thermal cutout. Courtesy Matsushita Electric Corp. of America

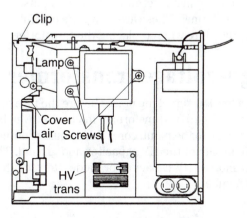

7-22 The filament leads and connection on the metal magnetron tube.

If there is no dc voltage applied to the magnetron, this might be caused by a shorted HV winding of the power transformer. First, discharge the HV capacitor. Then measure the resistance between the heater terminal on the magnetron and chassis ground. One side of the power transformer secondary winding is connected to the oven chassis ground, while the other end connects to the heater circuit.

In some ovens, you might find a high and low connection, providing a method to increase or decrease the amount of power produced by the magnetron. Normal resistance should be measured from the high side of the secondary winding to the chassis ground. In case a very low measurement is found, disconnect the HV transformer lead from the magnetron heater terminal. Then measure the transformer's HV winding. Compare the reading to those given in the manufacturer's literature. A lower-than-normal reading indicates shorted turns in the transformer, or the winding has arced over to the metal core area. If the transformer winding is normal, suspect a shorted or leaky magnetron tube.

You might find the power transformer has a loud buzz when the oven is operating. A low hum noise is normal. Although the oven operates correctly, the loud buzzing noise might be very disturbing to the operator. You might find the transformer buzz is intermittent and can only be heard when the oven is under a cooking load. Replace the noisy transformer with one having the original part number. Because these transformers are dipped after assembly, it's impossible to tighten them up to stop the noisy condition.

Typical high-voltage transformer replacement

1. Discharge high-voltage capacitor.
2. Disconnect all lead wires of the high-voltage transformer.
3. Mark down where each color-coded wire is attached.
4. Remove all mounting metal screws or bolts and nuts.
5. Pull out the transformer. These are heavy. Be careful.
6. Reverse the procedure when replacing the transformer.

Arcing high-voltage transformer

You might hear a few loud cracking noises before the fuse opens up, or you could smell something warm. The HV transformer might arc over between windings or from the high-voltage side to the metal core. Take a quick continuity resistance measurement between one side of the HV capacitor and ground. The symptoms can be similar to those of a shorted HV capacitor or diode. Replace the power transformer with the exact part number.

The HV capacitor

When the HV capacitor goes open, shorted, or leaky, no high voltage will be available at the magnetron. Before checking the capacitor or any component in the oven, discharge the HV capacitor. Be safe. Always discharge the capacitor. Use a couple of insulated screwdrivers for this purpose.

The HV capacitor is mounted close to the power transformer and magnetron circuits (figure 7-23). These oil-filled capacitors have a capacity rating from 0.64 to 1 µF with a 2- to 3-kilohm voltage rating. In early models, the physical size of these capacitors was quite large compared with those used today (figure 7-24). Many of these capacitors can be substituted from other microwave ovens.

In some circuits, you might find a bleeder resistor across the capacitor terminals. This resistor bleeds off the capacitor charges after the oven is turned off (figure 7-25). Before checking the capacitor for shorts or leakage, disconnect one side of the 10-meg resistor. Set the ohmmeter to the 10-k scale and preferably use a meter with a 6-V battery.

7-23 The HV diode is usually mounted close to the HV capacitor.

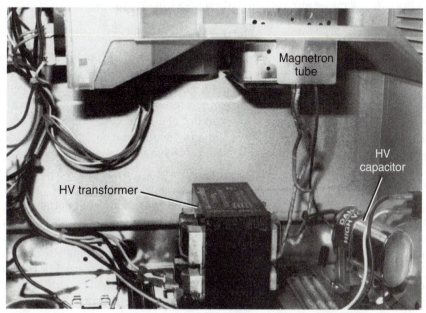

7-24 The HV capacitor is usually clamped to the metal side wall of oven.

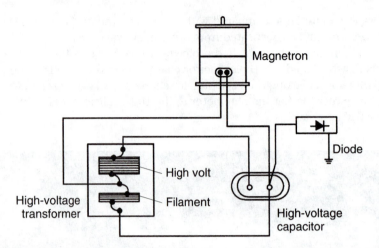

7-25 A pictorial diagram of the high-voltage circuits.

Remove connecting cables to one side of the capacitor and place the meter leads across the terminals. The capacitor will charge up with the meter hand going up the scale and then slowly discharge. Reverse the meter leads, and the same process should occur. Replace the HV capacitor if it will not charge.

You might find an HV capacitor with only a few ohms of resistance between the connecting terminals. Discard the capacitor because of leakage or shorted conditions. Test the capacitor for leakage between the terminals and the outside metal case. No continuity measurement should be shown between can and terminals. Any reading less than 100 k indicates a faulty capacitor. If in doubt, sub another oil-filled capacitor. The following are a few actual capacitor failure problems:

Dead/no cooking

The rest of the oven operated perfectly, but there was no cooking or heat. No dc high voltage was measured at the magnetron. When discharging the capacitor for tests, no arcing or snapping noise was heard. At first, the HV diode was checked and appeared normal. One terminal connection was removed from the capacitor with the ohmmeter clipped on. The capacitor would not charge in either direction. Another HV capacitor was clipped into the circuit and the oven came on. The open capacitor was replaced with the manufacturer's exact replacement part.

No cooking/no heat

Again, all functions of the oven except the water test appeared normal. The water temperature was cold. A dc high-voltage measurement of the magnetron indicated no high voltage, although 120 Vac was measured at the primary winding of the power transformer. Because the HV diode causes more trouble than anything else in the voltage-doubling circuit, it was checked right away. The diode was normal. A resistance measurement between the filament of the magnetron and ground measured 72 ohms. Either the magnetron or capacitor was leaky. The capacitor measured 1.3 ohms across the terminals and was replaced (figure 7-26).

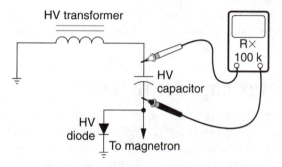

7-26 Check the resistance across HV capacitor with leads removed for a leakage test.

Popped and went out

The customer complained that she was cooking with the oven and it gave out a loud pop and then quit. A 15-amp fuse was replaced, but still no heat or cooking. Instead of measuring a high-dc voltage at the magnetron, low voltage was present. The HV diode appeared normal. The leads from one side of the capacitor were removed. Another HV capacitor was clipped into the circuit and the oven began to cook. A resistance measurement across the capacitor terminal indicated that the capacitor had internally broken down. It's best to remove the HV diode from the circuit when making high-resistance measurements.

Panasonic's HV capacitor tests

Before attempting to take resistance or voltage measurements in the oven, discharge the HV capacitor. An extremely low crack might indicate the capacitor or diode is open. Clip the meter across the HV capacitor terminals.

1. Check the continuity of the capacitor with the meter on its highest ohm scale.
2. A normal capacitor will show continuity for a short time and then indicate 9 megohms once the capacitor is charged.
3. A shorted capacitor will show continuous continuity.
4. An open capacitor will show a constant 9 megohms.
5. Resistance between each terminal and the chassis should be infinite.

Typical high-voltage capacitor tests

1. Pull the ac cord and discharge the HV capacitor.
2. Set ohmmeter to the highest range.
3. Remove the leads from HV capacitor.
4. A resistance measurement across the capacitor terminals will show continuity for a short time.
5. The open capacitor will show open or infinite resistance.
6. The shorted capacitor will show continuous resistance reading and might have a bulged or deformed body.

7. A leaky capacitor will show some signs of resistance.
8. No resistance measurements should be made between each capacitor terminal and outside metal case.

9-volt meter capacitor test

Several oven manufacturers recommend checking the HV capacitor with a meter having a 9-volt battery. Discharge the HV capacitor and remove the leads to the capacitor terminals. Set the ohmmeter on R×10,000 scale. The meter hand should momentarily deflect upward to show continuity, then return to show infinite ohms. Now, reverse the leads. The meter should give the same indications. The same should hold true with a DMM (figure 7-27). The numbers will go quite high and slowly discharge downward.

7-27 Discharge the HV capacitor before taking measurements in the oven.

If the meter hand goes to a low-ohm resistance, the capacitor is leaky. With an infinite measurement and no HV arc-over, the capacitor might be open. If all readings appear normal on the capacitor, check the HV diode, magnetron, or HV power transformer.

GE JEBC200 high-voltage capacitor test

Discharge the high-voltage capacitor before touching any components and wiring. If the capacitor is open, no high voltage will be available to the magnetron. Disconnect input leads and using the ohmmeter, check for a short or open between the terminals. Set the meter to the high-ohm scale. If the high-voltage capacitor is normal, the meter will indicate continuity for a short time and then should indicate an open circuit once the capacitor is charged. If that is not the case, use an ohmmeter to check the capacitor to see if it is shorted between either of the terminals and case. Replace the capacitor if it is shorted.

1. Disconnect the power and remove oven from wall to replace high-voltage capacitor.
2. Remove the top covers and discharge the capacitor.
3. Disconnect and mark the wires.

Samsung MW2500U HV capacitor test

1. Discharge high-voltage capacitor.
2. Check continuity of capacitor with meter at the highest ohm scale.
3. Once the capacitor is charged, a normal capacitor shows continuity for a short time, and then indicates 9 megohms (bleeder resistor resistance).
4. A shorted capacitor will show continuous continuity.
5. An open capacitor will show a constant 9 megohms.
6. Resistance between each terminal and chassis should be infinite.

Typical Samsung HV capacitor test

Discharge HV capacitor before taking continuity and resistance measurements. Remove leads from the HV capacitor. Check continuity of capacitor with meter at the highest ohm scale. The measurement should be above 10 megohms if no bleeder resistor is found across the capacitor terminals. A shorted capacitor will show low resistance across terminals. An open capacitor will not charge up or have a resistance above 10 megohms with reversed test leads. The resistance between any terminal and metal case should be infinite.

Replacing the high-voltage capacitor

The open or leaky high-voltage capacitor should be replaced with the original part number. In some Amana ovens, the HV capacitor has an internal fuse. The oil-filled capacitor must be replaced with the same type of capacitor. Make sure the 10-megohms bleeder resistor is removed for the open capacitor test. In the Goldstar microwave oven, you might find that the HV capacitor and diode have a built-in bleeder resistor within each component (figure 7-28).

Remove terminal leads from the high-voltage capacitor. In most ovens the capacitor leads are interchangeable. In the M-260, M-270, and MM-270 Hardwick microwave ovens, observe correct polarity—the negative terminal is connected to the HV diode. Remove two metal screws holding the capacitor to the metal cabinet. Replace the defective capacitor with a new one. To install, just reverse the procedure for removal of the capacitor. You might find that some capacitors have slip-on connections, while others are soldered.

Be careful when replacing the HV capacitor with universal replacements or capacitors from other manufacturers. Make sure the capacitor has the same capacity

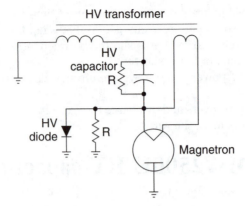

7-28 In Goldstar ovens, a bleeder resistor
is found across the HV capacitor and diode.

and working ac voltage. For instance, a 0.9-µF capacitor at 2000 WVac can be replaced with a 0.951 µF at 2100 WVac, but not reversed. You will find that different
microwave oven circuits have many different sizes of capacitors (Table 7-3.). The
working voltage could vary from 1700 WVac to 2100 WVac in the Panasonic ovens.
It's best to replace the high-voltage capacitor with the same capacitance and working voltage. Never replace the capacitor with a lower working voltage.

**Table 7-3. High-voltage capacitor
values with different oven wattages**

Manufacturer	Model	Capacity	Oven wattage
Goldstar	ER410M	0.6 µF	450 W
Goldstar	ER653M	0.76 µF	650 W
Goldstar	ER711	0.86 µF	700 W
Norelco	RR7000	1 µF	650 W
Panasonic	NN-9807	0.935 µF	700 W
Sharp	R-7350	0.80 µF	600 W
Sharp	SKR-9105	0.53 µF	500 W

Tappan high-voltage capacitor removal

1. Discharge the high-voltage capacitor (figure 7-29).
2. If you find the capacitor open or leaky, remove the high-voltage diode and
   capacitor leads.
3. Remove the screw in bracket.
4. Replace the HV capacitor with an original replacement part.

The HV diode

The silicon HV diode and magnetron are the most troublesome components in the
HV circuit. These diodes come in many sizes and shapes. They can be interchanged

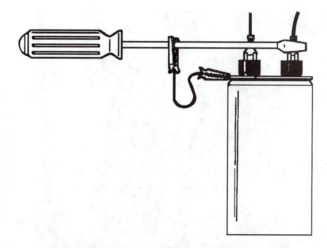

7-29 Discharge the high-voltage capacitor.

from other ovens if they can be mounted properly. Often, the HV diode is mounted quite close to the HV capacitor (figure 7-30).

A defective diode might keep blowing the 15-amp fuse or produce a no-heat, no-cook condition. Before attempting to check the diode, discharge the HV capacitor. If the body of the diode appears quite warm, replace it (figure 7-31). In most cases, the HV diode shorts or becomes leaky.

The suspected diode can be checked with the Rx10 k ohmmeter range. Preferably use an ohmmeter with a 6- or 9-volt battery. Do not use an ordinary VOM, or infinite resistance might be read in both directions. Isolate the diode by disconnecting

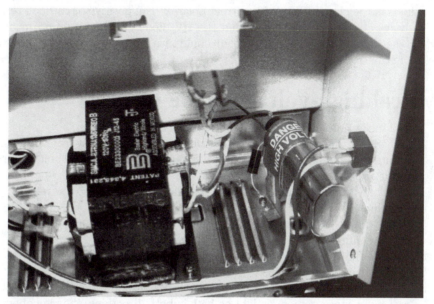

7-30 Here the HV diode is soldered in place with one side to the capacitor and the other to the chassis.

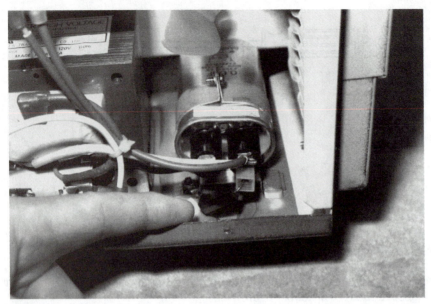

7-31 Replace HV diodes or capacitors if they feel warm.

one lead from the circuit. Clip the test leads across the suspected diode. The diode might read several hundred ohms. Now, reverse the test leads. No reading should be obtained (infinity).

Replace the diode when a low-resistance measurement or the meter shows continuity in both directions. If the diode measures below 150 kilohms, replace it. A VTVM or digital VOM is ideal to check for a leaky silicon rectifier. If in doubt, clip another diode in the circuit. Watch for correct polarity. The positive or cathode terminal is always at ground potential. When the HV diode cannot be found in older models, check the magnetron. The problem could turn out to be a defective diode inside the magnetron shield (figure 7-32).

Typical high-voltage diode tests

1. Pull the ac cord and discharge the HV capacitor.
2. Disconnect the HV diode leads. Remove bottom metal screw from chassis.
3. Rotate the ohmmeter to the highest range (preferably with a 9-volt battery inside the ohmmeter).
4. Check forward continuity; with a normal diode, it will indicate several megohms.
5. Check the reverse or backward direction; with a normal diode, it will indicate infinity.
6. Replace the HV diode if it has lower measurements in both directions.
7. Replace the diode if it runs warm to touch.
8. If the diode is measured with a FET or VTVM voltmeter, the HV diode resistance measurements might be high in both directions, showing open conditions—which is a false measurement.

7-32 In some early ovens, the HV diode is located inside the cage area of the magnetron.

GE JEBC200 high-voltage rectifier tests

Discharge the high-voltage capacitor before touching any oven components or wiring. Isolate the rectifier from the circuit (figure 7-33). Set the ohmmeter to the highest scale; read the resistance across the terminals and observe; reverse the leads to the rectifier terminals and observe the meter readings. If a short is indicated in both directions, or if an infinite resistance is read in both directions, the rectifier is probably defective and should be replaced.

1. Disconnect power and remove from the wall to replace diode.
2. Remove top covers and discharge high-voltage capacitor.
3. Remove one screw from case side of diode and pull connector off other side at the capacitor.

Samsung MW2500U high-voltage diode test

Isolate the diode from the circuit by disconnecting the wire leads. Set the ohmmeter on the highest resistance scale and measure the resistance across the diode terminals. Reverse the meter leads and observe the resistance reading again. Meters with 6-volt, 9-volt, or higher voltage batteries should be used to check front-to-back resistance of the diode; otherwise, an infinite resistance might be read in both directions. The resistance of a normal diode will be infinite in one direction and several hundred kilohms in the other direction.

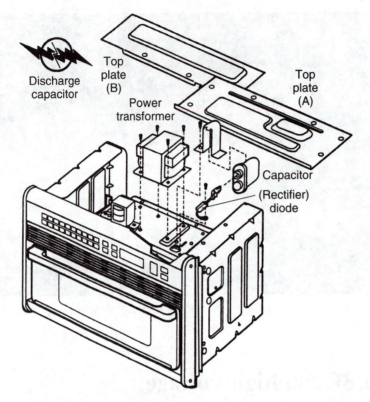

7-33 Remove top plates A and B to get at the HV diode and capacitor in a GE oven. Courtesy General Electric Co.

Norelco HV diode test

A meter with less than a 6- or 9-volt battery is usually not adequate for checking front-to-back resistance of the diode. The meter should be checked with a diode known to be good before judging a diode to be defective because of an infinite resistance reading in both directions.

Disconnect the two leads of the diode. With the ohmmeter set on the highest resistance scale, measure resistance across two diode terminals (figure 7-34). Reverse the meter leads and again observe the resistance readings. A normal diode will indicate infinite resistance in one direction; a deflection of the meter needle toward the 0 marking is indicated in the opposite direction. If continuity is indicated in both directions, or if an infinite resistance is read in both directions, the diode is probably defective and should be replaced. Suspect a leaky diode when the body appears warm.

Circuit Saver diode test

Pull the ac cord and discharge the high-voltage capacitor. If you suspect that the oven HV diode is defective or measurements are in doubt, sub the HV diode found in the Circuit Saver. A substitution of the HV diode can be made by clipping the two test leads of the tester into the high-voltage circuits (figure 7-35).

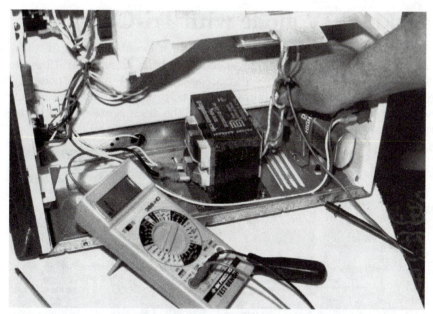

7-34 Check resistance across the diode for leakage.

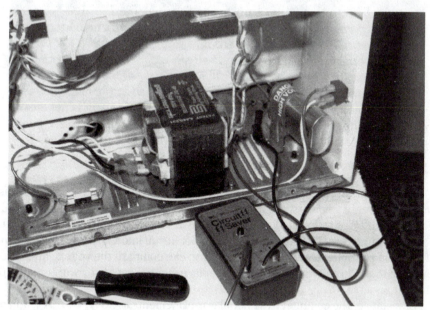

7-35 Remove one lead of the HV diode and test with the Circuit Saver.

Remove the high-voltage wire from the old diode (anode end) or the lead opposite the diode terminal that goes to chassis ground. Connect the test lead D (red) from the Circuit Saver to the high-voltage wire that was removed from the diode. Clip the black test lead (gnd) from G of tester to chassis of microwave oven. Fire the oven up and check high voltage and current with the Magnameter.

Checking HV diode with Tri-Check II

Pull the power cord. Discharge HV capacitor. Disconnect all wires to the HV diode. Turn on Tri-Check and middle light should blink. Connect Tri-Check to HV diode with black test lead (T1) to the anode and clear (T2) to the cathode of diode (figure 7-36). Do not use the test button for HV diode tests. Both lights will light with a shorted or leaky diode. No lights are on when the diode is open. Reverse the test leads and take another test. No lights should be on with a normal or open diode.

7-36 Checking the HV diode with Tri-Check II test instrument.

High-voltage diode replacement

In many cases, the high-voltage diode in the microwave oven can be replaced with universal replacements without any problems. Check the part number against the microwave HV diodes listed in the ECG, RCA, and NTE semiconductor manuals. When you do not know the capacitor part number, if you know the current of the magnetron (found with the Magnameter tester) you can compare the voltage and current with those found in the manuals. Always replace the HV capacitor with original part number if possible.

For instance, the common ECG popular replacements are 541, 542, 544, and 548. The ECG542 HV diode is interchangeable with the RCA SK9307 or an NTE542. Just check the interchangeable diode numbers with other universal replacements found in the semiconductor replacement manuals. The RCA semiconductor manual lists the SK9307 at 15 PRV (peak reverse voltage) at 15 kV average forward current at 350 mils, and VF (forward voltage drop) 14 V at 350 mils. The NTE542 silicon industrial rectifier rates PRV 16 kV at 350 mA -14 VF, which means both HV diodes are interchangeable.

Most HV diodes are enclosed in a plastic case with slip-on connections (figure 7-37). If the slip-on fittings do not match, solder diode wires to the holes in the clip terminals. The diode might be mounted with one or two screw holes. The Amana, Litton, Tappan, and others might have a diode with clip-on connections at one end and a ground eyelet on the other. Universal high-voltage diodes can be found at local electronics stores, mail order firms, or manufacturers' distributors and parts depot.

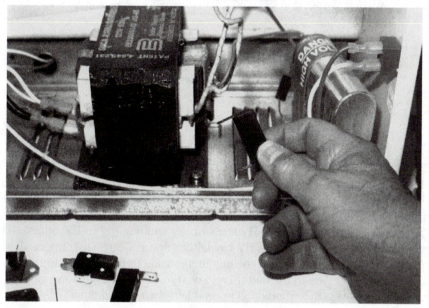

7-37 The HV diode comes in many sizes and shapes.

How to check the HV circuits

Only two test measurements are needed to check out the HV circuits. Clip the ac voltmeter across the primary winding of the power transformer (one and two). When 120 Vac is monitored at these connections, you know that all the low-voltage circuits are functioning (figure 7-38). If any other problems exist, the HV circuit and magnetron must be defective.

For the last test, measure the dc high voltage from the filament or heater connections to chassis ground. Because the heater terminal connections are covered with plastic insulation, the HV test can be made right across the HV diode. Use either an HV probe or meter such as the Magnameter. Usually, the proper high voltage across the HV diode indicates the HV circuits are functioning, except that the magnetron might be defective.

When the HV circuit and magnetron are working correctly, you will hear a loud hum from the power transformer. Sometimes the oven lights will dim a little when the magnetron is drawing current. No transformer hum might indicate a defective magnetron or no high voltage. Pull the power plug and discharge the HV capacitor.

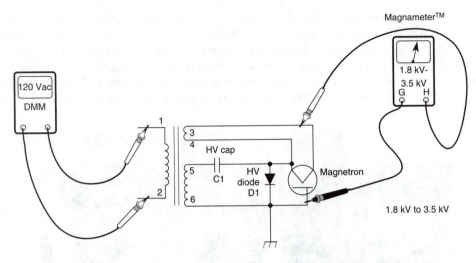

7-38 Monitor the low voltage on the secondary of transformer and high-voltage circuit with a Magnameter.

In case a dc HV meter is not available or you are afraid to take HV measurements, the HV circuits can be checked in the following manner (some oven manufacturers say it is not necessary to make HV measurements in the microwave oven circuits).

Measure the resistance of the HV transformer winding (50 to 100 ohms). If this measurement is normal, go to the HV capacitor (figure 7-39). Disconnect all leads from one side of the capacitor. Set the ohmmeter scale to R×10 k and apply the test lead to the capacitor. Watch the meter hand charge and discharge. Reverse the procedure. A normal capacitor should never read less than 100 kilohms.

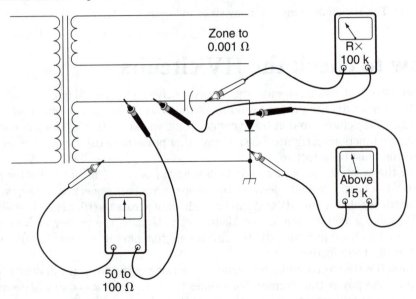

7-39 Checking the HV circuit with various resistance tests.

Check the HV diode with the ohmmeter. Disconnect one terminal so the resistance reading is accurate. You should be able to have a reading in one direction, infinite resistance with reverse test leads, and a normal diode. A normal diode can have a measurement of 150 kilohms. If the diode resistance is normal, measure the resistance from the heater terminal of the magnetron to the ground. Suspect a leaky magnetron if the ohmmeter reading is below 100 kilohms when all heater wires are pulled from the tube.

An open or poor heater terminal might prevent the magnetron from oscillating. Disconnect both heater terminals from the magnetron. Measure the resistance across these two cable ends. Most meters will read 0 ohms, while a digital VOM might indicate .001 ohms with a normal filament winding. Because the coil is made up of heavy copper wire, very seldom does the filament or heater winding go open.

Go a step further and measure the ac voltage across the heater terminals. Remove the heater cables from the magnetron. If not excessive, high voltage might be found on the heater terminals of the magnetron. Clip the ac VOM to the two heater cables of the power transformer. Keep the meter away from the metal oven area. Fire up the oven and measure the ac filament voltage. A normal filament reading is between 2.8 and 3.6 Vac. Although checking the HV circuits with the ohmmeter might take a few minutes longer, it is the safest method.

Caution: To prevent shock and injury, always be extremely careful while working around the HV circuits.

New Magnameter tests

The new Magnameter speeds up the servicing in microwave oven repair. If you service a lot of microwave ovens, this tester can make extra money. Besides safety in checking the high-voltage circuits, voltage, and current tests, this tester can indicate what part is defective. Low or improper current might indicate a defective magnetron or heater connections, with poor microwave oven cooking. No current indicates a defective magnetron if heaters light up and there is normal high voltage (figure 7-40).

The biggest advantage of the Magnameter is in measuring the high voltage at the magnetron. Most oven manufacturers discourage high-voltage tests in the high-voltage circuits. But this tester will indicate the exact high voltage applied to the magnetron in the yellow, green (normal), or red section. This tester measures up to 10 kV, way above the average oven.

If the magnetron goes open or if the heater voltage is missing at the tube, higher than normal voltage is found at the high-voltage diode. If this occurs in a commercial oven of 1000 to 1300 watts, the high voltage can increase above 5 kV. Of course, this voltage will damage all VOM or DMM test instruments. The Magnameter will indicate correct high voltage, current, and, with a red indicator light, whether the HV circuits are functioning.

When you have an oven that constantly blows the fuse, remove the fuse and connect the Circuit Saver to each fuse terminal. Now, each time the oven overloads, the

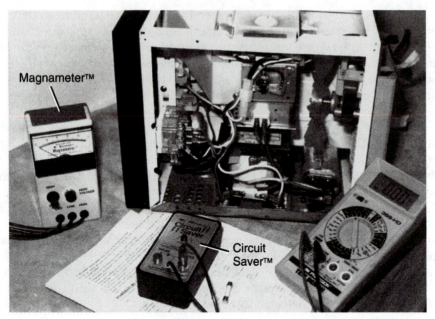

7-40 The Magnameter accurately checks magnetron, high-voltage, and oven operations.

circuit breaker within the tester will kick out; just push it in to reset the Circuit Saver. Try to locate the overloaded or shorted conditions before resetting the circuit breaker. Do not push and try to hold the circuit breaker when the short keeps kicking out the circuit breaker in the tester.

Caution: Do not touch or handle the Circuit Saver tester while the oven is operating. Always discharge the HV capacitor before connecting any test equipment to the microwave oven.

Panasonic basic microwave oven cavity construction

The microwave oven cavity is a multimade cavity resonator that is designed to resonate the microwaves emitted from the magnetron. A stainless steel, aluminum, or painted steel plate is used for the oven cavity.

The microwaves have several typical characteristics. They can be absorbed, reflected, or transmitted, depending on the materials they come in contact with. The oven cavity is made with metallic walls that reflect microwaves (figure 7-41).

Microwaves are emitted from the antenna of the magnetron and are transferred to the oven cavity through the waveguide, which is designed to transfer the microwaves to the oven cavity without any loss of the microwave energy. When the microwaves are

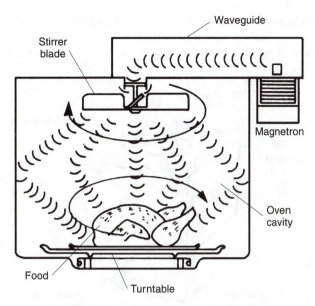

7-41 The RF energy passes through the waveguide into the oven and cooks food in the oven cavity.

fed into the oven cavity, they are either absorbed by the food or bounced around the oven cavity wall until they are eventually absorbed by the food. This difference in the energy mode will result in hot or cold spots.

In order to prevent the energy mode differences, microwave ovens might have a fan-like blade, called a stirrer blade, at the top of the oven cavity—or a turntable is provided on the bottom of the oven cavity to rotate the food. These devices prevent uneven cooking of the food by changing the energy path.

If the oven is operated empty or with large metal utensils inside, most of the microwaves are reflected back to the magnetron, where they will dissipate as heat. This excess heat might damage the magnetron or shorten its operating life. Although technological advances in the magnetron tube and in oven designs have reduced the likelihood of a catastrophic failure, this problem is the reason for the warnings in the customer instruction books against the use of large metal utensils or the empty operation of the oven.

Microwave oven hot spots

Excessive arcing and hot spots in the microwave oven can result from many different areas. Sparks or arcing occurring in the oven cavity might be caused by metal objects, bowls with metallic paint, or a temperature probe left lying in the oven or improperly inserted into the jack. Hot spots with the magnetron operation can result from:

1. Not replacing the embossed ring or mesh gasket around the neck of the antenna assembly when replacing the magnetron.

2. The radiating stirrer blade touching or being close to the oven cavity.
3. A loose or damaged waveguide assembly.
4. A defective magnetron firing from the antenna to the oven ground (figure 7-42).

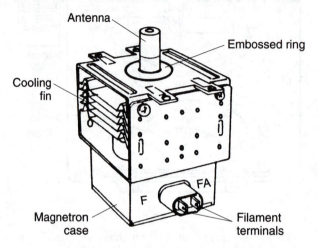

7-42 Check for cracked glass and arcing around the antenna and embossed ring.

Internal firing in the magnetron can be caused by an open filament, cracked antenna glass, or RF capacitors built right inside the magnetron. The magnetron must be replaced if arcing is found inside the tube.

Checking Tappan oven power wattage

There are two power ratings indicated on the model/serial plate of each microwave oven. For example, the following information may be printed below the model/serial number:

120 V, 60 Hz output 700 W
Frequency 2450 MHz
1500 W, 3 wire ac only

The 1500 W is the amount of power wattage needed to operate the oven during full power cooking cycle (input power rating). Output 700 W, on the other hand, is the amount of power wattage produced by the magnetron assembly during a full power cooking cycle (output power wattage).

When testing a microwave oven for power wattage, the output power wattage must be checked. To perform this test and obtain more accurate results, two 1-liter (microwave safe) containers and a glass Celsius scale thermometer should be used.

Fill the two 1-liter containers with tap water. Stir the water in each container with the thermometer and record each container's temperature. Find the average temperature of the two containers by adding the two temperatures and dividing the total by two. Place both containers in the center of oven. Set the oven power level to the highest setting and operate the oven for exactly two minutes. Note that when testing an oven with a timer, use your watch for better timing accuracy.

When the two minutes have elapsed, stir each container with the thermometer and record the temperatures. Again, find the average of both containers by adding the two temperatures and dividing the total by 2. Subtract the average temperature at the start and average temperature at the end to determine the rise in temperature.

Now multiply the temperature rise by a factor of 70 to determine the approximate output power wattage. Compare the results with that printed on the oven model/serial plate. If the result is within 10% of the oven output, the oven has sufficient cooking capabilities. When the results are lower than the 10% stated value, troubleshoot the oven. A weak magnetron or low high voltage can cause low output oven wattage.

8
CHAPTER

Locating and replacing defective magnetron

Magnetrons produce microwaves through the interaction of strong electric and magnetic fields (figure 8-1). When power is supplied to the magnetron from the high-voltage (HV) transformer, electrons are emitted from the cathode. They are accelerated toward the anode's 4000-volt potential. A magnetic field along the tube axis forces electrons into a rotating spoked-wheel pattern before they reach the anode. As electrons pass the anode cavity openings, resonant cavities produce electromagnetic oscillations. These microwave oscillations are trapped and radiated by an antenna (figure 8-2). The microwave electric field reverses itself 2450 million times a second. Polarized food or water molecules (separated electric charges) are vibrated back and forth in this field, and this tremendous molecular friction produces heat.

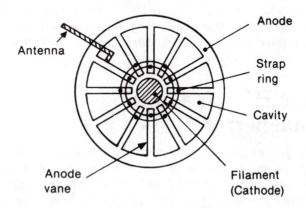

8-1 The anode of the magnetron is a hollow piece of iron with open areas that are constructed of an even number of anode vanes pointing toward the filament wall. Courtesy Matsushita Electric Corp. of America

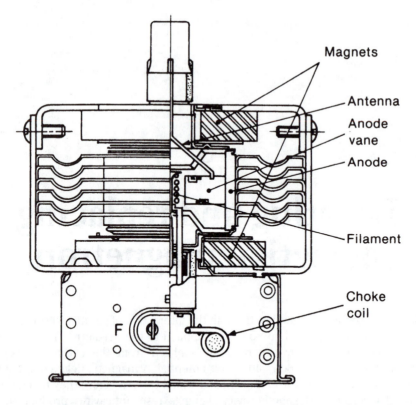

Magnets

Antenna

Anode vane

Anode

Filament

Choke coil

F

8-2 A magnet is located at the top and bottom of the internal part of a magnetron tube. Courtesy Matsushita Electric Corp. of America

You should perform several tests to determine if the magnetron is defective or if the trouble is in another circuit. Insufficient or no high voltage at the magnetron can be caused by a leaky magnetron or defective HV circuits. A leaky or shorted magnetron might lower the high voltage level (figure 8-3). Sometimes a shorted magnetron will keep blowing the 15-amp fuse. Improper high voltage or a defective magnetron produces a no-heat, no-cook symptom.

Discharge HV capacitor before magnetron tests

Before testing the magnetron or attaching test instruments to the tube, discharge that high-voltage capacitor. This cannot be stressed enough. Take time to discharge the HV capacitor with a screwdriver and test lead before connecting the magnameter and check the filament resistance (figure 8-4). Clip an alligator clip across capacitor lead while changing the magnetron. This capacitor can cause accidental shock which can be fatal.

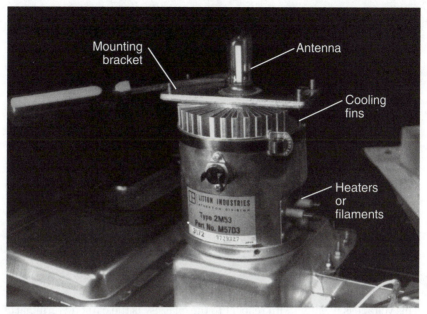

8-3 A leaky or shorted magnetron must be replaced.

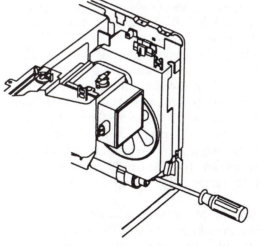

8-4
Discharge the HV capacitor before attempting to test and remove it.

How to test

A current and HV test will quickly determine if the HV and magnetron circuits are normal. Correct high voltage at the heater or filament terminals of the magnetron might indicate that the voltage-doubler circuits are functioning. High voltage can be monitored at the high side (negative) of the HV diode (figure 8-5).

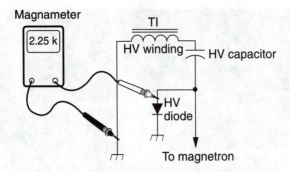

8-5 High voltage can be tested across the HV diode with Magnameter test instrument.

Caution: Be very careful. You are measuring up to 4500 Vdc in some microwave ovens.

This voltage is negative with respect to the chassis, as the cathode terminal of the diode is connected at ground potential. The high voltage can be measured with a high-voltage dc voltmeter, high-voltage probe, or a Magnameter.

When using a regular HV probe (found in the TV shop), the ground clips of the probe must be connected at the negative side of the HV diode. The probe tip must be clipped to the chassis ground. Lay the probe on a book or manual and clip the test leads to the probe tips. Do not hold the HV probe in your hands. Although the HV probe might not give a very accurate indication, at least you know high voltage is present.

An HV voltmeter or VTVM with an HV probe gives an accurate voltage reading because the correct voltage scale and polarity are available on a VTVM. Again, clip the meter into the circuit with test leads. Keep the voltmeter insulated from the metal oven area. Place the voltmeter on a service manual or book. Always discharge the HV capacitor before attempting to take any voltage measurements. Low-voltage measurements at the filament terminals and HV diode might indicate trouble in the HV circuits or a leaky magnetron. Excessively high voltage could indicate that the magnetron is open.

A new oven test instrument called the Magnameter is ideal to check voltage and current measurements within the magnetron circuit (figure 8-6). A correct negative voltage at the HV diode indicates the HV circuits are normal. Simply flip the toggle switch to the low reading and measure the current pulled by the magnetron. No current reading indicates the magnetron is open. Lower current than normal might indicate a low emission tube. Higher than normal current measurements could indicate a leaky magnetron.

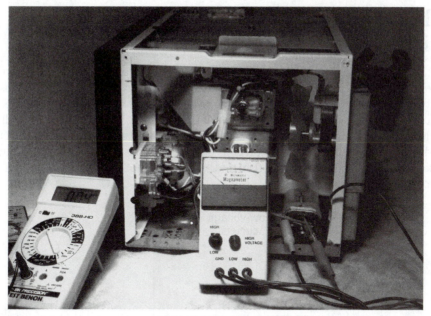

8-6 A Magnameter connected to the magnetron tube.

GE magnetron assembly test

Caution: High voltages are present during the cook cycle, so extreme caution should be observed. Disconnect oven from power source and discharge the high-voltage capacitor before touching any oven components or wiring.

To test for an open filament, isolate the magnetron from the high voltage circuit. Simply remove filament wires from heaters of magnetron. Check the resistance across the magnetron filament leads—the meter should indicate less than 1 ohm.

To test for a shorted magnetron, connect ohmmeter leads between the magnetron filament leads and chassis ground. This test should indicate an infinite resistance. If there is little or low resistance and the magnetron is grounded or leaky, then both must be replaced.

Power output of the magnetron can be measured by performing a water temperature rise test. This test should only be used if the above tests do not indicate a faulty magnetron and there is no defect in the silicon diode, high-voltage capacitor, and power transformer.

Checking the magnetron without the HV meter

You can check the operation of a magnetron with a regular pocket VOM and a 10-ohm resistor. Select a 10-watt and 10-ohm resistor, then connect alligator clips to each end. Pull the power cord and discharge the capacitor before attaching the resistor. Remove the cathode or ground end of the HV diode. This positive end will be soldered or bolted to the metal chassis. Insert the 10-watt resistor in series with the diode (figure 8-7). Connect one end to the diode and the other to chassis ground. Make sure these connections are firm and tight. Note: You might find a 10-ohm resistor in older microwave ovens.

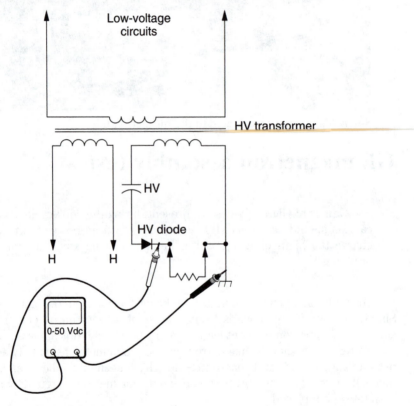

8-7 Measure the plate current by inserting a 10-ohm resistor in series with the high-voltage rectifier and common ground.

Clip the negative lead of the VOM to the chassis and the positive lead to the top side of the resistor, which is next to the diode. Switch the meter to the 100-Vdc scale. Prepare the oven for a cook test. Always start at or a little higher than dc voltage scale. If the voltage is lower, go down to the next meter range. With this method, you prevent the meter from hitting the peg.

With this setup, you are actually measuring the voltage across the 10-ohm resistor. If no voltage measurement is noted, you can assume that no high voltage is present at the magnetron or the tube is defective, since no current goes through the 10-ohm resistor. If there is a very high voltage drop across the resistor, this indicates that the magnetron is pulling excessive current and is running extremely hot. A low-voltage reading indicates the magnetron is operating as it should.

In one Sharp Model R-9314, a 10-ohm resistor was placed in series with the HV diode. With correctly applied high voltage and normal oven cooking conditions, the voltage measured across the 10-ohm resistor was 2.75 Vdc. The high voltage measured at the HV diode was 1800 volts, with the magnetron pulling 300 mils of current. You can assume that in most ovens with a 10-ohm series connected resistor, the voltage will vary from 2.5 to 5 volts under normal operating conditions. A simple voltage test with the low-voltage VOM might indicate that the oven is functioning properly. These test voltages should be marked on the schematic for future reference.

Firing around the neck of antenna

In the early magnetrons, the RF antenna was found inside a complete glass enclosure. Today, the antenna neck assembly consists of glass, ceramic, and copper ring at the top. The small antenna rod or wire lies inside and fits down into the metal cavity (figure 8-8). The RF energy generated by the magnetron occurs at a frequency of 2450 MHz.

Arcing around the neck of the magnetron can result from improper installation, defective mesh gasket, improperly secured, or defective magnetron. If the tube is not level and securely tightened down, arcing and sparks can occur. Replace the metal gasket if it is burned or torn. When installing a new magnetron, always install a new gasket. Recheck the placement of the wire mesh RF gasket around the antenna assembly. Make sure the magnetron is level and properly secured.

Checking the magnetron with the ohmmeter

Several tests of the magnetron can be made with the ohmmeter. Check for low resistance between heater terminals, with no reading to chassis ground (figure 8-9). You should measure infinite resistance or no reading at all to ground. Use the megohm scale for leakage tests. If a very low reading is obtained, suspect a leaky magnetron. Remove all connecting cables from the heater terminals for this test. You might find a reading from one side of the heater wires to chassis indicating a leaky HV diode. Suspect an internal resistor in the diode case if the resistance is 10 megohms (figure 8-10). Some newer ovens have a bleeder resistor across the diode to bleed off the charge from the HV capacitor.

Take another low-ohm reading across the heater terminals of the magnetron. Usually this reading is less than 1 ohm (figure 8-11). High resistance or no reading indicates the heater is open. A normal filament test made with the digital ohmmeter might show a fraction of 1 ohm. Always remove the filament transformer leads for this test, or you might measure the resistance of the transformer winding and still have an open filament inside the magnetron. Although these two resistance readings help to locate a defective magnetron, you could still have a tube with weak emission or no high voltage applied to it.

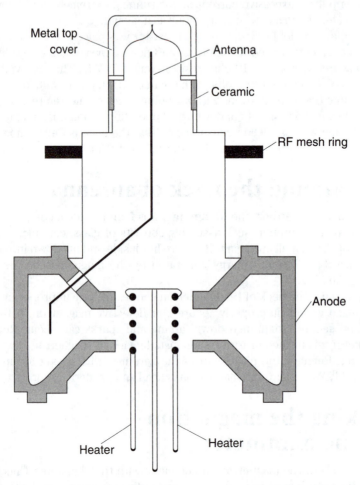

8-8 The small antenna rod or wire lays inside and fits down into the metal cavity.

Norelco magnetron filament

In normal oven cook operation, a visible glow from the heated tube filament can be seen through a small hole in the waveguide during a cook cycle.

If no filament glow is seen during a cook cycle, turn the oven off, unplug the power cord and discharge the capacitor with an insulated screwdriver. Disconnect the HV leads from the magnetron filament terminals. Measure the resistance across the two magnetron filament terminals with the ohmmeter set on R×1 (figure 8-12). If a normal resistance of less than 1 ohm between magnetron terminals is indicated

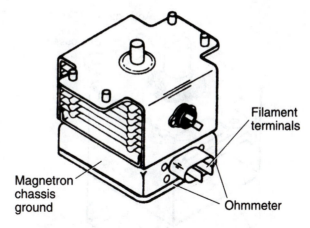

8-9 Check the magnetron with an ohmmeter with less than 1 ohm between heater terminals.

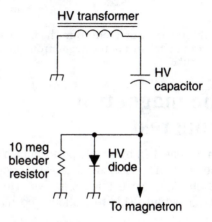

8-10 You might find a 10 megohms resistor across the HV diode to bleed high voltage off after oven is shut down.

on the meter, make a power transformer test. If high resistance or infinite resistance is indicated between magnetron terminals, replace the magnetron assembly.

Panasonic magnetron tests

Continuity checks can only indicate an open filament or a shorted magnetron. To diagnose for an open filament or a shorted magnetron:

1. Isolate the magnetron from the circuit by disconnecting leads.
2. A continuity check across the magnetron filament terminals should indicate 1 ohm or less.
3. A continuity check between each filament terminal and the magnetron case should read open.

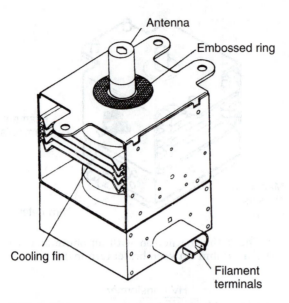

Antenna

Embossed ring

Cooling fin

Filament terminals

8-11 Infinite measurement from the filament terminal to a metal area of the magnetron indicates a normal tube.

Checking the magnetron with a cooking test

The magnetron can be checked by simply performing a water temperature-rise test. You will need a couple of 1-liter beakers, a glass thermometer, and a stopwatch. Before taking any tests, check for correct power line voltage. Low power line voltage will lower the magnetron output. This test should be made only with accurate test equipment.

Fill the two beakers with water and mark one (1) and the other (2). Stir the water in each beaker with the thermometer and record the temperature. Beaker temperature 1 is T1 and beaker 2 is T2. The average temperature of both beakers are as follows:

T = (record the average reading)

Place both beakers in the center of the oven cavity. In the Sharp oven, place the beakers on the revolving glass tray. Set the oven for high power for only two minutes. Close the oven door and begin the cook cycle.

After two minutes are up and the oven has turned off, remove the two beakers. Stir the water with the thermometer and measure the rise in the temperature. Be careful and do this rather quickly. Record each beaker's temperature. Now take the average temperature as before. Subtract these two average temperatures. You will have the cold water temperature and now the warm water temperature. You should have a temperature rise of 15°F to 20°F in a normal oven. If the temperature rise is

8-12 The DMM shows a 0.7-ohm filament measurement of a normal magnetron.

below 8°F, the magnetron is cooking very slowly. When no temperature rise is noted, either the high voltage is low, or there is a defective magnetron.

Now check for proper line voltage. Check the high voltage at the magnetron with an HV test instrument. When the high voltage is present at the magnetron and no temperature rise is observed in the glass beakers, suspect a defective magnetron tube. Replace the magnetron tube and take another water test.

GE microwave performance test

Measure the line voltage with oven operating. This test is based on normal voltage variations of 105 V to 130 V. Most power line voltage is around 120 Vac.

Place a WB64X0073 beaker containing exactly one liter of water between 59° and 75°F in the center of the shelf. Take the water temperature in beaker. Record the starting water temperature with an accurate glass thermometer. A good one is a Robinair #12084.

Set the oven to high power. Turn oven on and time exactly two minutes and three seconds. At the end of the time, record the water temperature. The difference between starting and ending temperature is the temperature rise. Depending on correct power line voltage, the normal minimum temperature rise should be 25°F at 105 V, and 28°F at 120 V.

Samsung MG5920T magnetron power measurement

The output power of the magnetron can be measured easily by performing a water temperature test. Select a 1-liter glass vessel and a glass thermometer with a mercury column. Take all temperatures and time tests with accurate equipment.

Fill the 1-liter glass vessel with water. (10 ±2 degrees Celsius.) Stir the water in the vessel with the thermometer and record the water temperature. The glass vessel temperature will be T1. Place the glass in the center of the cooking tray and set the oven on high power and set 52 seconds. Heat the water for exactly 52 seconds.

When heating is finished, stir the water again with the thermometer and measure the temperature rise and indicate as T2. Subtract T1 from T2. This will give you the rise in temperature. The output power is obtained as follows:

$$\text{Output power (W)} = \frac{4.187 \times 2000 \times \Delta T}{t\ (49.2)}$$

t is the heating time in seconds

4.187 is the coefficient for water

2000 represents 2000 cc of water

ΔT is the temperature rise (T2 – T1) (Celsius)

$$\text{Output power (W)} = 85 \times \Delta T$$

The normal temperature rise for this model is 11.5 to 13°C (52.7 to 55.4°F) at the high power setting. Remember that a variation or errors in the test procedure will cause variance on the temperature rise. Additional power tests should be made if the temperature rise is marginal.

GE JEBC200 magnetron test

Caution: High voltages are present during the cook cycle, so observe extreme caution. Disconnect the oven from the power supply and discharge the high-voltage capacitor before touching any oven component or wiring.

To test for an open filament, isolate the magnetron from the high-voltage circuits. A continuity check across the magnetron filament leads should indicate less than 1 ohm.

To test for a shorted magnetron, connect the ohmmeter leads between the magnetron filament leads and chassis ground. This test should indicate an infinite resistance (R×200 k). If there is little or low resistance, the magnetron is grounded and must be replaced.

The power output of the magnetron can be measured by performing a water temperature-rise test. This test should only be used if previous tests do not indicate a faulty magnetron and there is no defect in the high-voltage components, silicon rectifier, high-voltage capacitor, or power transformer.

GE JEBC200 magnetron and TCO replacement

The magnetron can be checked for an open or shorted tube. It is mounted to the waveguide by four screws. It also has a fuse assembly and damper fan assembly attached to the top of it. The magnetron temperature cutout (TCO) will open at a temperature of 302°F and will close at 140°F.

1. Disconnect the power; pull oven out to remove the top panel.
2. Discharge high-voltage capacitor.
3. Remove magnetron TCO leads and remove two screws that hold TCO (temperature cutout).
4. To replace magnetron, take off the rear top cover and right side cover.
5. Disconnect and mark high-voltage connectors to magnetron.
6. Remove metal damper duct (five screws) on side of oven.
7. Remove fuse assembly located on top of magnetron.
8. Remove one screw from damper door assembly that is located on top of magnetron.
9. Remove four magnetron mounting nuts (figure 8-13).
10. Slide magnetron sideways toward the middle to remove tube.

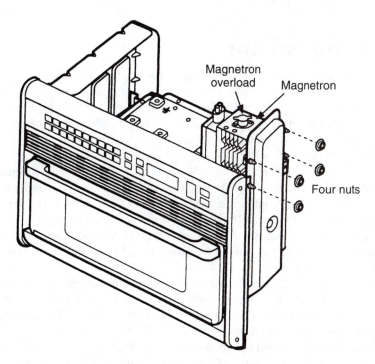

8-13 Remove four nuts at the outside to remove the magnetron from a GE oven. Courtesy General Electric Co.

Samsung magnetron tests

Check the magnetron by taking continuity tests for open filament or a shorted magnetron. Isolate the magnetron from the circuit by disconnecting leads. A continuity check across the magnetron filament terminal should indicate one ohm or less. A continuity test between each filament terminal and magnetron case should read open. Replace the leaky magnetron with a resistance measurement from filament to case (figure 8-14).

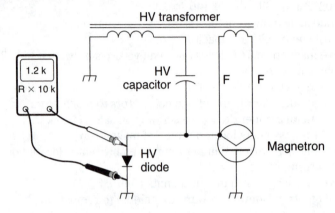

8-14 A 1.2 K ohm measurement indicates the magnetron or HV diode is leaky.

Magnetron failure

When you find the magnetron is arcing inside the top oven area, suspect a broken seal or that food particles are causing excessive arcing. Sometimes poor handling and striking the glass around the antenna area can let air inside, destroying the vacuum tube. This might occur while installing a new magnetron. Cracking can also be caused by thermal and mechanical stress.

A suck-in might occur, with the result of abnormally high power on the antenna glass, which softens the glass. The outside pressure pushes the glass inward toward the vacuum until a small hole is formed (figure 8-15). This can occur with improper use of the oven, such as using metal cooking utensils in the oven or operating the oven when it is empty. Often the breakdown occurs at the antenna glass to the metal seal. You can definitely hear any arcing that occurs inside the magnetron. After determining that the magnetron is defective, immediately shut off the oven to prevent HV component breakdown.

An open heater or filament can cause internal arcing. Discharge the HV capacitor, remove all cables to the heater terminals, and take a continuity test between the heater terminals. Very seldom does the heater short internally. Rough handling or shipping can cause the filament to go open.

The magnetron tube might have low emission. This can occur after many years of operation. Often, a tube suffering from low emission might take longer to heat up. When you receive a complaint of slow cooking, suspect a defective magnetron. A

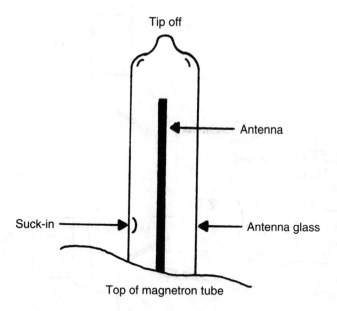

8-15 Be careful not to break the top seal where the antenna is located when you remove and replace the magnetron.

tube with low emissions might never reach the correct current requirements. The Magnameter is ideal for reading the current of any magnetron. Simply see what the current measurement is when the oven is turned on and check to see if it increases as the oven operates. It's nice to take voltage and current readings of each oven repaired. Mark these measurements on the oven schematic for future reference.

Quick magnetron resistance tests

The magnetron tube will fail if the vacuum is destroyed within the tube and air enters the magnetron. Internal arcing will occur, causing a shorted condition. Sometimes the RF capacitors or choke coils short against the chassis, producing a shorting high-voltage circuit. An open heater prevents RF energy and operation of magnetron. A tube with low cathode emission can take longer to get to the cooking point; tube current is too low for the magnetron to oscillate and produce power or cooking.

Three quick resistance tests on the magnetron can determine the condition of the tube (figure 8-16). Take a resistance measurement from the top of the high-voltage diode or anode side to chassis ground; this test might indicate a shorted, leaky, or defective high-voltage circuit.

Next, check the continuity of the heater terminals to determine if they are open. Remove the heater transformer cables and wires. Place the ohmmeter test probes across the heater (filament) terminals. Resistance should be less than 1 ohm in most ovens. Last, but not least, take a high-resistance measurement from the heater terminal to the metal shell of the magnetron. Any leakage indicates a defective tube. This should be an infinite measurement.

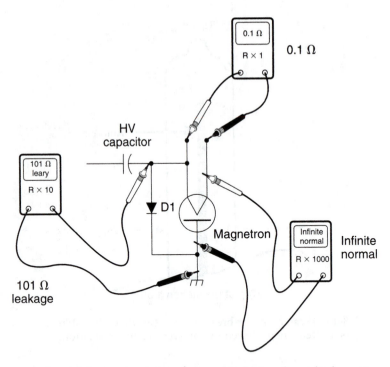

8-16 A quick low resistance test across the high voltage diode can indicate a leaky or shorted diode or magnetron. Infinite resistance across the diode shows both are normal.

Testing magnetron with the Magnameter

Always use the Magnameter test instrument when in doubt about a defective magnetron tube. Discharge high-voltage capacitor before attaching test leads. Connect the Magnameter to the magnetron. Remove screw holding the cathode terminal of diode from the base of the oven. Insert a 10-ohm resistor in series with the HV diode (figure 8-17).

Check the magnetron and Magnameter for a high voltage test. Then switch to the current test. A leaky magnetron will show low dc voltage and low current measurement of magnetron. An open magnetron or filaments will have a higher dc voltage measurement and no current reading. With a flip of a switch, you can tell if the magnetron or HV circuit is defective (figure 8-18).

Arcing in the oven

Excessive arcing in the oven can be caused by a shorted or leaky magnetron. You might find that the waveguide cover is burning. Remove the cover and fire up the oven. Sometimes grease behind the cover will cause arcing and burning of the cover. Replace the magnetron if the arcing continues. If excessive arcing occurs, turn the oven off at once.

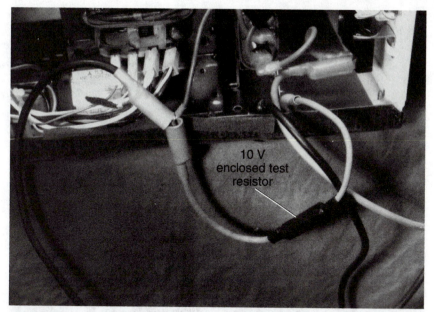

10 V
enclosed test
resistor

8-17 Insert a 10-ohm resistor in series with the HV diode for accurate current tests.

Sparks occurring in the oven cavity might be caused by excessive grease around the metal screw brackets or turntable shaft. Metal ties or objects in the oven cavity might produce sparks or arcing. Check the cup or bowl for gold or metallic paint applied. Leaving the temperature probe in the oven or not properly plugging into the jack might also cause sparks to fly. Check the radiating antenna blades for touching or being too close to the oven cavity.

Burns food

Suspect a defective magnetron when the complaint is that the oven runs hot and burns the food. Replace the magnetron if the oven appears extremely hot after operating for just a few minutes. The thermal protector switch will often intermittently shut down the oven when the magnetron appears too warm. Pull the power plug. Discharge the HV capacitor and touch the magnetron assembly. Replace the magnetron if it is too hot to touch. Take a current test and you will find if the hot magnetron is pulling excessive current.

Dead (No cooking)

A defective magnetron might cause the no-cooking symptom. First, take a voltage and current measurement of the HV circuits. Higher-than-normal high voltage indicates an open tube. No high voltage might indicate problems in the HV or low-voltage circuits. Poor tube emission will definitely show very little current measurement. A magnetron with no emission might have a defective cathode element. Discharge the HV capacitor. Take resistance measurements at the filament terminals. No resistance reading indicates an internal open filament.

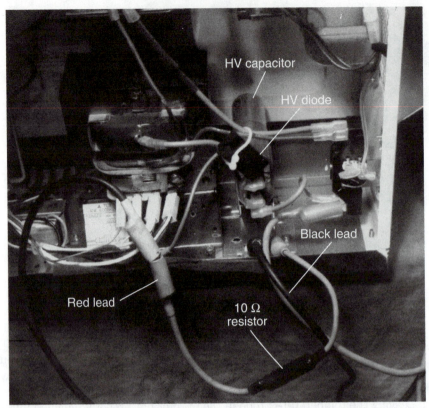

8-18 Connect the Magnameter to the magnetron and HV circuits to test voltage and current of the tube.

Erratic or slow cooking

A defective magnetron might produce erratic or slow cooking symptoms. With high voltage and current monitoring, you can quickly determine if the magnetron or other circuits are not functioning. The HV and current measurement should be read within two or three seconds. If the current reading is erratic and the high voltage fairly normal, you can assume that the magnetron is defective. Intermittent high voltage might indicate a defective HV circuit or magnetron. In some extreme cases, replacing the magnetron might be the only solution. Slow cooking might be caused by poor heater connections at the clip-on terminals. Check for burned and overheated connections.

Everything operates/no heat/no cooking

Sometimes the oven operates, but there is no heat and no cooking. This is a very common symptom in microwave ovens. All lights seem to be on, and the turntable is rotating, but there is no cooking inside the oven. Check with a water cook test. First, determine if voltage is applied to the magnetron. Check the magnetron current. When correct voltage is found at the heater terminals to ground, suspect a defective magnetron. No current reading indicates a defective magnetron. When an HV voltmeter and current meter are not available, check the heater continuity and resistance measurement between the heater and the chassis ground to determine if the magnetron is defective.

Cuts off after a few minutes

When the magnetron becomes quite warm after several minutes of operation, the thermal protector switch might open, removing the power line voltage from the primary winding of the HV transformer. After the thermal protector cools down, the oven begins to cook once again (figure 8-19). You can monitor the intermittent cooking with the ac voltmeter connected across the thermal switch. A pigtail bulb can be used here as a monitor (figure 8-20). When the bulb lights or the power line voltage is indicated on the ac meter, the oven is not cooking. Also, you might hear a loading-down noise from the power transformer under a cooking load that disappears when the voltage is disconnected.

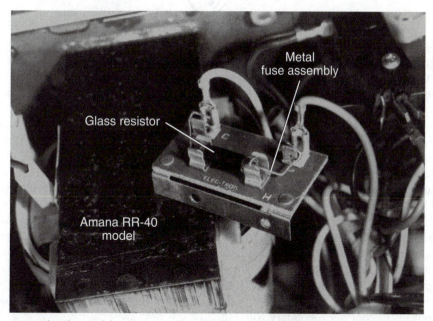

8-19 The thermal fuse is located in series with the regular fuse in some ovens.

Check the terminals of the thermal protector switch for burned areas. A defective thermal switch might cause the same symptoms. Replace the thermal switch if the magnetron is not running very warm. If the new switch is normal and the magnetron appears quite warm, install a new magnetron tube.

Intermittent cooking

Intermittent cooking in the high-voltage circuits can be caused by a defective magnetron, high-voltage capacitor and diode, or oven and magnetron cutouts. Monitor the primary winding of high-voltage transformer with a pigtail light or ac voltmeter.

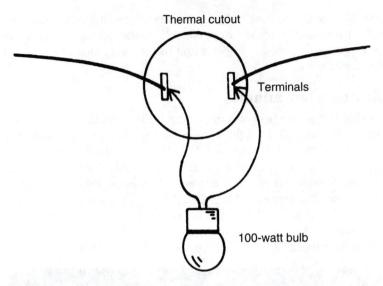

8-20 Clip a pigtail light across the thermal unit for a monitor indicator. If the light comes on, voltage is not applied to the magnetron.

Connect the Magnameter into the high-voltage circuit. The Magnameter will indicate if high voltage and current occur in the high-voltage circuits. Connect the pigtail light across the magnetron thermal unit.

If the oven becomes intermittent for a few minutes and then starts up, notice if the thermal cutout light comes on or if the thermal unit is open; this indicates that a defective thermal cutout or the magnetron is overheating. Do not overlook a broken belt on the radiating antenna or exhaust system. Replace the thermal unit and take another test. If the oven still comes on and off after a few minutes, replace the defective magnetron.

When the monitor light or ac meter indicates and the Magnameter shows adequate high voltage and low or no current, suspect a low-emission magnetron. Power output of the magnetron can be checked with a water test. If the high voltage is way above normal, suspect an open filament of the magnetron tube. If the oven operates a few minutes, then shuts down with the monitor light and ac meter normal, but there no high voltage or current, suspect a defective high-voltage capacitor, diode, or power transformer secondary winding.

Norelco magnetron thermal protector

The magnetron thermal protector, located on the magnetron assembly, is designed to prevent damage to the magnetron, if an overheated condition develops in the tube due to fan failure, obstructed air ducts, etc. (figure 8-21). The thermal protector opens at approximately 300°F and resets automatically at approximately 200°F.

Under normal operating conditions, the thermal protector remains closed. However, if abnormally high temperatures within the magnetron approach a critical level, the thermal protector will interrupt the circuit to the magnetron and coil of the surge relay. When the magnetron has cooled to a safe operating temperature, the thermal protector closes and a cook cycle can be resumed.

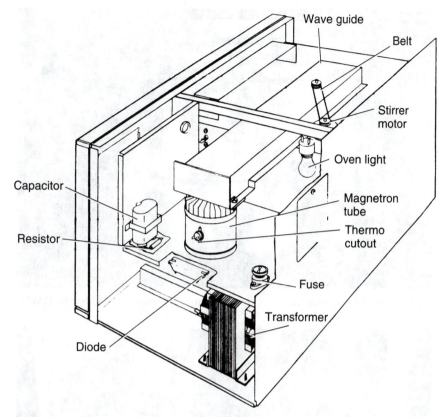

8-21 The location of a thermal unit on the magnetron of a microwave oven.

Intermittent thermal unit

It's possible to locate an erratic or intermittent thermal cutout when the oven cooks intermittently. Often the thermal switch element is weak or has burned contacts. Monitor the thermal cutout with an ac voltmeter or a 100-watt light bulb (figure 8-22). Under these conditions, check to see if the magnetron is overheating and quite warm. If the magnetron is fairly cool and the cutout is erratic in operation, replace it. Do not try to repair the thermal cutout component.

Removing the magnetron

Before attempting to remove the magnetron tube, once again discharge the HV capacitor. Take a visual inspection of the components surrounding the magnetron. You might have to remove a blower motor assembly before getting at the tube (figure 8-23). Disconnect wire leads from the thermal protector switch and heater terminals. Be careful when removing the wires so you do not damage other components. Mark down where the various cables are connected.

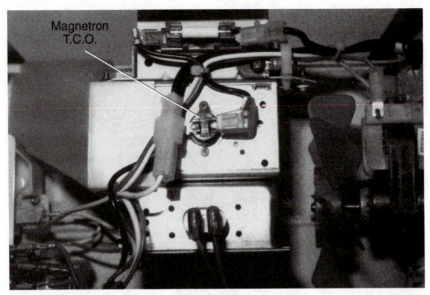

8-22 When the thermal cutout is closed, no measurement is read on the DMM.

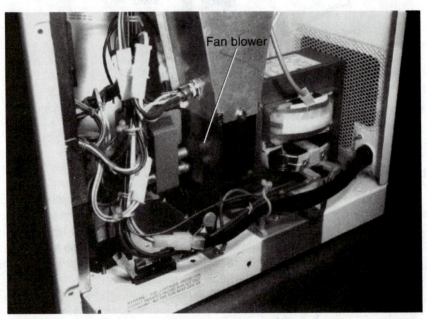

8-23 You might have to remove fan blower and magnetron cover to get at fan motor.

After the magnetron appears free of wires and shields, remove the four mounting nuts holding the magnetron to the waveguide (figure 8-24). Use socket wrenches instead of pliers to remove these nuts. Be very careful not to let the magnet pull the tool from your hands and break the glass seal of the magnetron. These magnets are very strong and might grab the socket wrench. Just loosen the nuts, then remove

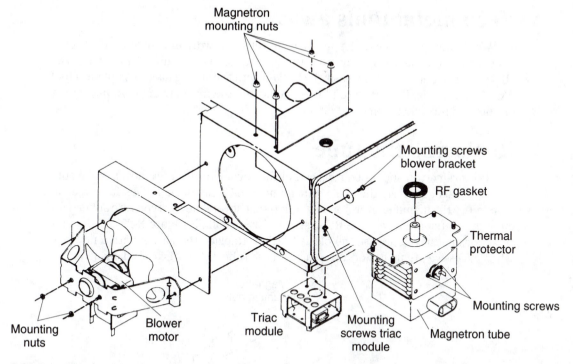

8-24 In a Norelco MC58100 model, the blower motor, metal brackets, and triac module must be replaced before removing the magnetron.

them by hand while holding the bottom of the magnetron so it will not drop. The magnetron assembly is quite heavy.

Now lower the magnetron until the tube is clear of the waveguide assembly. You might have to tip the tube to remove it from the oven. Be careful not to break the glass antenna assembly that is located on top of the magnetron. When mounted, the antenna assembly protrudes through the waveguide assembly. Place the magnetron on a firm surface and out of the way to prevent breakage. Here is the 10-step manufacturer's instruction for removing the magnetron:

1. Remove the blower motor assembly.
2. Remove the exhaust duct.
3. Remove the thermal protector switch assembly (two metal screws).
4. Remove the cable leads from the heater terminals (mark if needed).
5. Remove all metal flange or components that might be in the way of the magnetron.
6. Remove the four mounting nuts holding the magnetron to the waveguide assembly. (Be very careful with tools around the magnetron.)
7. Hold the bottom of the magnetron so it will not drop and damage the other components.
8. Lower the magnetron until the tube is clear of the waveguide assembly.
9. Remove the tube assembly completely from the oven.
10. Check to see if the RF gasket is still on this tube. This gasket might be needed for the new tube.

Keep metal tools away

When working around the magnetron with screwdrivers, pliers, and socket wrenches, keep the tools away from the glass antenna assembly. This unit sticks above the magnetron and can easily be broken with the strong magnets pulling a tool out of your hands. Be very careful when removing the magnetron mounting screws or nuts and replacing them.

Replacing the tube

The magnetron should be replaced in the reverse order of removal. Make sure the new magnetron has the same part number, or it might be a substitute recommended by the manufacturer. You might find that several newer types are subbed for an earlier version. Inspect the RF gasket before installing (figure 8-25). If the new tube does not have one, use the old RF gasket. Replace the gasket with a new one when excessive arcing has occurred around the tube or the gasket is damaged.

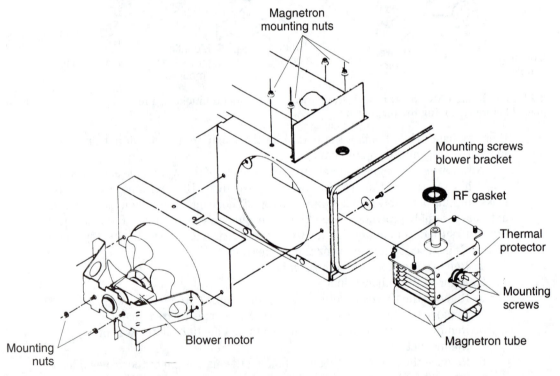

8-25 Inspect the RF gasket before replacing the magnetron. Replace the gasket if dented or damaged.

It's possible to replace one magnetron with one from another manufacturer (figure 8-26). You will find that some tubes are common in several ovens. The part numbers on the magnetron might even be the same. Of course, do not sub a different type of magnetron. You might find that some magnetrons are larger than others

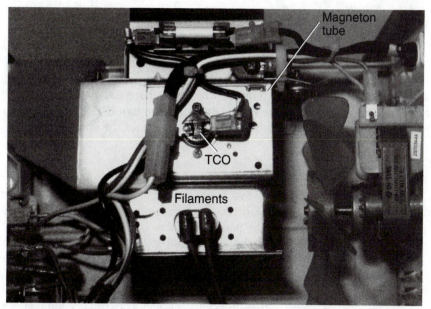

8-26 Some thermal cutout units can be replaced with a universal or other manufacturer's part.

and mount differently. The heater terminals might stick out at the wrong angle for correct mounting. Follow the manufacturer's replacement procedure.

Be sure to inspect the new tube before installation. Check for damage marks on the shipping box. When the bottom of the tube area is pushed to one side or up at an angle, the magnetron might arc internally. The glass sealed area might be cracked during shipment. All cooling fins should be equally spaced and not crushed.

Double check all mounting screws. Check the schematic or installation notes for correct hookup. Make sure the heater cables are in place. If the clips fit over the heater terminals rather easily, pinch the area together with pliers to make a greater contact. These contact clips will arc and burn if there is a poor contact, causing intermittent or erratic oven operation. Before replacing the back cover, check for leakage around the magnetron area and a normal cooking test.

Panasonic magnetron replacement

The positive lock connector is a specially designed locking connector that you will find in many lead wire connections. To remove this connector, pull the lead wire by pressing an extruded lever in the center of receptacle terminal.

1. Discharge the HV capacitor.
2. Remove the two screws holding magnetron thermal cutout.
3. Remove the three screws holding air guide and detach air guide.
4. Disconnect the two HV lead wires from the magnetron filament terminals.
5. Remove the four screws holding the magnetron. After replacement of the magnetron, tighten mounting screws properly, making sure there is no gap between waveguide and the magnetron. (This prevents microwave leakage.)

Costs too much to fix

Always provide a repair estimate for any repair involving magnetron replacement. These tubes are fairly expensive and, when added to normal labor costs, the repair could cost more than buying a new oven. When you provide a cost estimate, customers can make up their own minds whether to have the oven repaired or not. After making sure that the magnetron is defective through HV, current, resistance, and cooking tests, always provide an itemized estimate for the customer.

Too hot to handle

When working around microwave ovens, you must be careful to avoid burns from steamy hot dishes and food. Any time heat or cooking is involved, it is very easy for a person to receive burns on the hands and face. Although most burns are from the cooked food or utensils, most people are afraid of getting burned or injured with radiation from the oven. It is impossible to have an RF radiation problem if the oven has been checked for leakage with an approved leakage tester.

Excessive grease found in the oven cavity can cause fire or burning of plastic covers and waveguide covers. Often greasy food (such as bacon) accumulates after several years. The grease might run down behind the plastic shelf guide or behind the waveguide covers. Although the plastic waveguide covers are up at the top, they still collect grease on the top side. A lot of oven maintenance can be avoided if the plastic shelving and waveguide covers are removed and cleaned with a mild detergent.

In some early oven models, plastic shelf guides are found on each side of the oven cavity. These are held in place with metal screws. The grease collects behind the plastic and metal screws, causing the RF energy to burn the plastic guide assembly. Simply removing the plastic guide and wash it—you might prevent an oven fire and costly maintenance problem.

Burning of waveguide covers can be caused by excessive grease collected on top of the cover and by a defective magnetron. Often the defective magnetron might burn only a small section of the waveguide cover. Excessive grease on the cover might start to burn in several areas. Always replace the waveguide cover when it is burned or when replacing a defective magnetron. You might not find a waveguide cover in the early microwave ovens.

Two explosive examples

In this early microwave oven, a round hole was found in the aluminum hole cover. Undoubtedly, the owner was cooking whole eggs in the oven. Eggs can be cooked in this manner if a couple of pin-size holes are punched in the top of the shell. In this case, part of the egg and shell exploded, going through the 1-inch diameter hole and lodging against the magnetron. The magnetron began to arc between antenna and gasket area. The longer the oven cooked, the greater the arcing of the magnetron tube.

The magnetron was removed and inspected. In this particular case, the magnetron was not damaged. But if the customer continued to cook or to allow the oven to arc for several minutes, the magnetron might have become damaged, resulting in a very expensive repair job. Simply cleaning off the magnetron antenna and cavity area solved the oven arcing problem. Again, careful cooking methods must be observed when operating any microwave oven.

Some foods might become explosive dishes when cooked in microwave ovens. Most oven manufacturers warn that popcorn must be popped only in ready-made popcorn bags. In this particular oven, popcorn was placed in a brown paper bag without any steam holes, resulting in an open fire within the oven cavity. The paper bag caught on fire, melting down the top plastic waveguide cover (figure 8-27).

8-27 When popcorn was cooked in a brown paper bag, the plastic waveguide cover of an early microwave oven was melted down.

Of course, this plastic cover is inexpensive and easy to replace. If the fire had not been extinguished in time, the entire plastic front cover would have been burned. The inside plastic liner was burned, showing a burned area in the front door. The front door assembly was removed for inspection. Here, only the inside plastic area was burned. If the choked gaskets and whole door assembly had to be replaced, it would have resulted in a very expensive repair job. Always inspect the door area for possible choke damage, and double check when taking leakage tests.

Samsung MW2500U low- and high-voltage circuits

The flame sensor and magnetron cutout are found in each leg of the low-voltage circuit. A low-voltage transformer for the PCB and varistor are wired across the power line circuit. The main relay and primary interlock switch are found in one leg, with the power relay in the other power line side. The monitor interlock switch is located across the power line after the power relay (figure 8-28).

When the low-voltage circuits are normal, 120 volts ac is applied to the HV transformer primary winding. The filament secondary winding goes to F and FA of the magnetron. High voltage is generated with the HV winding fed to a voltage-doubler network. The HV rectifier provides dc voltage to one side of the filament or heater of the magnetron. Now the magnetron begins to draw current, and microwave cooking begins.

Samsung MG5920T magnetron operation

High voltage of 2120 volts ac is generated from the high-voltage transformer secondary, and the voltage is increased by the action of the HV diode and charging of the high-voltage capacitor. This resultant dc voltage is applied to the anode of the magnetron. The first half cycle of the high voltage produced in the high-voltage transformer secondary charges the high-voltage capacitor.

The dotted lines indicate the current flow. During operation of the second half cycle, the voltage produced by the secondary transformer and the charge of the high-voltage capacitor are combined and applied to the magnetron, as shown with solid lines; the magnetron begins to oscillate.

The interference wave generated from the magnetron is prevented by the choke coils of 1.6 mH, filter capacitors of 500 pF, and the magnetron shield case so that TV or radio signals are not interfered with.

Samsung MW5820T magnetron tests and replacement

Take filament checks on filaments and across the high-voltage diode to find if filaments are open and there is a shorted magnetron. Isolate the magnetron from the circuit by disconnecting several leads. A continuity test across the filament terminals (with transformer leads removed) should be less than 1 ohm. A continuity check between each filament terminal and magnetron case should read open or infinite.

Discharge the high-voltage capacitor before attempting to remove or test the magnetron. Remove the magnetron with the shielded case, permanent magnet, choke coils, and 500 pF capacitors, all of which are contained in one assembly. Disconnect all wire leads from the magnetron, motor assembly, and lamp. Remove the screws on the air cover. Remove the screws holding the thermal cutout switch.

Remove the screws holding the magnetron. Carefully remove magnetron. Remove the screw holding the high-voltage diode. Remove the screws and hooks securing the motor assembly. Remove the oven lamp by pulling outward on the two hooks that are placed on the air cover around the lamp holder.

Samsung magnetron removal

Pull the power cord. Discharge the HV capacitor. Remove the magnetron shield case, permanent magnet, choke coils, any capacitors, all of which is contained in one assembly (figure 8-29). Disconnect all lead wires from the magnetron. In some models, the fan motor assembly and lamp must also be removed. Remove a screw and clip securing the air cover.

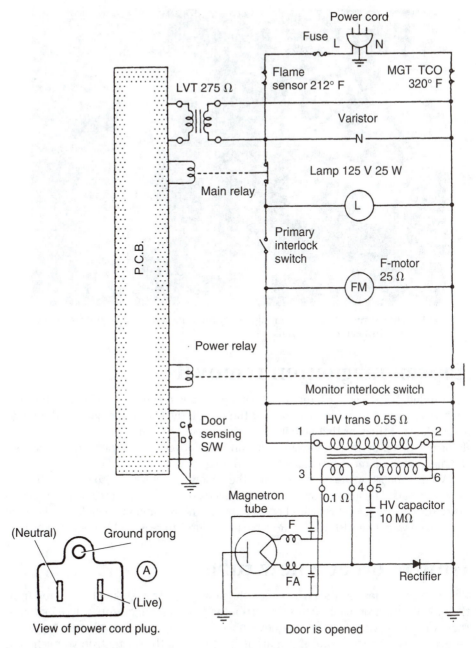

8-28 The low and high voltage circuits in Samsung's MW2500W microwave oven.
Courtesy Samsung Electronics America, Inc.

Remove screws around the thermal cutout switch. Remove screws securing the magnetron. Carefully remove the magnetron. Remove a screw securing the high-voltage diode. Remove screws and hooks holding the motor assembly. Remove the oven lamp by pulling the two hooks outward which are placed on the air cover around the lamp holder.

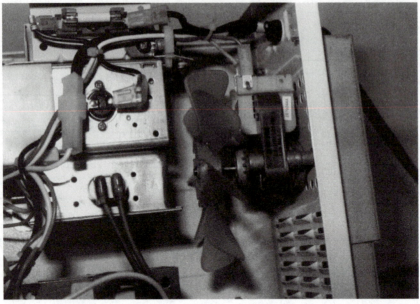

8-29 Many components on the magnetron must be removed before the defective magnetron can be removed.

Tappan magnetron removal

Unplug the power cord, remove the wrapper, and discharge the high-voltage capacitor. Remove the baffle by removing the plastic rivet from the back of the range and sliding out the tabs that hold it in position at back of unit. Either remove the baffle or position it so that the magnetron can be removed. Disconnect the leads from the magnetron filament terminals (figure 8-30).

While supporting the magnetron, use the hollow socket wrench to remove the four mounting nuts. The magnetron can now be pulled down and out. Care must be taken to make sure the RF gasket remains between the magnetron and waveguide. Replace the magnetron with a new RF gasket. Reverse procedures to reassemble the magnetron.

How to order a new tube

If you are a dealer or service dealer, the new magnetron should be ordered from the manufacturer or manufacturer's distributor. Most manufacturers will refer you to their various service centers. The new tube can be ordered from the dealer or microwave oven service center if you are going to install the magnetron yourself. You might find the magnetron in stock if it is a popular model. Otherwise, it might take from three days to several weeks to obtain a replacement.

Check the service manual for the correct magnetron part number. This same number might be found on the tube. In case the service manual is not available, use the numbers found on the magnetron. When ordering, don't forget to give both the tube part number and the model number of the microwave oven. When the tube

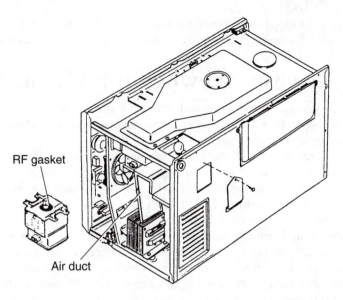

8-30 Remove the baffle before the magnetron can be removed in a Tappan oven. Courtesy Tappan, brand of Frigidaire Co.

arrives, the manufacturer or distributor might have subbed a new magnetron. The new tube will mount in the same place as the old one.

Check the tube over carefully when it arrives. Notice if the magnetron is packed in two separate boxes. Most manufacturers or parts distributors will ship the magnetron in a regular parts box, placed in a larger one for safe shipment. Check the boxes for broken areas. If one side of the metal flange assembly of the magnetron is bent out of line, refuse to install the tube. It might arc internally when fired up. Return the defective magnetron.

If the new tube appears intact, install the thermal protector switch in the matching screw holes. A new switch should be installed if it is defective or appears to have been operating quite warm. Now the magnetron is ready to be mounted in the waveguide assembly.

Small cabinet removal and replacement

To remove the back cover, loosen and remove all metal screws on the side and back area. Notice if screws around the sides have washers and should be replaced only on the side areas. Be careful when raising the metal cover upwards, as the two sides and top area might be slipped into a fold-back edge to hold the front edges into position.

Some covers are difficult to remove. Simply apply extra pressure at the rear lip and back metal cabinet with the blade of a large screwdriver and pry backwards. Do not lift the cover too high and pry upward, as you might damage the metal lip or edge and plastic front piece. Set the cover aside where it cannot be scratched or dented. To replace, make sure the front sides and top are in the groove before inserting metal screws. The top lip goes in easily, but the sides must be held together so all three edges slide into the metal areas. If not, the front sides will be loose and bulge outward at the front sides. So again, you have to remove the screws and try once again.

Cautions observed when troubleshooting

Check the ground wire in oven and outlet. Do not operate a 2-wire extension cord. The microwave oven is designed to be used when grounded. Make sure oven is grounded before servicing the oven.

Always discharge the high-voltage capacitor. As an electric charge in the high voltage capacitor remains for about 30 seconds after the operation stopped, short the current between the oven chassis and the negative terminal of the high voltage capacitor, by using a screwdriver when replacing or checking parts. When replacing parts, always pull the power cord from the outlet.

When the 15-amp fuse is blown by the operation of the interlock monitor switch, replace the primary interlock switch and monitor switch. This is mandatory. If not, the oven will come back in for repairs again. Check the fuse voltage and amperage rating before replacing a blown fuse. These 15-amp chemical fuses can be purchased at most hardware and appliance stores.

Avoid inserting any foreign material through any holes in the unit during operation. Do not insert a tool during operation. Never insert any foreign material or any other metal object through the lamp hole in the cavity or any holes or gaps. Such objects might act as an antenna and cause microwave leakage.

Be careful when wearing wristwatches and pacemakers around the oven. Consult the physician if the owner has a pacemaker. Write to the manufacturer of the pacemaker.

After repairing the oven, make sure that all screws of the oven are neither loose nor missing; this will prevent the possibility of microwave leakage. Make sure that all electrical connections are tight before inserting the ac plug into the outlet. Dress down and tie-up all wires as before servicing the oven. Last but not least, check the oven for radiation leakage.

How to replace while still under warranty

In case the old magnetron is in warranty, pack the tube in the same carton that the new one came in. Most magnetrons are warranted for a period of five or seven years. Some manufacturers issue a registration number when the oven is sold. Either the owner or dealer should register the oven with the manufacturer. If the oven has not been registered, a bill of sale must be attached to the warranty repair tag. Most oven manufacturers have their own warranty forms or will acknowledge NRA form number 317-515.

Make sure the magnetron is packed in its original cartons. Pack the tube well to prevent breakage. The warranty and work order form should be placed in the same carton. Fill out the forms completely. Usually, parts credit is issued and warranty labor paid the following month.

9

CHAPTER

Servicing oven motors

Up to seven different motor operations can be found in a microwave oven. The fan or blower motor keeps the magnetron cool and moves the air in and out of the vented areas. Two separate fan motors can be found in some ovens. The cooking time is controlled by a timer-clock motor, and a stirrer motor spreads RF energy out over the food in the oven cavity. Many ovens have a turntable motor that rotates the food for even cooking (figure 9-1). A vari-motor can be used for intermittent cooking or defrosting the frozen food before actual cooking begins. In the convection oven, you might have a heater or auxiliary blower motor to circulate the heat provided by the heating elements, and you might also have a small motor in the convection timer.

The rotation of the motors can help you locate a defective component in the oven. You can use each motor, which is tied into a different leg of the schematic diagram for trouble indications (figure 9-2). If the fan or blower motor does not operate, you know that the defective component is in the input of the power line circuits. No rotation of the turntable or stirrer motor might indicate a defective oven relay or contacts—as these motors are connected into the circuit after the relays. A rotation of the blower motor connected across the primary winding of the power transformer indicates that the low-voltage stages are normal.

Although the motors found in a microwave oven cause a variety of problems, a defective motor is easy to test and locate by voltage and continuity tests. With the oven in operation, measure the ac voltage across the motor terminals (figure 9-3). Remove the power plug and discharge the HV capacitor. Set the ohmmeter to the R×1 scale. Measure the resistance across the motor terminals. An open field will have no ohmmeter reading. Very low ohmmeter measurements might indicate a shorted field winding. Just clip the pigtail test light across the fan or blower motor terminals to make sure power is applied.

Small oven motors

A small, low-wattage oven might have only one fan blower motor. In other small ovens, you will find a blower and gear motor. The gear motor is also referred to as a

307

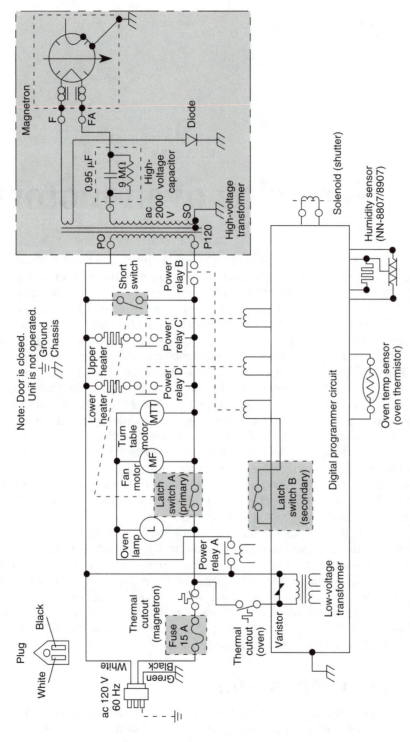

NN-8507/NN-8807/NN-8907 CPH

9-1 A typical oven schematic showing where the fan blower and turntable motor are located. Courtesy Matsushita Electric Corp. of America

Important safety note: The shaded area on this schematic diagram incorporates special features important for protection from microwave radiation, fire and electrical shock hazards. When servicing, it is essential that only manufacturer's specified parts be used for the critical components in the shaded areas of the schematic diagram.

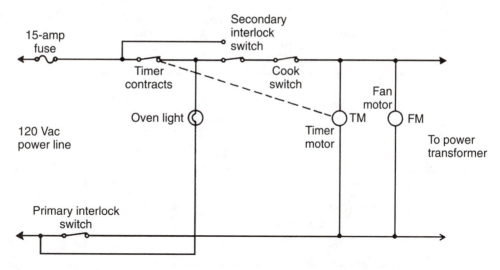

9-2 The oven light and blower motor can be used as indicators in the low-voltage circuits.

relay or triac. An exception is Sharp's small R-4580 (500-watt) oven—it has four different motors: the timer, cooking-fan motor, turntable, and vari-power motor.

GE magnetron fan motor

The magnetron fan motor drives a blade which draws in cool external air. The cool air is directed through the air vanes surrounding the magnetron and cools the magnetron assembly. Most of the air is then exhausted through the vents (figure 9-4).

The fan blower motor is found at the back of the oven with a fan cover. Remove three screws to remove the fan assembly. You must remove the fan blade before the motor can be removed. To remove motor assembly, remove two metal screws.

Fan motor problems

Fan or blower motor problems can produce many different symptoms in a microwave oven (figure 9-5). A dead or nonrotating blower motor can make the magnetron overheat, causing an intermittent cooking shutdown. When all other components seem to be in operation and the fan is not rotating, check the ac voltage across the motor terminals. If the power line voltage is present, pull the power plug and remove one motor terminal wire. Take a resistance measurement across the motor field terminals. An open field measurement might be caused by a poor or open motor socket connection. A quick method to see if ac voltage is applied to the motor field is to place a screwdriver blade against the metal motor area; the blade should vibrate.

Intermittent fan blower operation might be caused by a poor contact of the motor socket terminals. Sometimes these sockets will vibrate loose and make a poor contact (figure 9-6). Check for poor or old cable wires connecting to the motor field

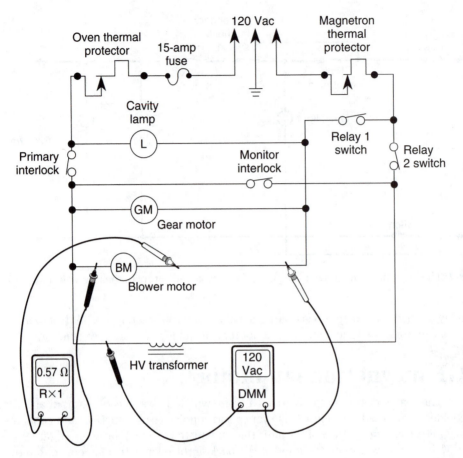

9-3 A defective motor can be located by voltage and continuity tests. Set the ohmmeter to the R×1 scale for a motor field continuity test.

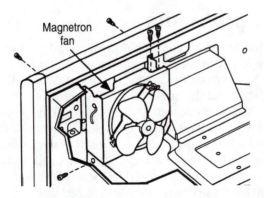

9-4 The magnetron blower fan cools the tube and exhaust on through vents in the back of oven.

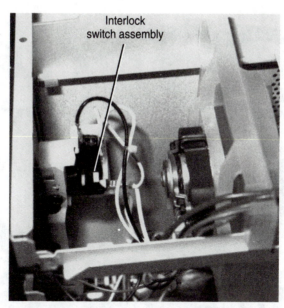

Interlock
switch assembly

9-5 The dead fan motor can cause the oven to overheat and shut down with thermal cutout.

coils. Use the R×1 ohmmeter range for all motor continuity tests. (The resistance of a fan motor in a Sharp R8310 oven was found to be 41 ohms with a digital VOM.)

An overheated motor might produce a shutdown after several hours of operation. Usually the field coil shorts the turns within the motor winding when the motor overheats. When the fan blower to the magnetron quits rotating, the magnetron becomes overheated and the thermal switch cuts off power to the magnetron. Most fan or blower motors operating at the correct speed produce quite a rush of air inside the oven. Compare the resistance reading of the motor field with that given by the manufacturer. A digital VOM takes a very accurate measurement of all motor field coils. Feel the motor body—if it is very warm, replace it.

When moisture appears around the door area, suspect poor ventilation. Check the fan motor for correct rotation. Clean out all vent areas. Don't overlook the clogged fin areas of the magnetron tube. Improper intake of cool air and venting of hot air might cause the magnetron to overheat. Check for correct vent and damper action, especially in the fan blower area in convection ovens. Also check to make sure there is enough air space around the sides of the oven vent areas.

A noisy fan blower motor can be caused by dry bearings or loose motor-mounting screws. Check the fan blade to see if it is bent out of line or loose on the motor shaft. A bent fan blade can cause the motor to vibrate, producing a noise inside the oven area. Often a high-pitched noise is a sign of dry motor bearings. Gummed-up motor bearings might cause the motor to overheat and run slowly. If the blade cannot be straightened properly, then replace it. Most fan motors are adequately lubricated. Don't overlook a dry fan motor bearing in the older ovens. Light motor oil placed on the motor bearings might solve the noisy motor-bearing problem.

9-6 If the fan is intermittently operating, check for poor fan wire connections

Various motor resistance

The fan motor resistance can vary from 2 to 50 ohms. Most fan motors are checked with a low ohm continuity measurement with one terminal lead removed. The timer motor resistance should be between 100–150 ohms while the turntable or stirrer motor can vary from 300 to 3.5 k. Some new turntable motors vary between 30–50 ohms. The convection motor resistance is between 10–25 ohms. The damper motor resistance can vary from 2 k to 3.5 k. (See figure 9-7.)

Replacing microwave oven belts

Improper heating or cooking can result from a broken or loose stirrer or antenna belt. The blower motor rotates, but the radiating antenna does not turn. Maybe the symptom is a broken drive belt or cotter key on the antenna wheel. The stirrer or antenna blade can be driven from a fan pulley or separate motor. Of course, some stirrer blades are directly driven units.

The drive belt could be flat, square, or have serrated teeth. Replace it with a new belt over motor, antenna, and tension assembly. Check the pulley bearing for lubrication if a squeaky noise is heard as it rotates. These belts can be obtained from microwave oven manufacturers, mail order firms, and local oven distributors of electronic parts.

Fan motor	35.8 ohms	
Convection motor	23.2 ohms	**9-7**
Damper motor	3.21 k ohms	GE motors and field resistance.
Stirrer motor	3.4 k ohms	

Fan starts after ten minutes

Suspect a dry or gummed-up fan motor when the blower does not come on for five or ten minutes (figure 9-8). This condition is likely to exist if the oven is five to ten years old. The fan blower motor might have gummed-up bearings that are caused by excess dirt and grease drawn into the fan area. Sometimes the thermal protector will shut down the magnetron if the fan motor takes a long time to come on. Clean up and lubricate the fan motor when this condition exists. If you don't, the magnetron tube might be damaged with erratic cooking problems.

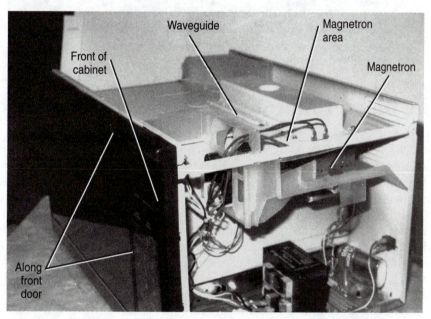

9-8 The fan blower is enclosed in the plastic framework around the magnetron in this oven.

Fan motor operates and stops

The fan motor might come on and operate a few minutes—or it might take a few seconds to start up. Often, this is a sign of gummed up motor bearings. A fan motor might stop after operating with dry bearings. Look for a poor connection to the motor with the erratic fan operation (figure 9-9). Just clean up the fan and wash out bearings with cleaning fluid, apply light oil and replace end bells of the motor.

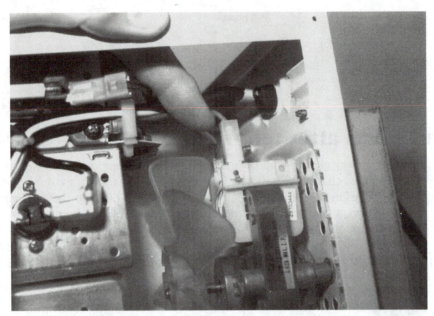

9-9 Check for poor wire connections and dry motor bearings if motor operates for a while and then stops.

The blower motor

The fan or blower motor rotates a blade that draws cool air from outside the oven. This cool air might be directed through different air vanes surrounding the magnetron. In some ovens, the fan motor is bolted to the bottom side of the magnetron. You will find two different fans in some ovens, one moving the air in and out, with the other fan cooling down the magnetron tube. Most of the air is exhausted directly through the back vents. In some ovens, the air is channeled through the cavity to remove steam or vapors given off by the food. Often the fan motor operates during microwave and convection cooking in the combination oven (figure 9-10).

The fan blade might be made of plastic or light metal. The blade can be held in place by an Allen set-screw or a nut mounted on the outside of the blade area. A bent fan blade might cause the fan to vibrate, thereby creating a vibrating noise. Some fan motors are mounted in rubber mounts for quiet operation.

In some fan or blower motors, the motor coil is fused internally. The fuse is embedded in the motor winding. If the fuse opens, the motor must be replaced. The motor field coil resistance might vary from 10 to 50 ohms. In the Hardwick ovens, the blower motor resistance varies from 16 to 22 ohms, while in the Amana ovens, the blower fan coil resistance might be from 5 to 10 ohms. The blower motor resistance in a small-wattage Goldstar microwave oven is 17 ohms (figure 9-11).

Remove one lead from the motor for accurate ohmmeter measurements. In some ovens, the fan motor connections unplug, and you could find the leads wired directly into the motor. Infinite resistance is measured from the metal motor frame

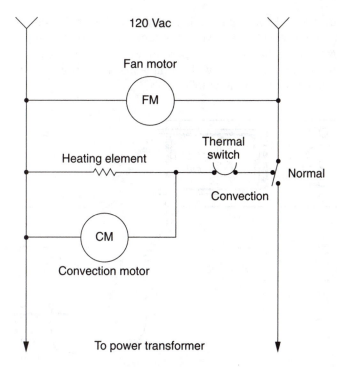

120 Vac

Fan motor

FM

Thermal switch

Heating element

Normal

Convection

CM

Convection motor

To power transformer

9-10 In convection cooking, both the fan blower and convection motor operate.

9-11 The fan blower resistance can be less than 1 ohm in the older ovens.

to one lead of the field coil (figure 9-12). Any resistance between the motor frame and field lead indicates a breakdown in the motor coil winding. Replace the blower motor to prevent shock hazard to the operator. Remember that the fan blower might also rotate the stirrer fan or blade with a pulley-and-belt arrangement. A typical fan motor replacement procedure is as follows:

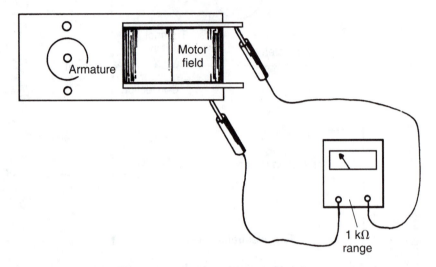

9-12 Infinite resistance should be noted between the metal frame and either side of motor terminals.

1. Remove the belt between the gear box and the fan motor pulley (only with combination fan and stirrer motor).
2. Disconnect the 2-wire plug or wire connectors of the fan motor.
3. Remove all screws holding the fan motor bracket to the oven.
4. If the motor is fastened to a separate bracket, remove these bolts and nuts.
5. Remove the fan blade (only if the motor cannot be removed).
6. Remove the ground strap.
7. On a combination stirrer-and-fan motor, remove the pulley to place it on the new motor.
8. Install the new fan motor by reversing the removal procedure.

Goldstar ER411M motor removal

Remove two screws holding the back cover. Unplug a 2-wire plug connector of the blower motor. Remove two screws holding the blower motor. Then remove the blower motor by turning the motor assembly to one side. Clean out fan area before installing the repaired or replaced blower motor.

Norelco RR7000 fan motor replacement

Disconnect the two leads from the fan motor. Remove the four blower assembly mounting screws secured to the air baffle (figure 9-13). Remove the blower motor. Remove the blower fan by pulling the fan from the motor shaft. The blower mounting brackets can be removed by taking out two mounting nuts and bolts that mount the brackets to the blower motor. To install a new motor, reverse the installation procedure.

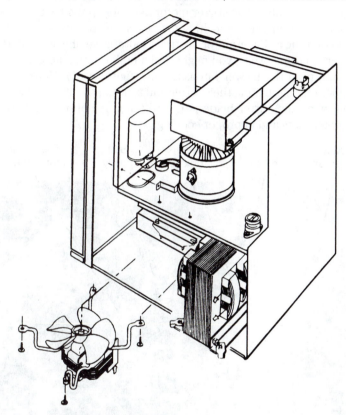

9-13 In a Norelco RR7000 model, the fan motor can be dismounted by removing four assembly screws. Courtesy Norelco Corp.

Tappan cooling fan removal:
1. Disconnect the oven from power source.
2. Disconnect the wires from the lamp socket, fan motor, and magnetron.
3. Remove the fan support by removing two plastic fasteners from bottom of oven.
4. Remove the nut from top of support and flex tabs enough to remove support from range.
5. The fan blade can be removed from the motor shaft by pulling it straight off, being careful not to distort plastic blades.

6. The fan motor can be removed by removing the two mounting nuts that secure the motor to the housing.
7. To assemble and install, reverse procedure.

The stirrer motor

The stirrer motor rotates or circulates the RF energy emitted from the magnetron waveguide assembly. The stirrer motor assembly provides even cooking and eliminates dead spots in the oven. In some Sharp microwave ovens, the food is turned instead of using a stirrer motor. A stirrer motor assembly is located at the top side of the oven. In some ovens, you might find a pulley with a long, flat belt driving the stirrer blade from the blower fan assembly (figure 9-14). Here the fan motor serves two functions: circulating the cool air and rotating the stirrer blade. Check for a broken belt when the stirrer blade is not rotating. A non-rotating stirrer motor might cause dead spots and improper cooking.

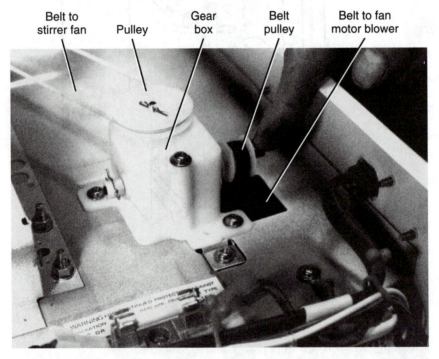

9-14 The fan blower circulates the air and the motor belt rotates gear box and stirrer blades in this microwave oven.

GE stirrer motor

The stirrer motor is the motor that turns the antenna located below the removable glass floor (figure 9-15). The stirrer motor is mounted on the bottom waveguide assembly and rotates a shaft that is fastened to the rotating antenna with two metal fasteners. This stirrer motor has a 3.4 k ohm resistance.

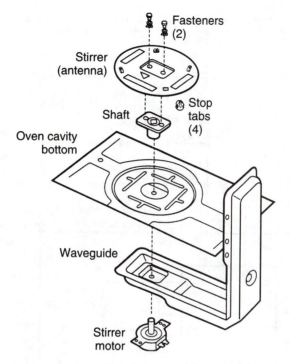

9-15 The GE stirrer motor is located under the glass bottom floor.

Norelco stirrer motor

The stirrer motor is mounted on a bracket. It operates a separate stirrer pulley via a belt. The blade revolves between 55 and 60 rpm and reflects the electromagnetic energy that is produced by the magnetron tube, back and forth, between the sides, top, and bottom of the metal cooking cavity (figure 9-16). This allows RF energy to penetrate food from all sides and gives a uniform heating pattern.

Because the stirrer motor has a relatively slower rotation, the field coil resistance ranges between 300 ohms and 3 kilohms. Check the manufacturer's literature when you suspect a defective stirrer motor. In a Norelco MCS 6100 model, the stirrer motor resistance is 2.7 k, while in a Hardwick EN 228 model, the motor resistance is 450 ohms. Check the continuity of the motor terminals with the low-ohm scale of the VOM.

Noise caused by the stirrer motor can originate when the blade comes loose on the motor shaft. The blade might rub or hit against the air chamber assembly. Check for a noisy pulley assembly at the hub or where the long belt goes over the oven edge (figure 9-17). This belt might be flat, with serrated teeth. Sometimes the pulleys will need lubrication or need to be tightened. Another source of noise is a dry stirrer-shaft bearing.

The stirrer motor or blade assembly might be difficult to remove. Usually the stirrer fan blade is located between the large waveguide cover and air guide assembly.

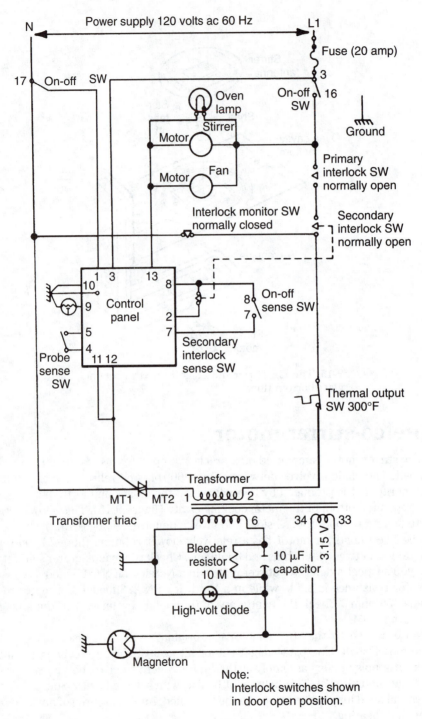

9-16 The stirrer motor and fan operate from control panel voltage.

9-17 The fan motor pulley also operates the belt which rotates the stirring antenna.

In most ovens, the blade or motor must be removed from the top side of the oven. Remove all waveguide metal screws holding the stirrer blade assembly. Some parts of the exhaust and air duct assembly must be removed before the stirrer assembly is free. Replace the stirrer motor and air duct assembly in the reverse order.

Panasonic stirrer motor replacement

Lay the oven on its top side. Remove the two screws holding the motor cover (figure 9-18). In this oven, the stirrer motor and antenna assembly is located in the bottom of the oven cavity area. Disconnect the two lead wires from the turntable/antenna motor. Remove the two screws holding the turntable/antenna motor (figure 9-19).

Replacing the Panasonic M8897BW antenna/turntable motor

1. Lay the oven on its top on a blanket or pad to prevent damage to the top finish.
2. Pull out the ac cord.
3. Remove the two screws holding the motor cover.
4. Disconnect the two lead wires from the antenna/turntable motor.
5. Remove the two screws holding the turntable/antenna motor.
6. Reverse procedure for correct replacement.

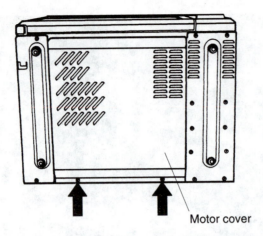

Motor cover

9-18 Lay the oven on its top side with padding underneath to prevent scratching. Remove cover to get at the motor. Courtesy Matsushita Electric Corp. of America

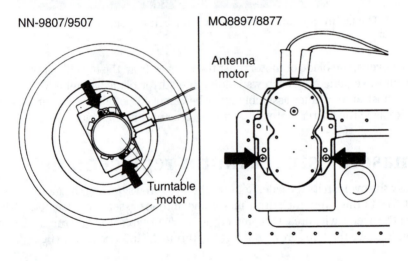

NN-9807/9507

MQ8897/8877

Antenna motor

Turntable motor

9-19 Disconnect the two lead wires from antenna/turntable motor and remove the two screws holding the turntable. Courtesy Matsushita Electric Corp. of America

Norelco stirrer motor replacement

Remove the stirrer belt. Remove the rubber pulley grommet by lifting off the shaft. Remove the two screws that secure the stirrer motor to the mounting bracket and remove the motor. In some cases, the wire must be cut and the replacement motor must be spliced and insulated with electrical tape or approved splice connectors. Reassemble the motor in reverse order.

The stirrer blade can be removed by stretching the belt above the rubber pulley grommets and lifting it off. Remove the shaft retainer fasteners by squeezing two tabs with needle-nose pliers and lifting them up. Remove the pulley from the shaft (figure 9-20), and remove the stirrer cover. The stirrer blade can be brought down and taken out of the oven. If a blade replacement is needed, remove the stirrer blade by removing the two screws that hold the blade to the shaft.

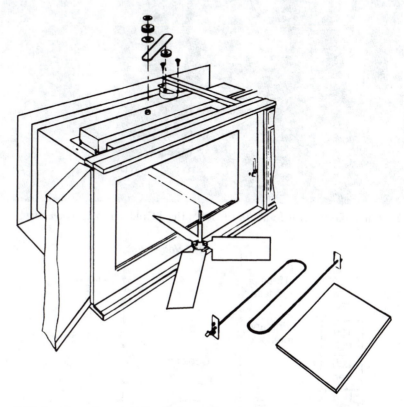

9-20 Remove the two tabs with needle-nose pliers and lift up to get at the stirrer motor.

The timer motor

The timer switch contacts are mechanically opened and closed by turning the dial knob located on the timer motor shaft. These contacts control the current path to the primary winding of the HV transformer and many other oven components. The correct cooking time is set on the timer unit, and the timer motor begins to rotate. When the timer reaches the 0 point on the scale, the timer opens the circuit and the cook cycle stops (figure 9-21).

The timer motor operates directly from the power line (120 Vac). Measure the power line voltage across the motor terminal (figure 9-22). A no ac voltage reading can

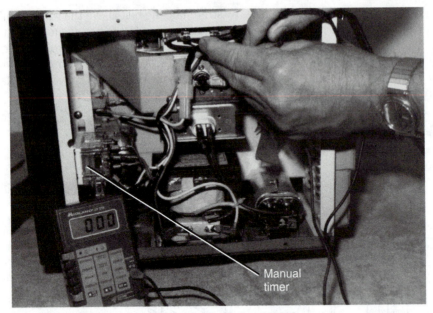

9-21 A manual timer is found in the low-priced Goldstar microwave oven.

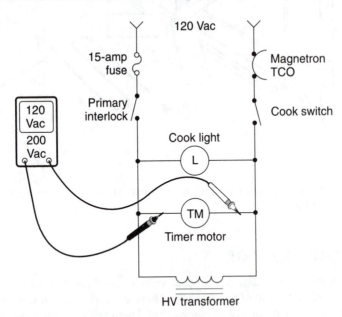

9-22 Check the timer motor with 120 Vac across motor terminals and a continuity test of less than 150 ohms.

be caused by defective interlocks or cook switches. If 120 Vac is measured across the motor terminals, and the motor doesn't operate, take a continuity test. Pull the power plug, discharge the HV capacitor, and remove one motor terminal lead. Set the VOM to the R×1 scale. Most timer-motor resistance tests should read from 100 to 150 ohms.

Replace the timer assembly if the motor winding is open or mechanically defective. A defective timer assembly might be erratic in operation and might never complete the correct cooking time. Sometimes the timer might stop without returning to 0.

Replace the defective timer assembly with one having the original part number. Even though you might have to move several components out of the way, most timer assemblies are easy to remove and replace. Remove all mounting screws to free the timer assembly. Replace the new timer assembly. Remount all other components and reconnect the wiring cables. Check the timer operation by rotating the timer knob to the 10-minute mark, then return it to 0. The bell should ring on most timers (figure 9-23). If the timer pointer does not correctly come to the 0 position, adjust the position of the timer assembly with the mounting screws loose. Then tighten the mounting screws and check once again.

Oven
timer

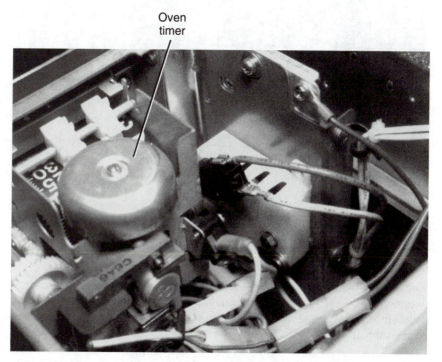

9-23 The manual oven timer has a metal bell that rings when cooking is finished.

Convection timer motor

In some ovens, a convection timer motor controls the cooking time and damper rod. The timer contacts are mechanically closed or opened by rotating the convection timing knob. The convection timer might cook up to two hours. The timer motor can be energized through the convection timer contacts. A CAM on the timer shaft operates to close the damper so no hot air escapes from the oven cavity. The damper is opened when the timer shuts off, letting hot air out of the convection oven.

Small gear or turntable motor

In newer model ovens, the small gear motor might be enclosed in one unit with step-down gear assembly (figure 9-24). This turntable motor is quite small physically, compared to the early Sharp turntable assemblies. The motor might start up in any direction. Usually the turntable motor is mounted in the bottom-center of the oven cavity.

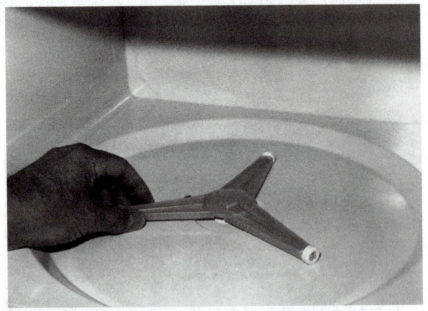

9-24 The turntable motor rotates a plastic tray holder in the bottom of a Sharp oven.

Remove the turntable motor wires and check the continuity with the low range of the ohmmeter (R×1). The resistance should have a low-ohm reading (30 to 50 ohms). Peak closely at the terminal wires for possible breaks in the wires. Replace the entire motor assembly if it is defective.

Turntable motor assembly

The turntable motor rotates the food for even cooking and prevents hot spots. The food is placed in a glass tray, with the tray rotating around on small plastic rollers. This turntable starts to operate when the oven relay contacts are made. The turntable rotation can be used as an indicator in troubleshooting the oven circuits. You know that all low-voltage circuits are normal when the turntable is rotating. Only some microwave ovens have a turntable assembly (figure 9-25).

The most common problems found with the turntable assembly are excessive noise while rotating, intermittent operation, and no rotation. Check the small plastic wheels found in the oven cavity for possible squeaky or dry-bearing noises. Remove

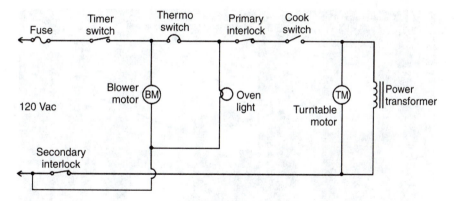

9-25 The new turntable motor can start up in any direction.

the glass turntable and let the motor rotate. If the noise has disappeared, lubricate the small wheels, holding up the glass tray. Only a drop of oil on each wheel will do. In case the noise is still present, check the plastic spindle bushing for dry or worn areas. You might hear a loud grinding noise with a broken or split spindle bushing. Check for a noisy and dry gearbox assembly.

Intermittent operation might be caused by a dry or defective plastic bushing (figure 9-26). Remove the turntable motor and inspect the bushing area. Check the bushing in operation. When found dry, lubricate the bushing with a light grease. Always replace a cracked or broken plastic bushing assembly. A jammed gearbox can cause the turntable to run slow or have intermittent operation. Check to make sure there is plenty of grease in the gearbox assembly. Worn or broken gears might cause the gearbox to freeze up and produce slow or intermittent rotation. For intermittent operation, also check for poor connections or improper voltage at the motor winding terminals.

A dead turntable motor could be caused by no voltage (120 Vac) at the motor terminals, a frozen bushing, or a jammed gearbox. Check the ac voltage at the motor terminals with the oven in operation. Usually the motor assembly is normal when voltage is found at the field coil. If a VOM is not handy, place a screwdriver blade against the motor frame—you should feel a magnetic pull or vibration on the motor assembly. If the motor will not rotate, suspect a jammed gearbox or open field coil if the correct ac voltage is found at the motor terminals (figure 9-27).

If no ac voltage is found at the turntable motor, suspect a defective primary or secondary interlock switch. Check the timer and cook switch terminals for poor contacts. Make sure that the timer and cook switch are operating correctly. In combination ovens, check for a defective connection timer assembly. Inspect the wire cables and connections to the turntable motor terminals and all the above components.

The motor armature might not rotate if the bearings are dry or excessively worn. A broken plastic bushing or gearbox can jam the motor so that the armature will not turn. You might want to remove the motor assembly to check the motor and gearbox. Remove the four mounting screws to separate the motor from the gearbox assembly. Check the motor for dry or worn bearings. Just lubricating motor bearings might solve the problem of slow rotation. Gummed-up motor bearings might cause the motor to run slow and cause the field coils to overheat.

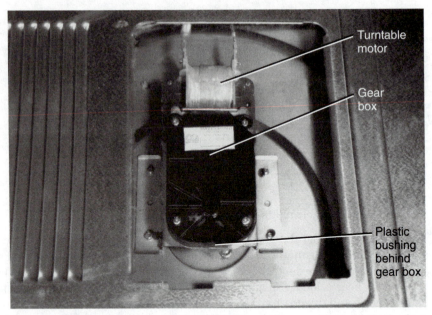

Turntable
motor

Gear
box

Plastic
bushing
behind
gear box

9-26 Intermittent and no rotation problems can be caused by dry or broken gears in old turntable motors.

9-27 Check the gear assembly of turntable motor for dry grease. Brush and clean out with cleaning fluid and apply new grease.

Wash and clean out the bearings when sticking or slow rotation is noted. It's best to remove one end bearing and pull the armature out. Wash off or spray cleaning fluid into the metal bearings and armature. Clean off with a rag and cotton swab tips. The gummed-up grease might be difficult to remove. After cleanup, lubricate the armature shaft and bearings with a light oil.

A continuity check should be made when correct voltage is found at the motor terminals and the motor will not rotate (figure 9-28). Most turntable motors have a resistance of from 30 to 50 ohms. With a digital VOM, 33.5 ohms was measured across a Sharp turntable's motor terminals. Short turns of the field coil might cause a lower resistance measurement. An open field coil will have no resistance reading. If the field coil is open, check for a coil break close to the motor terminals or coil assembly.

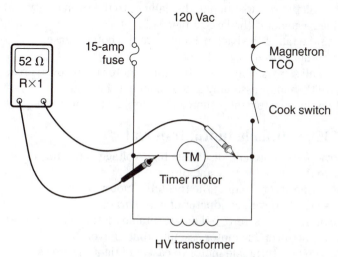

9-28 Take a continuity test across the turntable motor field coil to see if the defective motor winding is open.

An overheated motor will often show signs of heat or burned marks on the field coil assembly—you might find the area burned black from overheating. A motor suspected of running quite warm might slow down and stop. Touch the field coil assembly after the motor has operated for at least 30 minutes. A shorted coil can run extremely hot, while a normal winding will only get slightly warm, even after several hours of operation. Check the coil suspected of overheating with the low-ohm scale of the VOM for possible shorted turns. Discard and replace the turntable motor when the resistance reading is less than 25 ohms.

The turntable might stop or slow down if it has a dry or defective bushing bearing. Remove the turntable motor assembly and inspect the bushing. The plastic bearing might be cracked or enlarged, causing the turntable assembly to bind or stop rotating. Often a piece of the plastic is broken off. The plastic bushing can be replaced as a separate item. Replace the motor and gearbox assembly if a defective motor or broken gears are found. The turntable assembly is a special component that must be ordered directly from the manufacturer's part source.

In older ovens, the complete oven should be turned over on its side for turntable motor removal. Make sure the glass turntable tray is removed so it won't fall out of the oven and break. Remove the metal plate covering the turntable motor. Then disconnect the motor terminals. Remove the bolts securing the motor and gearbox assembly to the oven base plate. Pull the motor out to reveal a possible damaged plastic bushing bearing.

The following directions should be followed when removing a defective turntable motor from a Sharp convection microwave oven:

1. Pull the power plug and remove the back cover.
2. Discharge the HV capacitor.
3. Remove the one screw holding the door to the base cabinet.
4. Remove all the screws holding the cabinet to the oven cavity and back area (the back cabinet is the largest piece of metal).
5. Release the tab of the back cabinet from the bottom base.
6. Remove the bottom base cabinet.
7. Remove all screws holding the turntable motor to the oven area.
8. Remove the motor assembly and disconnect the leads.
9. Replace the new turntable motor by reversing the procedure.

Sharp R-7350 turntable motor removal

1. Remove power cord and discharge high-voltage capacitor.
2. Remove the turntable tray and disc.
3. Remove one screw holding the turntable motor cover.
4. Remove the two-wire conductor of wire harness.
5. Remove the three screws holding turntable motor to the cabinet.
6. Remove the turntable motor assembly out of oven.
7. Install the new turntable motor to case and tighten screws.

Goldstar ER-505M synchronous motor replacement

1. Remove the turntable and base in oven cavity.
2. Remove the three screws holding the motor cover to the base and gearbox.
3. Remove the six screws holding the gearbox to the lower plate.
4. Remove the gearbox cover.
5. Install the new motor with reverse procedure.

The vari-motor

The vari-motor provides variable cooking in the warm, defrost, simmer, and roast modes. The vari-motor rotates at two revolutions per minute, providing intermittent on and off operation of the power transformer and magnetron circuits (figure 9-29). For instance, in the defrost operation, the oven is on approximately 11 seconds and off 19 seconds, while in the roast mode the oven is on approximately 22 seconds and off 8 seconds. The vari-motor switches on the power transformer and HV circuits for variable cooking and then switches off for so many seconds. The vari-switch is activated by the vari-motor.

9-29 The vari-motor was used in early Sharp ovens.

The vari-motor rotates very slowly—about two revolutions per minute. Check for ac voltage (120 V) across the motor terminals. No voltage here could indicate problems ahead of the vari-motor assembly. If the micro cook light is on, suspect a defective vari-motor or open wiring connections. You might have a defective vari-switch if the oven seems to be operating, but little heat is produced. Of course, a defective magnetron power transformer, HV rectifier, or capacitor might also cause no heat or cooking.

Monitor the ac voltage at the power transformer. The 120 Vac will be present when the vari-switch is in the on position. Notice after a few seconds if the voltage drops off. Now the vari-switch contacts are open. When the ac voltage across the power transformer turns off and on every few seconds, you know the vari-motor circuits are operating correctly. If the ac voltage is present and there is no cooking, check the HV circuits.

The vari-motor and switch contacts can be checked with the Rx1 scale of the ohmmeter (figure 9-30). Always pull the power cord and discharge the HV capacitor before attempting continuity or resistance measurements. The switch contacts will be open in the off position. When in the on position, the vari-switch contacts are closed with a dead short reading. Remove one lead from the vari-motor terminals. Now check for continuity across these two terminals. Replace the vari-motor assembly when you find an open reading.

Because the vari-motor assembly is located on the front panel, the complete control panel must be removed before the screws holding the vari-motor can be removed. Most front control panels are fastened to the metal front panel with metal screws. In some ovens, removing the bottom screws will release the whole front control panel.

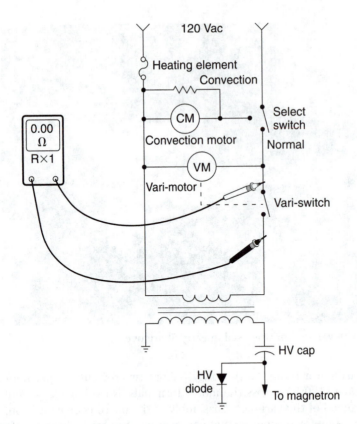

9-30 Check the vari-switch contacts with the R×1 scale of a DMM.

Disconnect the oven from the power plug and discharge the HV capacitor before attempting to remove any component.

After the front panel is removed, check for the mounting screws holding the vari-motor. Disconnect all wire leads from the motor and switch assembly. Pull off the control knob. Remove the screws holding the vari-motor assembly to the back plate panel. Remove and replace the vari-motor in the reverse procedure.

Auxiliary blower motor

In many convection ovens, another blower motor is turned on to provide cooling air to the microwave components. When the temperature reaches 150°F to 160°F, the thermal unit turns on the auxiliary motor, which remains on until the oven temperature falls below 125°F, with the thermal unit shutting off the blower motor.

In Panasonic convection ovens, the heater or convection fan motor moves the air around the heating element and over the food. The lower and upper heating elements are controlled by power relays D and C (figure 9-31). The auxiliary blower motor in the Hardwick MM-270 eye-level model is located in the top microwave oven

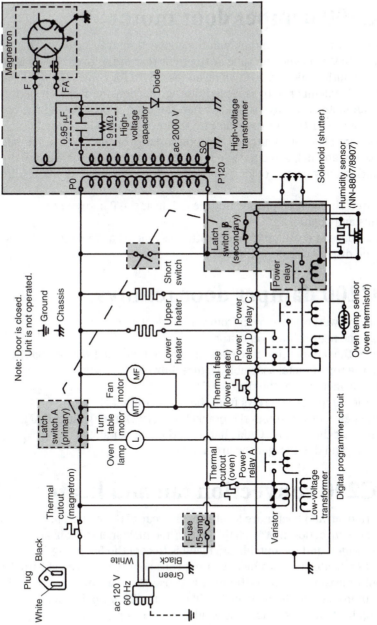

9-31 The turntable and fan motors are operating when a power relay closes in digital programmer circuit. Courtesy Matsushita Electric Corp. of America

Important Safety Note: The shaded area on this schematic diagram incorporates special features important for protection from microwave radiation, fire and electrical shock hazards. When servicing, it is essential that only manufacturer's specified parts be used for the critical components in the shaded areas of the schematic diagram.

section. Check the convection oven blower motors with the R×1 ohmmeter scale. Set the ohmmeter to the R×1 scale to check the thermal unit. This thermal element should read infinite ohms at less than 160°F.

GE JEBC200 damper door motor

The damper door motor is found only in the convection oven models. This motor is located next to the magnetron tube at the right rear of the oven. The damper door remains open during microwave cooking and is closed for convection and combination cooking. The damper door is closed during this cooking function because the magnetron fan motor is always running when the oven is on. Leaving the door open would allow more air (cooler room air) to be blown into the oven cavity.

Closing this door also contributes to the pressurizing of the cavity. This minimizes the air mixture and helps to cut down on moisture buildup. If the damper door did not close during convection cooking or combination cooking, these two things could happen.

1. More heated air could be pushed out the exhaust; the flame sensor could detect this and open.
2. If the oven doesn't come up to temperature as quickly as the micon thinks it should, the display will show ERROR.

GE JEBC200 damper door removal

1. Disconnect power; remove oven from wall to remove cover.
2. Discharge the high-voltage capacitor.
3. Disconnect the damper motor leads and remove two small screws that hold motor to the top of metal assembly (figure 9-32).
4. Separate the motor from the damper CAM.
5. Remove the three larger screws from the damper bracket.
6. Install the new motor through the bracket and attach the damper CAM. Make sure that the damper CAM shaft is in the damper slot.
7. Reassemble, and don't forget the ground wire.

GE JEBC200 convection fan and heater

The convection fan and heater are located at the rear of the oven. There are two fan blades on the convection motor shaft. One is for movement of air across the heater (internal blade), and the outside fan blade is to cool the fan motor.

The convection heater has two helix heating coils contained in the sheath. The resistance should measure about 9–10 ohms because they are in parallel. If they measure 18–20 ohms, one of the helix coils will open up. This would cause a shortage of convection heat, and the helix coils would have to be replaced.

1. To remove the convection heater: disconnect the power, discharge the HV capacitor, and remove the top panel. Remove the screws that hold the panels to back panel. Remove the outer plate, and remove the back panel.

2. Remove two nuts from each side of the burner box, located in the middle (figure 9-33).
3. Remove the four screws across the top of the burner box and five screws across the bottom of the burner box.
4. Remove the four screws holding the heating element in place. The heater is now free.
5. To remove the convection fan motor, follow steps 1–3 (previously discussed) for convection heater removal.
6. Remove the left threaded nut holding the fan blade in place.
7. Remove the four screws that are located inside of the heating element.
8. Remove the three screws at the other end of the assembly.
9. Remove the two screws holding the motor to the motor bracket. The spacer on the motor shaft goes between the outside fan blade and the burner box.

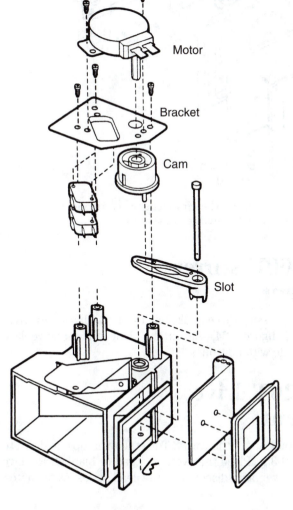

Motor

Bracket

Cam

Slot

9-32
Remove the motor leads and two small screws that hold the top panel in a GE JEBC200 oven. Courtesy General Electric Co.

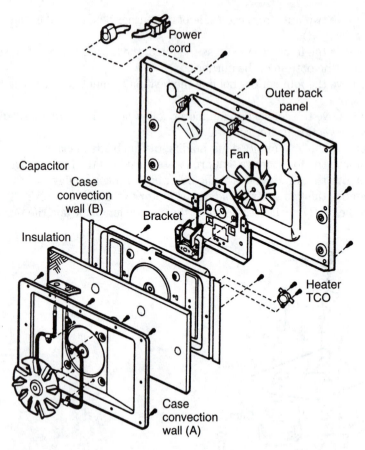

9-33 Check the resistance of the two heating elements (9–10 ohms each) before removing suspected heaters. Courtesy General Electric Co.

Samsung MW2500U stirrer motor replacement

Remove the stirrer cover by pressing three hooks one by one with a screwdriver and by pulling the cover forward (figure 9-34). Lift up the blade stirrer and replace the blade stirrer and housing blade with new ones.

Samsung MG5920T drive motor replacement

Take the glass tray and coupler out of the cavity. Turn the oven upside down to replace the drive motor (figure 9-35). Place the oven on a pad or blanket. Remove screws from the drive motor cover and disconnect the lead wires from motor.

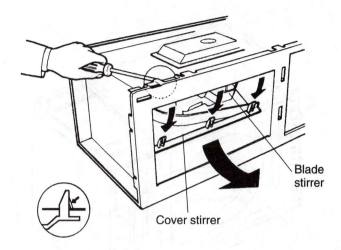

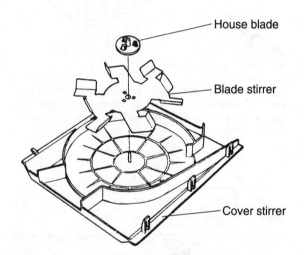

9-34 Remove three hooks with a screwdriver to remove the stirrer cover in a Samsung MW2500U oven. Courtesy Samsung Electronics America, Inc.

Remove the screws securing the drive motor. Remove the defective motor. Reverse the procedure to replace the drive or turntable motor.

Tappan stirrer cover removal

In browner models, the stirrer cover is held in place by guides along each side of the oven. To remove the cover, push up on the locking tabs and pull the cover forward (figure 9-36). Do not place the cover directly on top of the browning element.

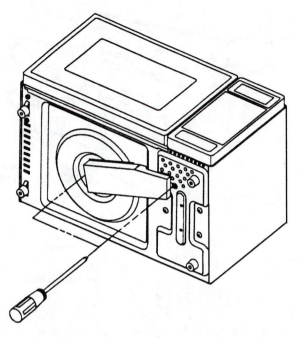

9-35 Lay the oven on its side on top of a pad or blanket and remove screws holding cover to the drive motor in a Samsung MG5920T oven. Courtesy Samsung Electronics America, Inc.

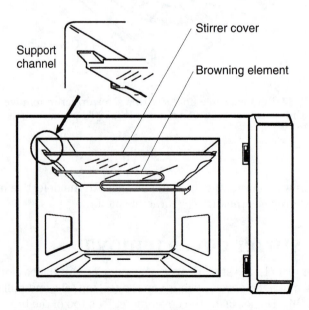

9-36 The stirrer cover is found above the browning element in some Tappan ovens. Courtesy Tappan, brand of Frigidaire Co.

Tappan turntable motor removal

1. Disconnect the power by pulling the cord.
2. Remove any loose items from inside the oven cavity and set the oven on the hinge side.
3. Remove the two screws from the rear of the bottom plate, then slide the plate toward the rear and out.
4. Unplug the connector to the turntable motor.
5. From the inside of the oven, remove the two screws that hold the turntable motor to the bottom of the oven.
6. Lift the square drive from the motor shaft and pull the motor out.

Motor problems

The following are various motor problems related to the fan, blower, turntable, and timer motor assemblies.

Noisy fan

Check for a loose fan blade. A vibrating noise might be caused by loose mounting-assembly bolts. Inspect for fan blade hitting against the fan assembly.

Fan slows down

Suspect dry or gummed-up motor bearings when the fan slows down after operating for several minutes. A squeaky motor noise might be due to dry motor bearings. Overloading the HV circuit can also cause the fan to slow down.

Intermittent fan operation

Check for a poor wire connection or cable plug to the motor. A jammed fan blade can also produce intermittent fan operation.

Oven shuts down after a few minutes

Suspect an overheated magnetron and open thermal switch if the fan blade is not rotating. Usually the magnetron overheats, opening up the thermal switch and closing down the cooking process. A leaky or shorted magnetron tube can cause the same symptoms.

Intermittent turntable

Check for a broken or cracked plastic coupling between the gearbox assembly and the turntable assembly. Also inspect the motor cable connections for intermittent turntable motor operation. The turntable motor rotation and fan operation can be used as low-voltage circuit indicators.

Longer oven cooking time

If the oven seems to take longer than it should to cook food, check the cooking time, power transformer, magnetron, and radiating antenna. The radiating antenna or blade might not be moving at all—or be moving rather slowly. (Remember that stirrer motors rotate the blade slower than the blower motor.) The radiating antenna blade might have stopped, the belt might be off or broken, or the stirrer motor might be dead. Inspect for a missing cotter key on the antenna wheel and shaft bearing. Check the stirrer motor for proper rotation and check the belt drive assembly on the blower motor for improper deflecting blade operation.

Exhaust cleanup

Dirty or plugged exhaust areas can cause the microwave oven to overheat, intermittently cook, and produce oven shutdown. Periodically clean the exhaust vents of grease, dirt, and dust. Remove the fan duct on small models to get down inside the vent areas. Clean up with soapy water, a cloth, and a brush.

10
CHAPTER

Servicing control circuits

The advantage of the control board over manual controls is easy and accurate operation. Simply tap or push the desired buttons or pads on the control panel, and the oven automatically takes over all the cooking process. Most microwave oven control boards are operated in a similar manner. For correct operation, always check the manufacturer's operating guide or service literature on how the oven operates (figure 10-1).

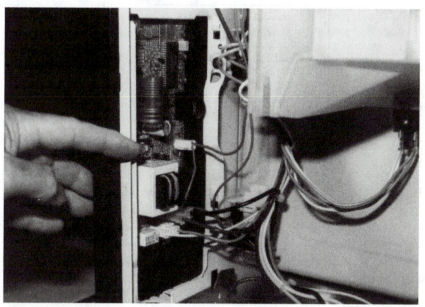

10-1 The back side of the control board panel in a Sharp oven.

Each oven has its own method of operating, so you should always read the operational directions before attempting to use it. The following is a general operational description of how a Norelco Model MCS 8100 functions.

Cooking with time or temperature

Microwave cooking is controlled by either time or temperature. Cook with temperature when internal temperature is the indication required. Use time when visual appearance is the indication required (figure 10-2).

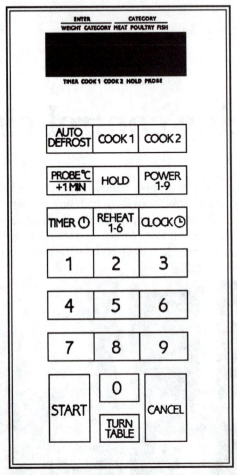

10-2 The electronic control board in a Tappan microwave oven. Courtesy Tappan, brand of Frigidaire Co.

Power level

The power level control provides total flexibility in choosing the speed (or power) of cooking that gives the best results for each type of food. Power level (used when cooking with time or temperature) is the name for variable power and includes high, saute, reheat, roast, bake, simmer, braise, defrost, low, and warm. The high setting provides the greatest speed; settings between high and warm represent decreasing amounts of microwave power.

Cooking with memories

Model MCS 8100 has a memory, which can be programmed with any combination of a variable power setting and a time or temperature setting. The memory automatically changes the power setting, cooking time, or temperature. A tone can be heard between memories.

Temperature control and automatic hold warm

"Temperature control" is a feature name for automatic food temperature control. It is used as a guide when cooking, reheating, or warming food by temperature. When the temperature-control food sensor is in place in the food and the temperature-control plug is inserted in the oven receptacle, the oven cooks by judging the internal temperature of the food. After final set temperature is reached, the oven automatically goes into a hold-warm setting.

Automatic hold warm

"Automatic hold warm" allows food to be kept warm for an extended period of time. Hold warm will operate only if the temperature is in the set program. It will continue to operate until food and probe are removed from the oven and the timer has been erased, or until 60 minutes have elapsed. Prolonged holding can result in overcooking.

Program check

The oven is designed so you can check or recall any cooking step programmed before or after cooking has started. Pushing the memory pad brings the set time or temperature of each step into the display; pushing the power-level pad displays the power-level setting.

Change or cancel

A cooking step can be changed or canceled at any time.

Delay cook

The "delay cook" can be used to postpone the beginning of the cook cycle by entering the delay time desired. It can also be used as a conventional timer (no cooking involved). The oven light will illuminate, but the oven is not operating. The timer is simply counting down.

Samsung MW2500U control pad

The Samsung oven control pad contains the display at the top that indicates numbers and letters for cooking. The UN2003A fluorescent tube is supplied filament and VCC voltage from low voltage power supply. The defrost pad is pushed to defrost frozen food. The number pads begin with 1 through 9 and 0 to set the hours, minutes and seconds of cooking (figure 10-3).

A clock set pad is pushed to set the time of day on the display. (If the electricity has been shut off for a period, as during a power failure, the display will flash off and

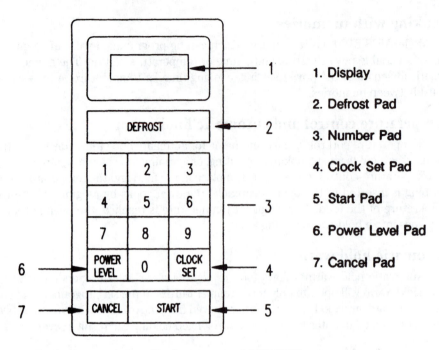

1. Display

2. Defrost Pad

3. Number Pad

4. Clock Set Pad

5. Start Pad

6. Power Level Pad

7. Cancel Pad

10-3 The front touch pads of a Samsung MW2500U oven. Courtesy Samsung Electronics America, Inc.

on until the clock time has been reset.) The start pad begins the cooking process. The 120 Vac is applied to the primary winding of power transformer. A power level pad sets the amount of power needed for cooking. If mistakes are made in setting up the time for cooking, you can cancel any settings by pressing the cancel pad.

Time and power-level operation

1. Open the door. Oven interior light comes on. Colors appear in display. Place food in oven. Close door.
2. Push the time pad.
3. Set the desired time. Push the appropriate number pads. The time will appear in display. Example: five and one-half minutes, push number pads 5, 3, and 0.
4. Push the power level pad and HI appears in display window. This is a reminder that oven will always operate at full power unless the power level setting is changed.
5. Select desired power level. Push the number pad with selected word. Example: number 5 is simmer. The display will read 50. This means the oven is operating at 50% power.
6. Push the start pad. The blower will come on and time will start to count down. A tone sounds three times when the time is up. The oven shuts off automatically. The display colors and oven light will remain on for about one minute after cooking has stopped.

7. Power-level check.
   a. Push the power level pad. The power level setting will appear in the display.
   b. Push the start pad to return to the time countdown.
8. Cancel the time and power level. Push the change/cancel pad twice.
9. Change the time or power level.
   a. Push the time pad (or power level pad). TIME or POWER will appear in the display.
   b. Push the change/cancel pad once. The display will blink for a time and show that HI is the power-level setting.
   c. Enter the new time or power-level setting.
   d. Push the start pad.

Temperature operation

1. Open the door. The oven interior light comes on and colors appear in display. Place food in the oven; insert probe plug into oven receptacle. F appears in the display. Close the door.
2. Push the temperature pad.
3. Set desired temperature. Push the appropriate number pad. Temperature from 90°F to 195°F can be entered. Example: to set 160°F, push number pads 1, 6, and 0. (If temperature below 90°F or above 200°F are entered, a tone will sound and display will go blank except for 6.)
4. Push power level pad; HI appears in display. This is a reminder that the oven will always operate at full power, unless setting is changed.
5. Select the desired power level. Push the number pad with the correct power level word. Example: The roast setting is the number 7 pad. The display will read 70. This means oven is operating at 70% power.
6. Push the start pad. 90 will appear in display (or actual temperature if greater than 90°). As the food cooks, the temperature display will increase in 5-degree increments. When the set temperature is reached, a tone will sound three times and the oven will automatically reduce power to 10% (hold down); display will show time counting down from 60 minutes to indicate how long the oven is in hold warm. The oven will continue to cook in the hold warm setting for 60 minutes or until change/cancel pad is pushed or door is opened (interrupts but does not erase without change/cancel pad).
7. Temperature or power-level check.
   a. Push the temperature pad (or power level pad). Set the temperature or power level. The setting will appear in the display.
   b. Push the start pad to return to the actual temperature.
8. Cancel the temperature and power level. Push the change/cancel pad twice.
9. Change the time or power level.
   a. Push the temperature pad (or power level pad). Set temperature or power level setting will appear in display.
   b. Push the change/cancel pad once. Display will blink for temperature and show a HI power-level setting.

    c. Enter in the new temperature on power-level setting.

    d. Push start pad.

Memory operation

1. Open the door. The oven interior light comes on. Colors appear in the display. Place food in oven. (If cooking with temperature probe, plug it into oven receptacle. F appears in display.) Close door.
2. Push the time (or temp) pad.
3. Set the desired time (or temperature). Push the appropriate number pads. The time or temperature will appear in display.
4. Push the power level pad. The letter A appears in display. This is a reminder that oven will always operate at full power unless the power-level setting is changed.
5. Select a power-level setting. Push the number pad next to the correct power-level words. Example: The simmer setting is number pad 5. The display will read 50, indicating that oven is operating at 50% power.
6. To enter next the memory, push the time or temperature pad. The display blinks and the color (or F) will blink, indicating the second winding.
7. Set the desired time. Push the appropriate number pads. The time or temperature will appear in the display (colors or F will continue to blink).
8. Push the power-level pad. HI appears in display. This is a reminder that oven will always operate at full power unless the power-level setting is changed.
9. Push the number pad next to the power-level words selected. Example: The roasting is pad number 7. The display will read 70, indicating that the oven is operating at 70% power.
10. Push the start pad. Time for first memory will start, counting down (90 will appear in display for temperature and will increase with actual temperature). At end of first memory, a tone will sound once and the oven will automatically shift to next memory. Colors (or F) will blink in the display as a reminder that oven is cooking in the second memory. At end of cooking, a tone sounds three times. The oven will automatically stop cooking (if cooking with time) or reduce to automatic hold warm (if cooking with temperature). The oven will continue to hold warm until the food and probe are removed from the oven and the timer has been canceled—or until 60 minutes have elapsed.
11. Set the time or temperature.
    a. Push the memory pad. Set time or set temperature will be displayed.
    b. Push the memory pad again and the next memory will be displayed.
    c. Push the start pad to return to time counting or actual temperature.
12. Check the power-level setting.
    a. Push the memory pad until the memory to be checked is in display.
    b. Push the power-level pad.
    c. Push the start pad to return to time counting or actual temperature.
13. Cancel all time, temperature, or power levels. Push the change/cancel pad twice.

14. Change the set time or temperature.
    a. Push the memory pad until the memory to be changed is in display.
    b. Push the change/cancel pad once.
    c. Enter the new time or temperature by touching the number pads.
    d. Push start.
15. Change the power level while cooking.
    a. Push the memory pad until the memory to be changed is in display.
    b. Push the power-level pad.
    c. Push the number pad for the new power-level setting.
    d. Push start.

Samsung MG5920T PCB circuit diagram

The circuit diagram of the control panel circuits includes the low-voltage power supply, reset, power relay, reset, door switch, sensor, main relay, buzzer, display, and key pad circuits (figure 10-4). The low-voltage power supply furnishes voltage to all of the circuits with a –5 V, –12 V, and –31 volts. The –5 volt source is fed to the most circuits of the Micom IC1. A –12 volt source operates the power, browner and main relays, and buzzer circuit. The –31 volts is applied to the VCC source of display "DIGITRON" with ac filament voltage to the filaments of display tube.

The display "DIGITRON" tube receives a voltage signal from Micom IC1 and the key pads. The different key pads operate with a –5 volts and are tied into pins 10, 11, 12, and 21. Pin 22 of IC1 is the voltage supply terminal of –5 volts.

The power relay circuits operate out of pins 36 and 39, while the relay operates from pin 17. Pin 13 of IC1 provides signal to the browner relay circuits. The crystal pulse oscillator circuit, which generates the oscillating frequency that controls the steps of various input and output signals, is connected to pins 33 and 34.

The control board or DPC

The basic control board might have a keyboard, key harness or membrane, external small ac voltage power transformer, and control circuit. Most control circuits are powered by a small ac power transformer (figure 10-5). A temperature probe thermistor operates from the control circuits, and the control board operates a power cooking relay or triac assembly to control the oven cooking.

In the Panasonic ovens, the digital programmer (DPC) can have a microcomputer circuit with a pulse oscillator, buzzer, display, humidity sensor, temperature control, clock pulse, initializer, door check, variable-power relay control, power relay control, membrane switch keyboard, and low-voltage power supply (figure 10-6). The different functions of each operation are shown in Table 10-1.

Internal control board problems can result from improper operation, improper temperature control, erratic programming, improper sequence operation, improper display numbers, and no cooking or improper cooking. Always double check for proper operation of the control board. Refer to the manufacturer's service literature.

P.C.B Circuit Diagram

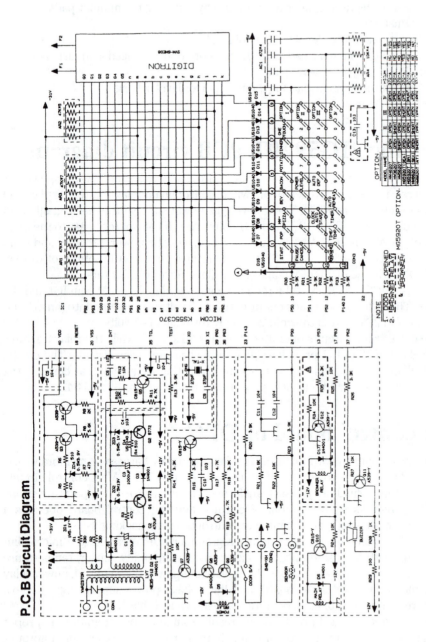

10-4 The PCB circuit diagram of control circuits in a Samsung oven. Courtesy Samsung Electronics America, Inc.

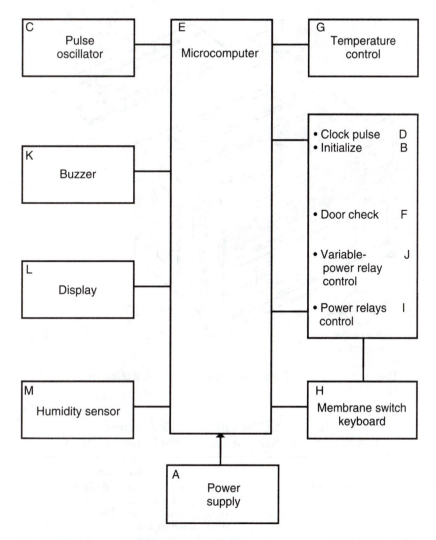

10-5 The block diagram of the DPC in the Panasonic oven. Courtesy Matsushita Electric Corp. of America

When an oven with a touch-control panel malfunctions, try operating the controls. If you press the proper sequence of keys or buttons and do not hear audible signals, suspect a defective keyboard. Check for proper illumination of the lighted display. Are there two or more figures on the display at the same time? If so, the control panel is defective.

Check the power transformer input voltage when you have no display lights or the control board appears dead. Measure the output voltage from the controller when the display and board seem normal (figure 10-7). Improper input and output voltages from the controller can indicate external or internal control problems.

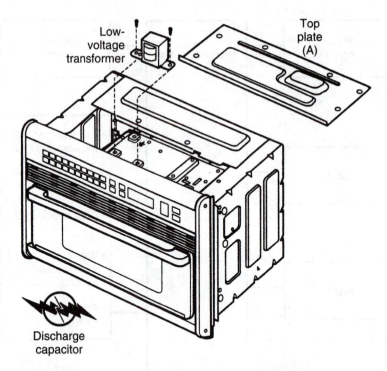

Low-
voltage
transformer

Top
plate
(A)

Discharge
capacitor

10-6 Location of LVT transformer in the GE JEBC200 oven. _{Courtesy}
General Electric Co.

Table 10-1. The various block diagram
components and their functions in the Panasonic DPCs

	Block	Functions
A	Power supply circuit	Rectifies the ac voltages supplied from the low voltage transformer, regulates the dc voltages and supplies them to the respective circuits
B	Initialization circuit	Resets the microcomputer when the oven is first plugged in
C	Pulse oscillator	Oscillates at a specific frequency to control the processing speed of the microcomputer
D	Clock pulse generation circuit	Generates a clock pulse at the power supply ac frequency for synchronizing the clock timing
E	Microcomputer	Determines and processes various information and generates the appropriate output signals in the sequence that was preprogrammed
F	Door signal check circuit	Inputs a signal to the microcomputer to indicate whether the door is open or closed by detecting the signal from the door signal switch

	Block	**Functions**
G	Oven temperature control circuit	Reads current temperature within the oven cavity and feeds this information to the microcomputer
H	Membrane keyboard	Inputs various data and instructions into the microcomputer through the mechanical contacts of the membrane switch keyboard
I	Power relay control circuit	Controls on-off timing of power relay to turn the microwave oven on-off with signals from the microcomputer (magnetron control)
J	Variable power relay control circuit	Controls on-off timing of variable power relay within a specific duty cycle to vary the output power of the microwave oven with signals from the microcomputer
K	Buzzer control circuit	Generates beep tones indicating the completion of a programmed operation in the microwave oven
L	Display circuit	Turns on necessary segments of the display tube to indicate the appropriate figures, symbols, letters, etc.
M	Humidity sensor control circuit	Reads current humidity conditions within the oven cavity and feeds this information into the microcomputer

Courtesy Matsushita Electric Corporation of America

10-7 The low-voltage power transformer is mounted on the control panel in the latest ovens.

Samsung MW5820T PCB connections

The main PCB panel of this Samsung oven has many components mounted on the board. The secondary interlock switch circuit is connected to connector 0-1 and 0-2 (figure 10-8). The gas sensor connects to R, W, R, B, and B of PCB. The secondary interlock relay and main relay has separate connectors. A varistor connects across the 120 Vac input voltage.

The primary winding (WW) of the low-voltage power transformer connects to the varistor power source with the secondary voltages out of the Y-Y and R-R connectors. The LVT transformer supplies low ac voltage to the –5 V, –12 V, and –31 volt sources. This transformer has a primary winding resistance of 270–285 ohms.

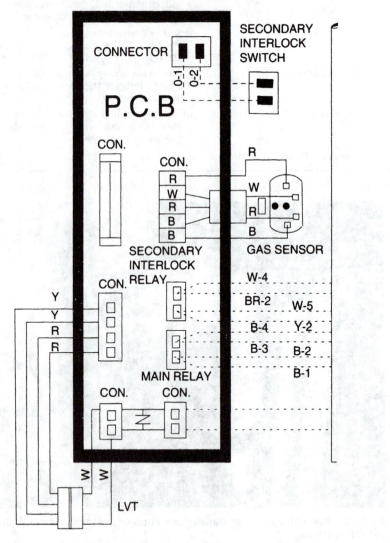

10-8 A pectoral diagram of components tied to a Samsung MW5820T oven. <small>Courtesy Samsung Electronics America, Inc.</small>

Samsung MW2500U touch-control circuits

In the Samsung oven, the keyboard and PCB assembly have 20 keys. The basic wiring circuit of the selector pad key matrix is shown in figure 10-9. The relevance between terminals and key matrix termination order is shown in figure 10-10. The keyboard continuity measurement can be taken between D10 and D17 to pin 5 (K) and 6 (J). R28 and R29 can also be checked with the ohmmeter. Check D18 and D19 with the diode test of a digital multimeter (DMM).

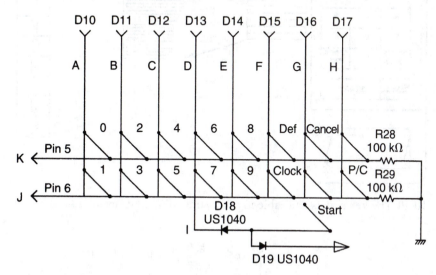

10-9 The wiring diagram of selector pad keys in a Samsung oven. Courtesy Samsung Electronics America, Inc.

Samsung MG5920T touch-control circuits

This Samsung oven has 27 keys with the function wiring of the selector pad key matrix shown in figure 10-11. Check each key pad with a continuity ohmmeter reading between each key pad from A to 1 on terminals M, K, J, and 12. For instance: if the Clock Auto Start button is erratic or does not function when pressed, take a low ohmmeter measurement between 2 of IC (B) and pin 12 of IC at 10.

Tappan oven touch-control circuits

The Tappan oven touch-control pad pins on the flex tail are numbered from the top of the board to the bottom. When a defective switch or pad does not function, check the touch pad with two pin numbers. Take a resistance measurement between the pin numbers (figure 10-12).

Control board circuits

Before replacing the suspected control board, check all input voltages. The low ac voltage from the small power transformer might be missing, causing improper control-board functioning. This low-voltage power transformer is fed directly from

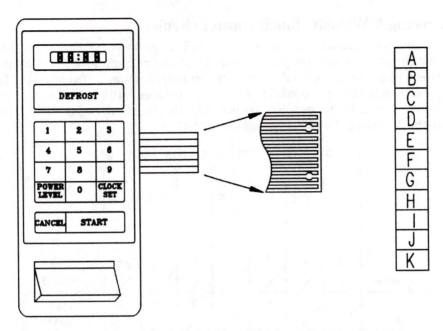

10-10 The key matrix and cable in a Samsung MW2500U oven. Courtesy Samsung Electronics America, Inc.

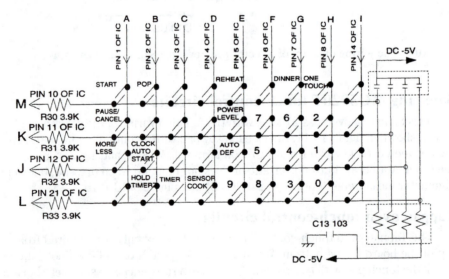

10-11 The key touch pad connection in a Samsung oven. Courtesy Samsung Electronics America, Inc.

the power line. You'll find this transformer on the control board of some ovens (figure 10-13). Check the low ac voltage right at the terminals where voltages tie into the control board. Only a few manufacturers provide these transformer voltages, so it's a good idea to mark them on the schematic for future reference (figure 10-14).

Note: THE PINS ON THE FLEXTAIL ARE NUMBERED
FROM THE TOP OF THE BOARD TO THE BOT-
TOM.

TOUCH PAD	PINS
AUTO DEF.	5 & 3
COOK 1	3 & 4
COOK 2	4 & 5
PROBE	8 & 7
BROWN	9 & 10
POWER	10 & 11
TIMER	8 & 10
REHEAT TIME	9 & 8
CLOCK	12 & 13
1	10 & 7
2	11 & 12
3	10 & 13
4	12 & 10
5	14 & 13
6	14 & 15
7	7 & 3
8	13 & 15
9	13 & 7
0	7 & 16
START	1 & 2
TURNTABLE	3 & 16
CANCEL	15 & 16

Pin Number 1

10-12 The flexible cable pin connection in a Tappan oven. Courtesy Tappan, brand of Frigidaire Co.

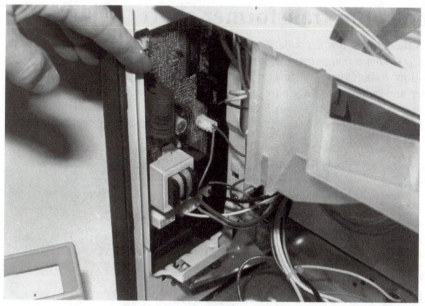

10-13 The LVT transformer operates from the ac power line.

Test set-ups	Test points	Normal voltage (approximate)
Attach meter leads to wire harness test points A-B, apply power to oven and open oven door.	A-B	120 Vac
Disconnect power and remove low voltage transformer connector attached to control circuit board and attach meter leads into harness side of connector at test points shown in chart. Apply power to oven and open oven door.	2(blue)-3(yellow)	2.5 Vac
	1(red-6(brown)	20 Vac
	1(red)-4(orange)	20 Vac

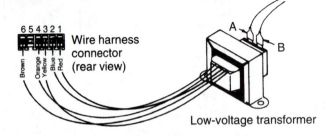

6 5 4 3 2 1

Brown Orange Yellow Blue Red

Wire harness connector (rear view)

A
B

Low-voltage transformer

10-14 You might not find LVT ac voltages in some ovens.

Controller transformer chart

Although the low secondary voltages and color coded wires are different, the chart provides a comparison when repairing other program controllers. The regulated low voltages provide power to the processors and solid-state circuits, while higher and negative voltages are applied to the display circuits. An open primary winding of the low-voltage transformer might result in no oven control, and an open secondary winding might prevent operation of controlling circuits (Table 10-2).

Table 10-2. Controller transformer chart with secondary leads and voltage

Oven	Color-code	Ac voltage
General Electric	R-R	13 V
JEBC200	Y-Y	3 V
	BL-BL	8.6 V
Goldstar	R-R	2.4 V
ER-505	Y-Y	4.8 V
	Y-BRN	24 V
	Y-BRN	24 V

Oven	Color-code	Ac voltage
Hardwick M-250	R-R	21.5 V
	G-G	2.5 V
Norelco MCS8100	BU-Y	2.5 V
	R-BRN	20 V
	R-ORG	20 V
Samsung MW-4630U	R-R	13 V
	BLU-BLU	2.5 V

General Electric JEBC200 touch-control panel and board test

The touch-control panel is divided into two units—a touch-control pad and a smart board. To repair, replace each unit. The following symptoms indicate a defective control pad:

1. When touching the pads, a certain pad produces no signal.
2. When touching a number pad, two figures or more are displayed.
3. When touching the pads, sometimes a pad produces a signal. Each touch pad can be checked with a R×1 scale of the ohmmeter.

The following symptoms indicate a defective smart board:

1. When touching the pads, a certain group of pads do not produce a signal.
2. When touching the pads, no pads produce a signal.
3. At a certain digit, all or some segments do not light up.
4. At a certain digit, brightness is low.
5. Only one indicator does not light up.
6. The corresponding segments of all digits do not light up, or they continue to light up.
7. Wrong figures appear.
8. A certain group of indicators do not light up.
9. The figures of all digits flicker.
10. The buzzer does not sound or continue to sound.
11. The clock does not operate properly.
12. Cooking is not possible.
13. Proper temperature measurements are not obtained.

GE JEBC200 key panel test

The key panel pad connections can be tested by checking continuity between pad connectors at the end of the ribbon. When a certain pad does not work, take an ohmmeter test. Set the ohmmeter to the R×10 scale. A shorted indication should occur between the connector connection numbers on the ribbon (Table 10-3).

Check the pad numbers against the connector numbers to take continuity ohmmeter readings when a pad is defective.

Table 10-3. Check the pad numbers against the connector number to take continuity ohmmeter readings when a pad is defective

Ribbon Pad	Connectors	Pad	Connectors
TIME COOK	2-8	1	5-7
TIME DEFROST	6-9	2	4-7
TEMP. COOK HOLD	3-9	3	3-7
MIN/SEC. TIMER	6-11	4	2-7
POWER LEVEL	4-10	5	1-7
ADD 30 SECONDS	1-8	6	6-8
AUTO COOK	5-10	7	5-8
AUTO ROAST	2-9	8	4-8
AUTO DEFROST	5-9	9	3-8
AUTO REHEAT	5-11	0	6-7
AUTO START	2-10	CLOCK	4-11
START	1-10	COMB. COOK	6-10
CLEAR OFF	3-11	CONV. COOK	1-9
POPCORN	3-10		

Courtesy General Electric Co.

Samsung MW5820T touch-control transformer circuits

A small ac transformer power supply provides a –5 V, –12 V, and –31 Vdc source. The –5 V and –12 V sources are regulated by transistors Q1 and Q2 (figure 10-15). These dc voltages are fed to the microcomputer IC, pulse oscillator, secondary detect switch, timer pulse shaping, and display circuits. The –31 volts is applied to the fluorescent display tube. A separate low-ac winding on the transformer supplies filament voltage for the display tube.

Check the power transformer winding with the ohmmeter when low or no ac voltage is found at the small transformer. Always discharge the high-voltage (HV) ca-

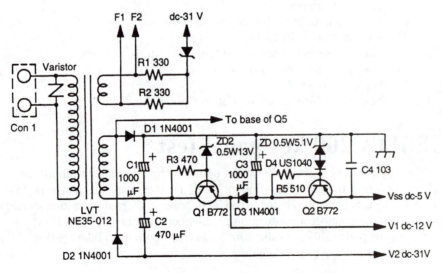

10-15 The control panel LVT circuits in a Samsung MW5820T oven. Courtesy Samsung Electronics America, Inc.

pacitor and unplug the power cord when taking ohmmeter measurements in the oven. Remove one ac input lead from the power transformer and take a resistance reading across the primary winding. Replace the transformer when you find an open reading. Sometimes these primary windings open after lightning, power-outage damage, or an overloading of the control board circuits (figure 10-16). Low continuity resistance of the secondary winding can indicate that the windings are normal.

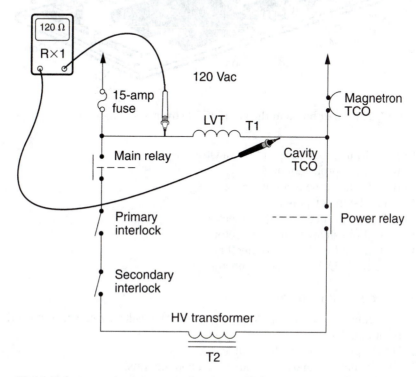

10-16 Take a continuity test to determine if the primary winding of LVT is open with no dc voltage source.

The ac primary voltage for the small power transformer operates directly from the power line. When the oven is plugged in, the color numbers should appear in the digital display in most ovens. If the ac voltages are normal from the transformer but the display does not light up, suspect poor board connections. Measure the transformer ac voltage right at the control board terminals. Poor connections can cause erratic oven operation.

GE JEBC200 smart board

The smart board is held in place by six plastic standoffs. When standoffs are squeezed together, the board can be pulled over them for board removal. Many diagnostic circuit tests can be made at the disconnect plugs (figure 10-17).

Con 1. 6-pin LVT secondary connector

Con 2. 4-pin damper and heater motor relay

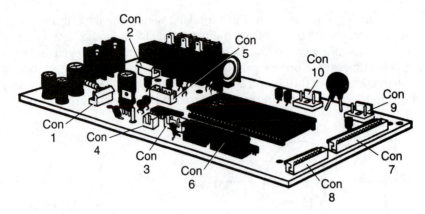

10-17 The smart board in the GE JEBC200 oven. Courtesy General Electric Co.

Con 3. 4-pin door sense and thermistor
Con 4. 2-pin probe connector
Con 5. 5-pin gas sensor connector
Con 6. 11-pin key panel connector
Con 7. 13-pin smart board connector
Con 8. 9-pin smart board connector
Con 9. 2-pin fusible link connector
Con 10. 2-pin LVT primary connector

To replace sensor board

1. Disconnect the power; pull the oven from the wall so you can remove the front panel.
2. Discharge the high-voltage capacitor.
3. Disconnect and mark all smart board connections.
4. Squeeze plastic standoffs (6) and lift up board.

Replacing GE JEBC200 touch-control pad

The touch-control pad can be replaced as one assembly. Check the key pad panel with a continuity test of the ohmmeter.

1. Disconnect the power and discharge the high-voltage capacitor.
2. Pull the oven out from wall far enough to remove the top panel.
3. Disconnect the ribbon from the connector and connectors 1 and 2 on touch control panel (figure 10-18).
4. Remove top trim—two screws.
5. Remove grille—two screws.
6. Remove fuse bracket (on top of magnetron)—two screws.
7. Remove four screws (two on each side) that hold the control panel to the frame. The screws are horizontal, with heads facing the rear of the unit. A stubby screwdriver is helpful.

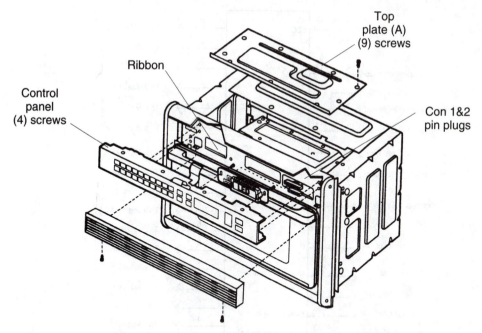

Note: Disconnect ribbon and con 1&2 pin plugs

10-18 Remove the GE touch-control pad in one assembly. _{Courtesy General Electric Co.}

Samsung MW5820T microcomputer circuit

The function of the microcomputer (IC) is to receive the information from other circuits, store the information, process the information according to the predetermined sequences, and send signals to the other circuits (figure 10-19). Take critical voltage and resistance measurements on the IC to determine if the IC or other components are defective. A low supply voltage (VDD) might indicate a leaky IC.

Removal of Samsung MW58207 control circuit board

Make sure no static electricity has formed on your body. It's best to use a grounded wrist strap when working around or upon the control circuit board. Remove the wire leads and disconnect the connectors from the control circuit board. Remove the screws that secure the control box assembly.

The control panel will come off the front of the oven. Remove the screw that holds the ground tail keyboard. Pull the lever end of the plastic fastener and remove the flexible circuit (FPC) connector from the circuit. This is the flat flexible cable. Remove the

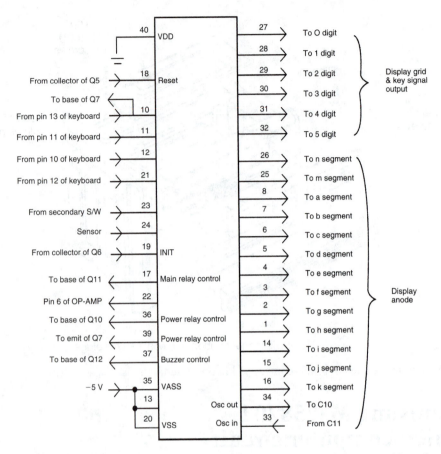

10-19 Microcomputer IC1 controls the many oven circuits of the control board.
Courtesy Samsung Electronics America, Inc.

screws securing the control circuit board. Lift up the control circuit board from the base assembly. When you reconnect the FPC connector, make sure that the holes in the connector are properly engaged with the hooks on the plastic fastener.

GE JEBC200 sensor cooking

When using the Auto Cook, Popcorn, or Auto Heat function, the foods are cooked without calculating time, power, level, or quantity. When the oven senses enough steam from the food, it relays the information to the microprocessor, which calculates the remaining cooking time and power level needed for best results. When the food is cooked, water vapor is developed. The sensor "senses" the vapor, and resistance increases gradually. When the resistance reaches the value set according to the menu, supplementary cooking is started.

The time of supplementary cooking is determined by experimenting with each food category and is entered into the microprocessor. For example: Auto Cook 2 (figure 10-20).

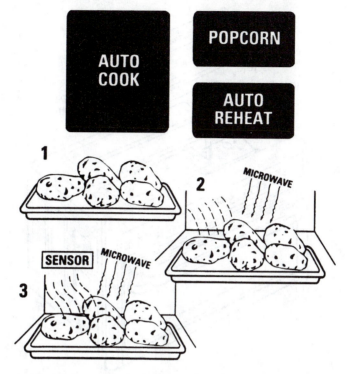

10-20 Sensor cooking determines what cooking time and power level settings are needed in GE oven. Courtesy General Electric Co.

1. Potatoes at room temperature. Vapor is emitted slowly.
2. Heated potatoes. Moisture and humidity are emitted rapidly.
3. The sensor detects the moisture and humidity and calculates cooking time and variable power.

GE JEBC200 gas sensor tests

Microwave sensor cooking uses a special gas sensor that detects both humidity (steam) and hydrocarbons (food odors) during the cooking process. The sensor is located on the exhaust duct at the top left corner of the cavity, behind the grille (figure 10-21).

Checking the initial sensor cooking condition

1. The oven should be plugged in at least 5 minutes before sensor cooking.
2. Room temperature should not exceed 95°F (35°C).
3. The unit should not be installed in any area where heat and steam are generated (for example, next to a conventional unit).
4. Exhaust units are provided on the back of the unit for proper cooling and air flow in the cavity. To permit adequate ventilation, do not block the small vents. Leave space for circulation.

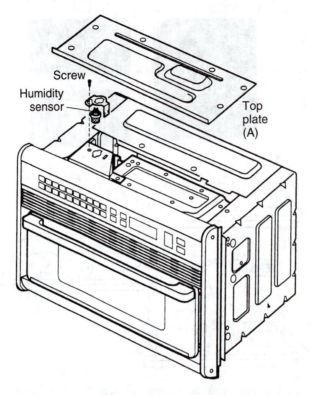

10-21 The Gas Sensor detects humidity and food odors during the cooking process. Courtesy General Electric Co.

5. Be sure the exterior of the cooking container and the interior of oven are dry. Wipe off any moisture with a dry cloth or paper towel.

6. The sensor works with food at normal storage temperature. For example, chicken pieces would be at refrigerator temperature and canned food at room temperature.

7. Avoid using aerosol sprays or cleaning solvents near the oven while using sensor setups. The sensor will detect the vapor given off by the spray and turn off the oven before the food is properly cooked.

8. After about two to nine minutes, if the sensor has not detected the vapor of the food, ERROR will appear and the oven will shut off.

Quick test sensor

Using three fingers, touch and hold pads 6, 7, and 8 at the same time. Observe diagnostic numbers in display (numbers are approximate).

    a. 40–20 normal (can verify with detection test).

    b. 255 or higher (sensor failed to open, unplugged, wiring problem, or smart board problem).

Do not check the sensor terminals with the ohmmeter, as this might damage sensor. Only the heater terminals should be checked with the ohmmeter (30 ohms).

To perform the sensor detection test:

1. Place ⅓ cup of water in oven. Do not use styrofoam cup.
2. Touch Auto Cook-10 start.
3. After 1½–2½ minutes, control should beep and display EN4.
4. Touch Clear/OF6. If test is normal, the sensor is okay. If test fails, check the sensor.

GE JEBC200 oven sensor

The oven sensor is a negative coefficient thermistor; when the temperature goes up, the resistance goes down. The oven sensor should measure 60 kilohms when cold.

1. To remove oven sensor, disconnect the power plug, pull the oven from the wall, and remove the top panel (figure 10-22).
2. Remove screw inside cavity (next to probe).
3. Lift sensor out and unplug connector.

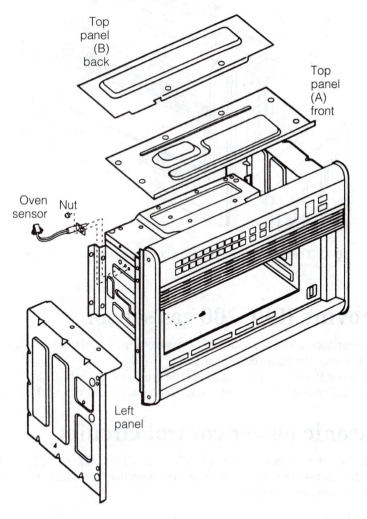

10-22 The location of the oven sensor in the GE oven. Courtesy General Electric Co.

GE JEBC200 flame sensor

The oven thermal cutout is located on the exhaust duct. The temperature rating is 248°C. If open, find the cause and replace the defective component.

1. To replace flame sensor, disconnect the power plug, remove the oven from wall, and take off the front cover.
2. Discharge the HV capacitor.
3. Remove the two screws holding the flame sensor (figure 10-23).

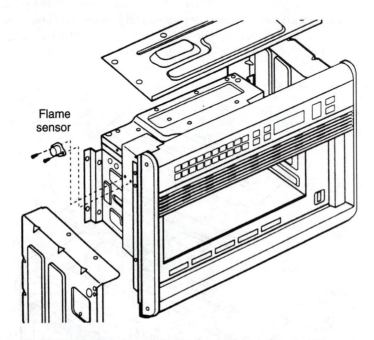

10-23 The flame sensor is located on exhaust duct in GE oven.
Courtesy General Electric Co.

Removing JEBC200 gas sensor

1. Disconnect the power and pull the oven out from wall to remove the top cover.
2. Discharge the high-voltage capacitor.
3. Remove the one screw that holds the bracket (figure 10-24).
4. Lift the sensor and connector out. Unplug the sensor from the connector.

Panasonic power control circuits

In the power relay control circuits, when pin 13 of IC2 goes high and the power relay control transistor turns on, its collector goes low, energizing relay RY1, which turns on the oven lamp (figure 10-25).

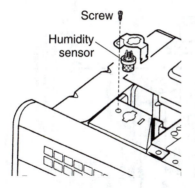

Screw

Humidity
sensor

10-24
How to remove gas sensor in
GE JEBC200 oven. Courtesy General
Electric Co.

Within the magnetron relay control circuit, an initialization E25 of the microprocessor goes low. When you press the start key, pin 6 will go low and the latch circuit output will go high, turning on the power relay drive transistor. At the same time, R11 of the microprocessor unit will go high, turning on Q4 and energizing the magnetron via relay RY2.

In the initialized circuit, the output pin 14 of IC2 goes to the microprocessor unit reset. Under normal conditions, pin 14 goes high. If power is interrupted, pin 14 to the microprocessor unit reset goes low, and the microprocessor unit will lock up.

With the clock pulse generator circuit, pin 16 of IC2 receives 18 Vac from the secondary winding of the low-voltage transformer. The inverter clips and inverts the input ac and generates a 7-V P-P output square wave to IR Q of the microprocessor.

Other circuits

Pulse oscillator circuit

The pulse oscillator circuit functions from pins 25 and 26 of the microprocessor (figure 10-26). The pulse oscillator circuit is 4.19 MHz. The oscillator's peak-to-peak voltage is 5 volts. This signal controls the processing speed of the microprocessor unit.

Buzzer control

When the appropriate membrane key is pressed and accepted by the microprocessor, the microprocessor unit generates a single burst of 2.1-kHz pulse at E24, which activates the buzzer circuit (figure 10-27). When the cooking program is completed, the microprocessor unit will generate a series of five 2.1-kHz bursts, creating five audible beeps.

Samsung MG5820T buzzer circuit

The buzzer control circuit is found on pin 37 of Micom IC1. A –12 volt source is fed to the buzzer which is in parallel with R26 (1 kilohm) resistor (figure 10-28). The emitter terminal of Q11 is at ground potential while the base circuit is fed signal from pin 37 of IC1. The buzzer sounds off indicating cooking process is finished or when the pad is recognized by the microcomputer IC1.

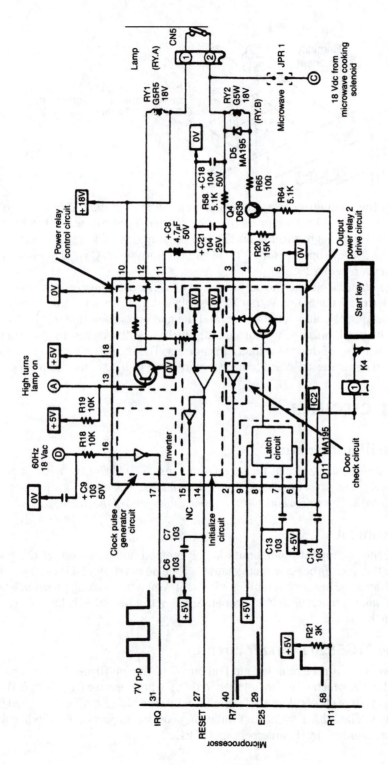

10-25 The power control circuits in a Panasonic oven. *Courtesy Matsushita Electric Corp. of America*

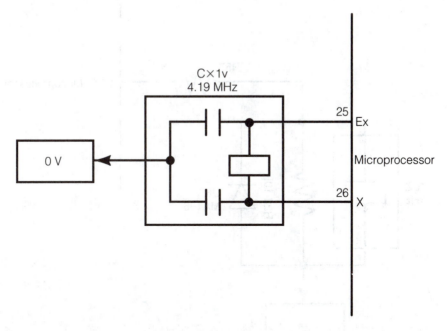

10-26 The 4.19-MHz signal controls the processing speed of the microprocessor.
Courtesy Matsushita Electric Corp. of America

Samsung MW2500U display circuits

The LED circuit is controlled by pins 11 through 17 of the microcomputer I3. With push-button pads, the microcomputer chip controls the numbers upon the fluorescent display tube (figure 10-29). The filament ac voltage is found on pins 1, 2, and 16 of ULN2003A and negative dc voltage (–18.5 V) at pin 9 (VCC). The pin assignment of the display is shown in figure 10-30.

Samsung MW2500U pulse oscillator circuit

The pulse oscillator circuit generates an oscillating frequency that controls the various input and output signals and measures the time of data-processing speed (figure 10-31). The crystal provides oscillation on pins 18 and 19 of IC1.

Samsung MW2500U timer pulse-shaping circuit

The timer pulse-shaping circuit converts the sinusoidal waveform of 60 Hz to a shaped waveform and becomes the time base in the microcomputer, supplying zero-crossing signal, which turns the power relay 2 on (figure 10-32).

Samsung MW5820T door-sensing switch circuit

The door-sensing switch circuit detects whether the door is open or closed and sends this information to the microcomputer IC (figure 10-33, Table 10-4). A –5 volt dc voltage is fed to the door-sensing switch and is applied to pin 23 of the IC. The detecting circuit is off when the door is opened and on when the door is closed.

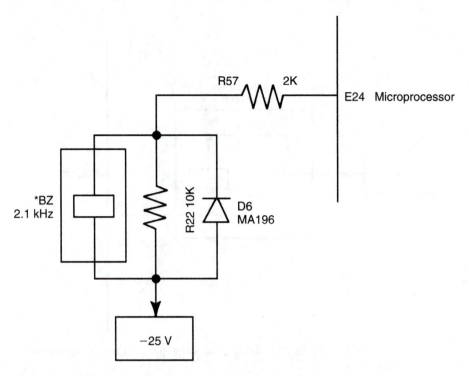

10-27 The piezo 2.1 kHz buzzer indicates when cooking process is done in a Panasonic oven. Courtesy Matsushita Electric Corp. of America

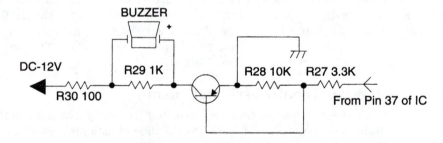

10-28 The buzzer circuit in a Samsung MG5820T oven. Courtesy Samsung Electronics America, Inc.

Troubleshooting guide for touch control circuit

A defective control circuit in the Samsung MW5820T oven can be indicated by segments missing, partial segments missing, digit flickering, and colon does not turn on. Suspect a problem with the display when you see a distinct change in the brightness of one or more numbers (figure 10-34). Check the display if one or more digits in the display does not light up when it should and no indicator lights for the selected

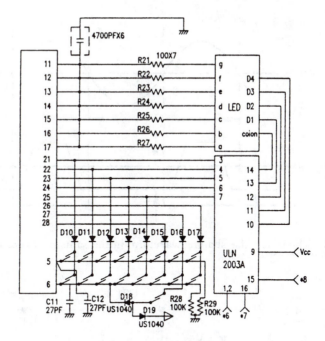

10-29 The fluorescent display circuits in a Samsung MW2500U oven. Courtesy Samsung Electronics America, Inc.

Pin no	1	2	3	4	5	6	7	8	9	10	11	12
Con-nection	A	B	C	D	E	F	G	D4	D3	D2	D1	Colon

10-30 The pin numbers and connections on display in Samsung MW2500U oven. Courtesy Samsung Electronics America, Inc.

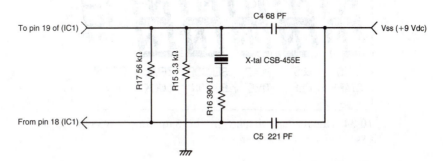

10-31 The crystal provides oscillations on pin 18 and 19 of IC1. Courtesy Samsung Electronics America, Inc.

cooking function. Also suspect the display circuits when indication of a number is different from the one touched. For instance, if pad 7 was touched and 8 appears in the display.

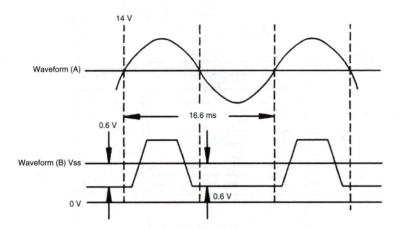

10-32 The time pulse-shaping waveform in the Samsung oven. Courtesy Samsung Electronics America, Inc.

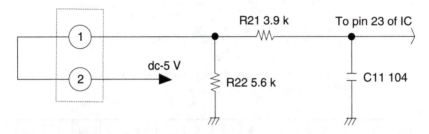

10-33 The door sensing switch detects if door is opened or closed in a Samsung MW5820T oven. Courtesy Samsung Electronics America, Inc.

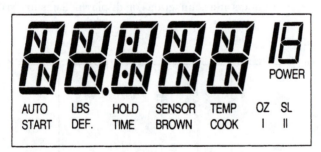

10-34 The readout display in the Samsung MW5820T oven. Courtesy Samsung Electronics America, Inc.

Check the display circuits when the display does not count down, the clock does not work, or the display obviously skips periods of time while counting down. If the display counts down too fast while cooking and display can not shift from the first stage cooking to the second stage cooking while in 2 phase cooking, suspect the control board circuits.

When the display continues counting down time cooking when the door-open button is pressed, suspect board circuits. Check to see if the door sensing switch shorts when the door-open button is pressed. If the door sensing switch is normal, replace touch control circuit board.

Refer to Table 10-4 if the digital readout display does not show programming when the membrane key board is programmed by touching the proper pads. But before you follow the above steps for the membrane keyboard's failure, check for the continuity of each wire harness between the membrane keyboard and touch control circuit.

When the microwave oven does not operate properly with time cooking, check the ac output voltages of the LVT transformer (figure 10-35). The primary winding L and N or 1 and 2 connects to the 120 Vac power line. Secondary voltage 1 at terminals 3 and 4 supply 2.5 volts ac and terminals 5 and 6 have a 13-volt ac source.

If the oven does not operate at all, although the START pad is touched, notice if oven lamp is on at low brightness and the motor fan operates slowly. If so, suspect low voltage circuits. Also, the touch control circuit might not be detecting the key input signal. Check the primary and secondary voltage of the low-voltage (LVT) transformer.

Replace the transformer if it has improper output ac low voltages. If no voltage is measured on the primary side, check the components in the power line circuits. Do not overlook an overloaded low-voltage source.

If the oven operates normally only at high power settings, but does not properly at other power settings, measure the time periods of the line voltage being applied to the two terminals of the relay. Check the foil patterns connected to the power and main relay circuits.

If the oven lamp and fan motor operate normally but the microwave does not oscillate or cooking time takes too long when compared to the cooking time described in the cookbook, suspect the power level off and on system. If the oven operates with a water load in the oven and power level at a high selection for only a few minutes, suspect power control relay circuits and high voltage. Measure the ac voltage continuity between terminals A and B of the power control relay with the oven started. Be very cautious of high voltage. Always use water in beakers when operating the oven in service tests.

Suspect improper timing periods of the line voltage being applied to each power setting compared to Table 10-5. When the timing period of the power control relay is other than those listed in the table, replace PCB assembly. Replace touch control panel when defective components are inside the touch control circuits. All outside parts can be replaced as individual components.

Temperature control

The oven-temperature sensor monitors the temperature changes in the oven cavity. This sensing device is called a thermistor, and its resistance change is converted into a voltage variation across the resistor network R14 and R15 (figure 10-36). This voltage change is fed back to the microprocessor unit port AN6 to control the cycle on time of the magnetron or the heaters.

The temperature-control sensor monitors the heat produced by the heater and establishes the oven temperature the user has selected. In a kitchen, normal room temperature can vary anywhere from approximately 50° to 90°F. The resistance reading across the oven thermistor should be within 380 kilohms at 50°F and 800 kilohms at 90°F.

Table 10-4. Samsung's MW5820T troubleshooting guide control circuits

CONDITION	CHECK	RESULT	CAUSE	REMEDY
Display does not show programming at all, even if keyboard is touched	Check continuity of each pad of membrane keyboard	Normal	Malfunction of touch-control circuit of control box assembly	Replace Ass'y PCB main
		Abnormal	Malfunction of the membrane keyboard	Replace the membrane keyboard
	Check the voltage of low-voltage trans secondary windings	Normal	Malfunction of touch-control circuit of control box assembly	Replace Ass'y PCB main
		Abnormal	Malfunction of the L.V. transformer	Replace the L.V. Trans

Courtesy Samsung Electronics America, Inc.

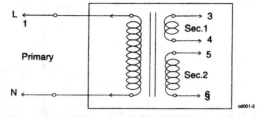

Primary Voltage	(1-2)	120V
Sec.1 Voltage	(3-4)	2.5V
Sec. 2 Voltage	(5-6)	13V

10-35 The low-voltage transformer (LVT) in the Samsung oven. <small>Courtesy Samsung Electronics America, Inc.</small>

Table 10-5. Timing period of power control settings in Samsung MW5820T oven

Power level %	Relay turn on time (sec)	Relay turn off time (sec)
10	5	25
20	8	22
30	11	19
40	14	16
50	17	13
60	20	10
70	23	7
80	25	5
90	29	1
100	30	0

Courtesy Samsung Electronics Corp. of America

Tappan temperature-probe test

Remove the roast probe test, if installed. Connect the ohmmeter prods between the tip of roast probe and the barrel of the roast plug (figure 10-37). Measure the resistance of the probe at a room temperature of 77°F. The probe is normal at 55 kilohms of resistance.

Panasonic's variable-power circuit

The variable-power circuit controls on-off timing of the variable-power relay within a specific duty cycle to vary the output power of the microwave oven with signals from the microcomputer. The variable-power circuit incorporates a phase-control circuit, a multivibrator circuit, and a switching circuit. In some advanced models, these circuits are combined into one piece of microprocessor (LSI). However, we will study the analog circuit (commonly used as the variable-power circuit) in this section (figure 10-38).

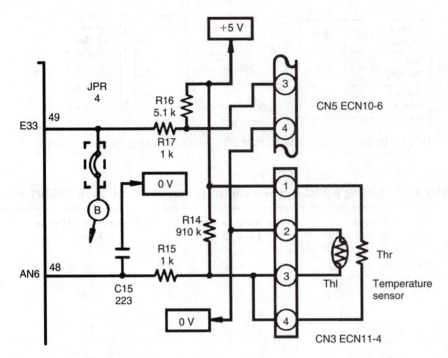

10-36 The temperature sensor is around 380 kilohms in a Panasonic oven.
Courtesy Matsushita Electric Corp. of America

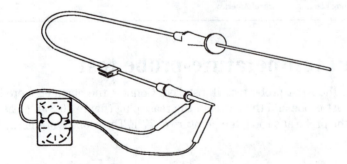

10-37 The temperature probe in a Tappan oven should be around 55 kilohms. Courtesy Tappan, brand of Frigidaire Co.

Phase-control circuit

The phase-control circuit functions to control the open/close timing of the variable-power switch to prevent possible damage to the switching device and possible interference to TV and radio as the result of arcing generated across the contacts of the switching device.

In the HV circuit, the high voltage is applied to the magnetron only during the negative half-cycle, and no high voltage is developed across the variable-power

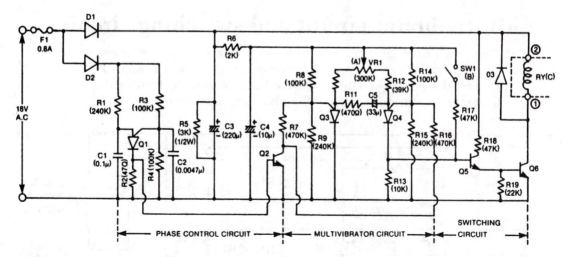

10-38 The Panasonic variable-power control circuits with phase control, multivibrator, and switching circuits. Courtesy Matsushita Electric Corp. of America

switch during the positive half-cycle, as explained in the HV circuit in this chapter. The basic idea of the phase-control circuit is to open or close the high-voltage circuit when no voltage is applied to the magnetron. In the following section, you will see how this is done.

1. When the start button is pressed with door closed and timer set, approximately 18 volts is generated in the tertiary winding of HV transformer and supplied to the variable-power circuit.

2. The 18 Vac is converted to 24 volts half wave by the diode D2 and this half-cycle voltage is divided by the R3(100 K) - R4(100 K) combination, and the resulting 12-volts half wave is applied to the gate of transistor Q1.

3. Capacitor C1 is charged through R1 and when the voltage across C1, which is equivalent to anode voltage of Q1, exceeds the gate voltage of Q1, it causes Q1 to conduct.

4. When Q1 conducts, a voltage is developed across R2 and will forward-bias the base-emitter junction of transistor Q2, allowing it to conduct. Note: Conduction of Q1 discharges C1, and this will cause Q1 to cut off when C1's voltage drops to approximately 0 volts. The Q1 remains cut off until the next positive half-cycle charges C1 to 4.2 volts.

5. When Q2 conducts, it causes the gate voltages of Q3 and Q4 to drop from 14 volts to 12 volts through R7 and R16. Note: The transistor Q2 conducts only when Q1 conducts, which means that Q2 conducts only during the positive half-cycle and does not function when the high voltage is across the variable-power switch.

6. As long as Q2 remains cut off, the gate voltages of Q3 and Q4 remain 14 volts, and this will result in no conduction of both Q3 and Q4, as the anode voltages of both transistors never exceed 14 volts.

In this way, the variable-power switch is controlled to open or close only when no high voltage is across the variable-power switch contacts.

Multivibrator circuit and switching circuit

This circuit functions to control the open/close time period of the variable-power switch. As shown in figure 10-39, the programmable unijunction transistors Q3 and Q4 are connected in parallel with the voltage divider resistors R8-R9 for Q3 and R14-R15 for Q4. In the following section, you will see how this circuit functions.

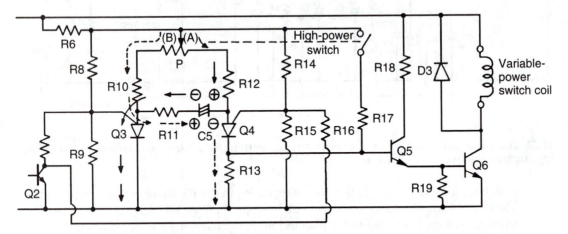

10-39 The switching and multivibrator circuits in a Panasonic variable-control circuit. Courtesy Matsushita Electric Corp. of America

1. 24 Vdc generated by diode D1 is smoothed by capacitors C3 and C4 and is then applied to the multivibrator circuit. The gate voltages of Q3 and Q4 are determined by the R7-R8-R9 combination for Q3 and R6-R14-R15 combination, plus the switching effect of Q2 on both combinations. When the circuit is first energized, the transistor with the higher anode potential conducts first.

2. Assuming that the anode voltage of Q3 is higher than Q4, when the circuit is first energized Q3 conducts and 12 Vdc (as the result of Q2 conduction) on the gate of Q3 becomes approximately 0 volts. This causes current to flow (as shown by the solid line) and causes capacitor C5 to charge.

3. When the anode voltage of Q4 exceeds the gate voltage as the result of the charging effect of C5, it causes Q4 to conduct and the current flows as shown by the dashed line; this causes Q3 to cut off. The charging time of C5 is determined by the time constant of the series combination of the variable resistor (portion A) and R12-C5-R11.

4. The voltage across R13 forward-biases the base-emitter junction of Q5, causing it to conduct. The conduction of Q5 instantaneously causes Q6 to conduct, allowing 24 Vdc to be applied to the variable-power switch coil, thereby closing the HV circuit to start microwave oscillation.

5. As soon as Q4 conducts, the capacitor C5 starts to charge in the opposite direction through R10 and R11. When the anode voltage of Q3 exceeds its gate voltage (12 volts), Q3 conducts and Q4 cuts off. The complete time cycle of one transistor conduction, through cutoff and back into conduction, is approximately 22 seconds. This on/off cycle of Q3 and Q4 is repeated as long as the circuit is in operation.

The length of time within the 22-second period that Q4 conducts is determined by the setting of the variable-power resistor mounted on the front panel of each microwave oven. When the variable-power resistor is set at medium-low power, the multivibrator circuit is balanced and each transistor will have equal conduction time periods of 11 seconds each, resulting in 50% of full output power.

As the variable-power resistor moves toward the B portion end (low power), Q4 conducts for a shorter period; when moved toward A portion, it conducts longer. By adjusting the external variable power resistor controller, the output power of the microwave oven can be varied to any power setting desired. When the variable-power switch is moved toward the extreme portion A end, the high-power switch closes. This will keep Q5 and Q6 conducting constantly to maintain full power oscillation. Note: It is possible to check the time duration of one complete cycle of the variable-power circuit. To check the on/off cycles of the magnetron, simply observe dimming of the oven light. If the oven light dims, the secondary circuit is supplying high voltage to the magnetron, and when the oven light goes bright, it indicates no voltage supply to the magnetron.

As previously explained, one complete cycle is 22 seconds. However, it can vary approximately 7 seconds due to the tolerances of resistors and capacitor C5. This tolerance of the duty cycle will not affect the actual output power because the proportion of on/off time period remains the same. For example, if the duty cycle is measured as 26 seconds, the duty on time of the medium-low power setting should be 13 seconds, resulting in 50% of full power output.

Troubleshooting the rest

To determine if the rest of the oven circuits are operating, clip a wire across the power relay terminals (figure 10-40). Always pull the power plug and discharge the HV capacitor before attempting to connect the clip wire. Now plug the oven in—if the HV and magnetron circuits are normal, the oven will begin to cook without tapping in any time on the control board. Replace the control board when no voltage or low voltage is found at the oven relay and the rest of the oven circuits are normal.

In ovens using a triac assembly controlled by the control board, check the ac voltage applied to the gate terminal of the triac. No voltage here indicates a defective control-board assembly with proper ac input voltage. To test the rest of the HV and magnetron circuits, clip a wire across MT1 and MT2 of the triac assembly (figure 10-41). Again, discharge the HV capacitor and pull the power cord. When the oven begins to cook, you know the rest of the circuits are normal.

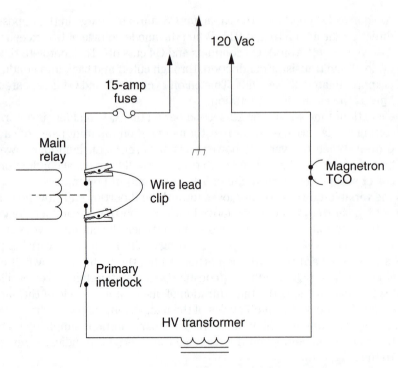

10-40 Clip a wire across relay contacts to determine if oven circuits are operating.

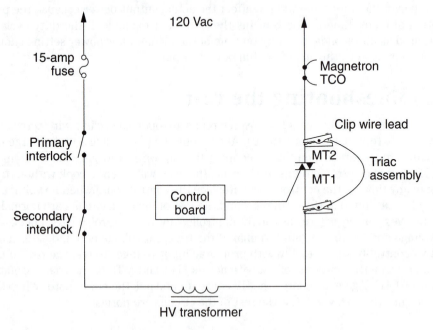

10-41 Discharge the HV capacitor and clip across triac wires MT1 and MT2 to determine if triac defective or control circuits.

Replace the defective control board with proper low ac input voltages when abnormal voltages are found at the power relay or triac assembly. When the digital display does not come on with proper input ac voltage, replace the control board. If after programming the control board at least three times according to the manufacturer's directions the oven does not start to count down, suspect a defective control board. Make sure the temperature probe is out of the socket when making these tests. If the display numbers do not register properly after several attempts, suspect a defective controller assembly (Table 10-6).

Table 10-6. A control board troubleshooting chart

Symptoms	Possible problems
No display light or flashing	1. Check fuse or power line. 2. Check low ac voltage from small power transformer to controller. 3. Make sure temperature probe is out of socket. 4. Suspect controller board when low voltage is found at control board terminals. 5. Check circuit breaker. 6. Check defective varistor.
Display light with flashing	1. Low ac voltage. Transformer normal and applied voltage to control board. 2. When time will not register, suspect control board. 3. Go over time and temperature setting at least three times before replacing control board.
Improper number display and incomplete segments	1. Double-check the oven operation from manufacturer's service manual. Make sure you are tapping proper numbers in sequence. 2. Check each pad membrane keyboard continuity. 3. Defective control board. 4. When some of the numbers light up and not others, suspect defective control board. 5. Defective touch pad.
Improper countdown	1. To check the timer functions, touch pads and set desired time. 2. The display should start to count down. 3. Replace board when display shows correct time and no countdown. 4. Replace control board if display jumps or too fast countdown.
Time and temperature set. Time counting down. No blower motor or cooking operation	1. Suspect no control voltage or defective cook relay or triac. 2. Measure voltage to controlling component. If voltage is present check defective relay or triac. 3. Use test clip method to isolate defective relay or triac. 4. Replace control board with no control voltage. 5. Check door interlock. 6. Check the microprocessor in control board.
To check power display and stop operation	1. Set time and temperature. Push switch. 2. Oven begins to cook. Then stops switch or off button. 3. Oven should stop. If not, check triac assembly.

Table 10-6. Continued

Symptoms	Possible problems
Time and temperature pushed. Time counting down. Blower motor operating with no heat.	1. Check 120 Vac at the primary winding of the high-voltage transformer. 2. If power line voltage present, problems within high-voltage circuits. 3. No power line voltage at the primary winding of the high-voltage, suspect defective oven relay, triac or control board. 4. Check safety concealed switches.
A faint display number is seen in background of display	1. Although the oven counts down and cooks properly, sometimes you can see a faint number in the background at all times. 2. Replace defective control board. This might occur after power outage or lightning damage.
Control panel defective a few months after replacement	1. Check for improper fusebox ground. 2. Intermittent gate terminal or triac. 3. Magnetron tube.
Temperature probe will not key in . . . no indicator light.	1. Check continuity of probe. 2. Defective sensor probe. 3. Check jack terminal leads. 4. Defective control board or microprocessor.
Oven stops . . . soon as start button pushed in with temperature probe in operation.	1. Check continuity of probe. 2. Defective sensor probe. 3. Check continuity of probe jack and wires. 4. Check for shorted probe. 5. Defective control board and microprocessor.

Lightning damage

Often the control board is damaged when the oven is struck by lightning or affected by a storm-caused power outage. Before replacing the control board, check for a burned or damaged varistor on or near the control board. In some ovens, the line-voltage varistor is outside the control board. Check the varistor for burned marks (figure 10-42). Replacing the varistor might solve the control board problem. These small varistors are added across the ac line to protect the oven or control board circuit.

Remove the control board and check for damaged pc wiring. Some ovens have a test cord kit to connect the board to the oven for easy repairs. Sometimes the wiring might be damaged around the defective varistor (figure 10-43). Bridging the burned pc wiring with regular hookup wire might solve the dead-oven symptom. If the control board is excessively damaged or appears erratic after being struck by lightning, replace the entire control board.

Check the control board contacts and plug-in connections. Arcing around the contact pins will show melted areas around the plastic. Inspect both male and female connections. If dark or arcing marks are found around the pin connections, remove the board and inspect for badly soldered pins. Clean up the bad contact connections.

When these connections are very bad, try and replace them. For temporary repairs, solder a flexible lead wire around the two connections.

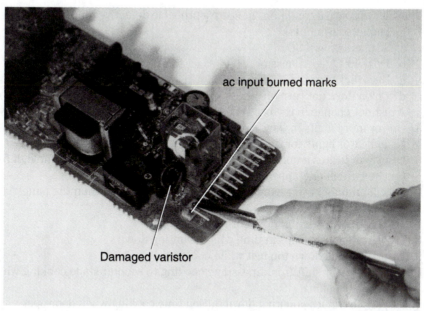

10-42 Lightning damaged this control board panel.

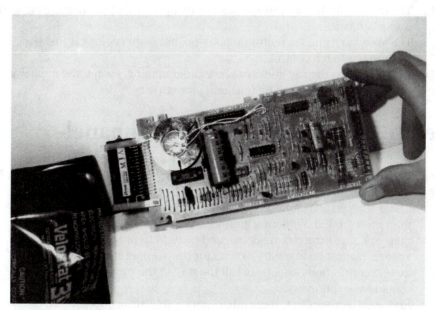

10-43 Check the PC wiring for possible damaged or stripped areas of control board. Also remember to handle the control board by the edges and wear static wrist-strap in handling these boards.

Samsung MW5920T troubleshooting of touch control circuit

The following visual conditions indicate a probable defective touch control circuit:

1. Incomplete segments.
    a. Segments missing.
    b. Partial segments missing.
    c. Digit flickering.
    d. Colon does not turn on.
2. A distinct change in the brightness of one or more numbers in the display.
3. One or more digits in the display does not light when they should.
4. No indicator lights for the selected cooking function.
5. Display indicates a number different than one touched. For example, if 5 is touched, number 8 appears in display.
6. Specific numbers (for example 2 or 3) will not display when the panel is touched.
7. Display does not count down or no clock works.
8. Display obviously skips in time while counting down.
9. Display counts down too fast while cooking.
10. Display cannot shift from first-stage cooking to second-stage cooking while two-phase cooking.
11. Display continues counting down during time cooking when door-open button is pressed. Check if poor door-sensing switch shorts when door button is pressed. If door sensing switch is normal, replace touch control panel.
12. The time of day in the display does not reappear when the pause/clear pad is touched twice.
13. The oven lamp and fan motor do not stop, although cooking is finished. Replace the touch control.
14. Digital readout display does not show programming, even if the membrane keyboard is programmed by touching proper pads.

Before you replace the control panel

The following is a list of things to check before you replace the control panel in an MCS 8100 oven. It will also help you to service any other panels in the future. Do not assume the board is bad on every call you get.

1. Check the customer's electrical system for correct polarity and especially for a good ground. Without a ground, the system will act erratically, skip numbers, give incorrect readouts, and be hard to program. Check line voltage. It must be at least 110 volts for proper operation. Always check around connections from board all the way to the round pin plug. You should have continuity.
2. Recheck the complaint. Remember, some problems can be caused by the customer not really knowing how to operate the unit. A prime example is

if the customer waits too long between touching the clock-set pad and number; the unit will not enter a number. There is a three-second time limit. Become completely familiar with the operation of the touch panel. Always have the customer operate the unit for you. Use the check procedure in the service manual.

3. Make sure all door interlocks are operating properly. An ohmmeter check of these interlocks should show (with door open) interlock monitor closed and primary interlock open, secondary interlock open, door interlock sense switch closed. With door closed, all switches will check opposite of above. Remember, if the door interlock sense switch is not working, you will be able to program the unit, but the unit will not start when start is touched. An improperly adjusted door-interlock sense switch will cause the symptom of not shutting countdown off when door is opened.

4. Check the on/off sense switch, which is a contact between pin 7 and 8 of on/off switch. An open on/off switch will not allow the light and fan and stirrer to come on. The only thing that will happen is you will get all 8888s on display.

5. If you have a unit that programs, but the light, fan, and stirrer will not work, the browner will not heat, and the oven will not operate, check pins 3 to 16 on on/off switch. With switch on, these contacts will be closed.

6. Many times a control panel is changed because the unit starts operating as soon as the on/off switch is pushed. This could very well be a bad triac. To check, pull the wire off the gate terminal of the triac in question. If the unit still operates as soon as the on/off is pushed, you have a shorted triac. If it does not, you either have a bad board or bad disconnect block. If you notice a burned spot on the board edge connector, it indicates that you have had a loose terminal; order part number 355T498S02.

7. If unit shorts as soon as on/off switch is turned on, this problem could be caused by a short in either the magnetron fan or stirrer motors. To isolate which one it is, make a resistance check of these motors. With motor disconnected, read the following: stirrer motor 2000 ohms, magnetron fan motor about 25 ohms.

8. Always check to see if the printed circuit board is loose in the frame. You should not be able to move the board up and down at all. If board moves, bend up the tabs on bracket, which the board sets on, until the board no longer moves.

Installing a new control board

Be careful when removing the control panel from the packed carton. Most control boards are enclosed in static-type bags. Always hold the control board by the edges. Keep fingers and metal material away from the printed wiring and board components (figure 10-44). Touch the metal oven cabinet to discharge any static electricity before mounting a control board.

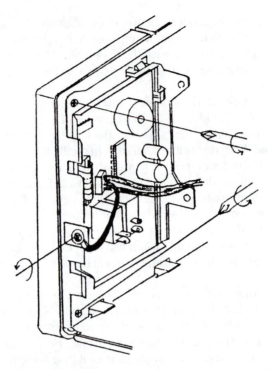

10-44 Disconnect all socket, cables, and wire from control board before removing.

In some ovens, the control board is easily removed from the backside—while in others, the front control panel assembly must be removed. Install the new board in reverse order. Make sure that all board plugs are tight and snug. In ovens with flexible rubber cable, make sure the cable is pushed in tight. Inspect the plug as you plug it in. Tighten all metal screws holding the board to the metal frame.

Before replacing any control board or panel, make sure it is defective. Run through the operation procedure several times. Take input and output voltage measurements of the control board. Determine if the control board or other circuits are at fault. Be careful in handling and mounting the control board. Control boards are fairly expensive and take several minutes to install. If in doubt, follow the manufacturer's service manual for control-board troubleshooting procedures.

Samsung MW2500U board disassembly

Before removing or replacing a control board, put a ground wrist strap or make sure your body does not touch the control circuitry. Disconnect the connectors from the control circuit board and wire leads. Remove three screws securing the control panel attached to the oven circuitry (figure 10-45). Remove one screw securing the ground tail of keyboard. Push up the lever end of the plastic fastener and remove the flexible printed circuit (FPC) connector from the circuit.

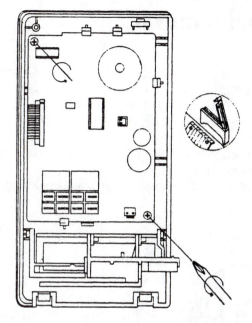

10-45
Remove four screws to remove
the control board. Courtesy Samsung
Electronics America, Inc.

Remove the four screws holding the control board circuit (figure 10-46). Lift up
the control circuit board from the right side, and take off the hooks of the control circuit board from the box assembly.

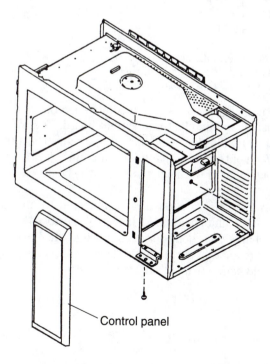

10-46
Remove the screws from
bottom of control board to
take it out of a Tappan oven.
Courtesy Tappan, brand of Frigidaire Co.

Control panel

Tappan control panel removal and assembly

In Tappan ovens, the plastic control panel housing containing the electronic control mounts to the oven chassis, using one screw at the bottom center and one plastic hook at the top. Remove the one screw at the bottom, being careful not to let the panel drop, possibly damaging a component. Always use static electricity protection when handling the circuit board.

1. Disconnect the power, remove wrapper, and discharge high-voltage capacitor.
2. Tag and disconnect wires from control board.
3. Remove screws from bottom of the control panel assembly and detach panel (figure 10-47).

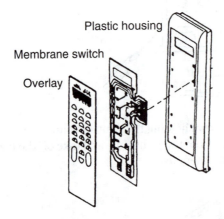

Plastic housing

Membrane switch

Overlay

10-47 Remove self-adhesive to separate overlay and membrane in the Tappan oven. Courtesy Tappan, brand of Frigidaire Co.

The electronic control panel assembly consists of a decorative overlay, a membrane touch switch, a plastic housing, and an electronic circuit board. The overlay and membrane switch attach with self-adhesive. To remove the circuit board from the plastic housing, gently press each of the three snaps (one at a time) just enough to release the board from the engagement of the snap, then slide the board out from under the hooks (figure 10-48).

When handling electronic controls, use a wrist strap connected to an earth ground source to protect against static buildup and avoid damaging sensitive circuits. If a wrist strap is not available, discharge your body (prior to touching the board) by touching an earth-ground appliance or another earth source. This should be repeated after any body movements that could generate additional static electricity.

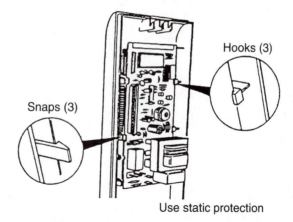

Hooks (3)

Snaps (3)

Use static protection

10-48 Remove the straps and hooks to take the circuit panel out of the oven. Courtesy Tappan, brand of Frigidaire Co.

11
CHAPTER

Microwave leakage tests

If a customer is afraid that a microwave oven is leaking radiation, you might be called upon to just check the oven for RF leakage. Most people are deathly afraid of becoming burned or receiving RF radiation. After every repair, always check the microwave oven for energy leakage. If the oven is still in the warranty period, you are required to take radiation leakage tests after every oven repair. Be safe. Always take a radiation test to prevent possible damage or even a lawsuit (figure 11-1).

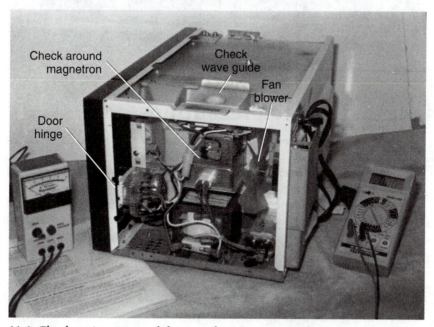

11-1 Check various areas of the oven for microwave radiation.

For a microwave oven in a customer's home, the U.S. government leakage standard is 5 mW/cm². The power density of the microwave radiation emitted by a microwave oven should not exceed 1 mW/cm^2 at any point 5 cm or more from the external surface of the oven, measured prior to acquisition by the purchaser. Throughout the useful life of the oven at any point, 5 cm or more from the external surface of the oven should not exceed 5 mW/cm^2. Should the leakage be more than 2 mW/cm^2, some microwave manufacturers require that they be notified at once. A defective oven should not be used until the radiation leakage has been corrected according to the manufacturer's instructions (figure 11-2).

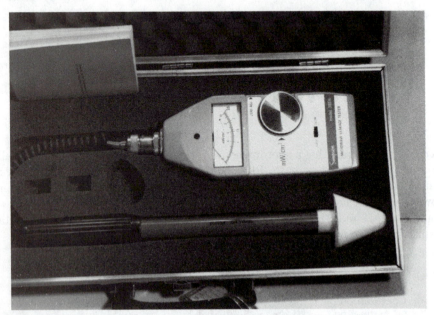

11-2 Test for radiation leakage with a certified tester after each repair.

Important safety precautions

Before the oven is returned to the customer and installed in the home, check the following safety features:

1. Do not operate the oven on a two-wire extension cord. Always operate the oven from a properly grounded outlet.
2. The oven should never be operated with a defective door seal or warped door. Check the choke and gasket area. Keep the door area clean and free of foreign material. Readjust the door for proper closing.
3. Do not defeat the interlock switches. If the oven operates with the door open, shut it down and check all interlock switches.
4. Check the action of all interlock switches before and after repairing the oven.
5. Before returning the oven to the customer, make sure the door is adjusted properly (tight and snug without any play).
6. Never operate the oven if the safety interlocks are found to be defective.
7. To meet the Department of Health and Human Services (HHS) guidelines, check the oven for microwave leakage with a recommended microwave radiation meter.

Required test equipment

Besides an electromagnetic energy leakage monitor or other test equipment, a glass beaker and thermometer are required for servicing microwave ovens. Select a 600- or 1000-cc glass beaker or equivalent. Check the manufacturer's literature for their required leakage tests. Most manufacturers recommend that the glass beaker should hold from 8 to 10 ounces of tap water and should be made of an electrically nonconductive material such as glass or plastic. A 100°C or 212°F glass thermometer is ideal for these tests. The water test is used when taking radiation leakage and normal temperature-rise tests.

Place the water container in the center of the oven cavity. The placing of this standard oven load not only protects the oven, but also ensures that any leakage is measured with accuracy. Set the microwave oven to high or full power. Close the door and set the timer to about three minutes. If the water boils before all leakage tests are completed, replace it with cool tap water. Turn the oven on for accurate leakage tests.

Radiation leakage monitor

There are several survey instruments on the market that comply with the required leakage procedure tests prescribed for microwave ovens. Those recommended by most oven manufacturers are: Holaday H1 1501, H1 1500, and H1 1800; Narda 8100, Narda 8200; Simpson 380M. Any one of the suggested leakage testers are accurate and will do the job in leakage tests. Some manufacturers request a certain leakage tester in their service literature. One thing to remember: The microwave leakage tester is no good unless you really know how to use it. Double-check the meter leakage operation several times. Take several leakage tests to get the hang of it. The leakage test is the last regular job in finishing the oven repair.

The Narda 8100 and the Simpson Model 380M are described thoroughly in Chapter 2. The operation and description of the Holaday H1 1500 is given here. Although each model has its own method of operation, radiation leakage tests are the same. All of these leakage monitors operate on the 2450-MHz band.

Holaday model H1-1500 survey meter

This survey meter contains a 1-mA meter with an operational amplifier. It is powered by two 9-volt batteries. You will find a fast and slow position switch. In the fast operation, the time for the meter to reach 90% of final value is less than one second. It takes just under three seconds with the slow switch position. A temperature compensation network for the diodes (within the probe) is located inside the meter box.

Eight carrier diodes are housed in the end of the teflon probe. This antenna array and diode detection has the unique feature of being able to sum the microwave electric field in a place perpendicular to the probe axis. The antenna is very broad and makes measurement easy when taking microwave leakage around the oven door and components. A 5-cm cone spacer attaches to the tip of the probe. The probe is attached to the amplifier and meter box with a 3 1/2-foot shielded cable.

Before you put the monitor into service, check the batteries and probe. Turn the selection knob to the battery test position. Make certain the meter hand reads above the green marker line on the meter. If not, replace the batteries.

To test the probe, turn the selector knob to probe test with the zero adjustment knob in the center position. The meter should read in the "okay" probe test area between the green lines. If not, the probe might have been damaged and should not be used. Return the defective probe to the manufacturer. Do not try to fix it.

To operate the Holaday monitor, turn the selector knob to the desired meter range (2, 10, or 100 mW/cm^2). When the microwave oven is suspected of leakage, try the highest scale. Remember, the maximum allowable leakage in the field is 5 mW/cm^2. In most cases, the low 2 mW/cm^2 scale will be used, because the leakage found in most microwave ovens is very low. Allow the meter to warm up if used under extreme temperature conditions.

Place the tip of the cone against the equipment or component to be measured. The spacer cone keeps the detecting probe five centimeters from the equipment. Read the leakage level on the meter. Switch the meter to slow response when ovens with stirrer equipment are being tested. This makes the measurement easier to read. Always turn the selector knob off when the survey meter is not in use.

Norelco RF radiation leakage tests

Check for leakage before and after servicing. If leakage is more than 5 mW/cm^2, immediately inform North American Phillips Corporation—attention product assurance department. After repairing or replacing any radiation material, keep a written record for future reference, as required by HHS regulation. This requirement must be strictly observed. An electromagnetic radiation monitor, Holaday, Narda, or HHS-accepted equivalent, 1000-cc glass beaker or equivalent, and a glass thermometer 100°C or 212°F are required test equipment.

Start by pouring 275 ±15 cc (9 oz. ±^{1}/$_2$ oz.) of 20 ±5°C (68° to 90°F) water in a beaker. Set the radiation monitor to 2450 MHz and use it, following the manufacturer's recommended test procedure to ensure correct results. When measuring leakage, always use the 2-inch spacer supplied with the probe. Place the unit in a cooking mode and measure leakage (using an electromagnetic radiation monitor) by holding the probe perpendicular to the surface being measured. Move slowly, no more than one inch per second.

Whenever the magnetron is replaced, measure for radiation leakage before the outer panel is replaced and after all the necessary components are replaced or adjusted. Special care should be taken in measuring around the magnetron and waveguide. These tests should be made with the outer panel removed.

With the oven is fully assembled, including outer panels, measure all components for radiation leakage around the door periphery, door viewing window, exhaust openings, and air inlet openings. Radiation leakage should not exceed the values prescribed in the following:

1. Maximum allowable RF leakage level is 5 mW/cm^2.
2. If the reading is over 5 mW/cm^2, refer to door and switch assembly adjustments section. The door is designed for leakage of less than

1 mW/cm^2, and adjustment should be attempted to bring all readings to below 1 mW/cm^2.

Measuring precautions

1. Do not exceed full-scale deflection.
2. The test probe must be moved no faster than one inch per second along test areas; otherwise a false reading might result.
3. The test probe must be held by the grip portion of handle. A false reading might result if operator's hand is between the handle and probe.
4. When testing near a corner of the door, keep the probe perpendicular to surface, making sure the probe end at base of cone does not get closer than two inches from any metal. If it does, erroneous readings might result.
5. When high leakage is suspected, do not move the probe horizontally along oven surface; it might cause probe damage.

After adjustment and repair of any radiation-interrupting or radiation-blocking device, make a repair report, record measured values, and keep the data. Apply this measurement to the service invoice. If the radiation leakage is more than 5 mW/cm^2, first determine that all parts are in good condition and functioning properly, and that genuine replacement parts as listed in the manufacturer's manual have been used before notifying the manufacturer of the radiation leakage in the oven. Have the radiation monitor checked for accuracy by the manufacturer at least once a year. Check the monitor batteries at least every six months or once a year.

Where to check for leakage

Microwave radiation leakage tests should be made around the front door and gap area (figure 11-3). When the magnetron tube or waveguide assemblies have been replaced, leakage tests should be made around these components. Always be careful while taking leakage tests around the magnetron with the oven in operation. Keep your hands on the handle area. Just let the plastic cone touch the various components.

Before the outer panels are installed (figure 11-4), check around the magnetron tube for microwave energy leakage. Go slowly around the seam area where the magnetron bolts into the channel cavity. Check around the heater or filament terminals for possible leakage. Double-check around the magnetron if you find dented or bent areas in any of the metal areas.

Take a leakage test around the waveguide area seams (figure 11-5). Check for leakage on top of the waveguide cover. In a convection oven, if the shielded metal areas have to be removed to repair the convection heating elements or to get access to them, check for leakage after these components have been replaced. Try to avoid the belt-driven pulley and fan-rotating components while taking these tests. You should not encounter HV terminals on the top area of the oven, but do be careful of possible low-voltage connections.

Always take a leakage reading around the front door area after the oven has been repaired and cleaned up. When testing near a corner of the door, keep the probe perpendicular to the surface, making sure the probe end at the base of the cone does not get closer than two inches from any metal. Slightly pull on the door while the oven is operating and notice if there is any leakage. The door should have no play that lets the microwave energy escape. This test might show a possible damaged rubber or

11-3 Check for leakage around the front door seam, latches, and window area.

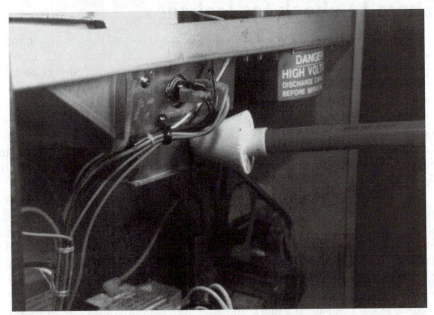

11-4 Check for leakage around the magnetron and waveguide assembly with the oven operating.

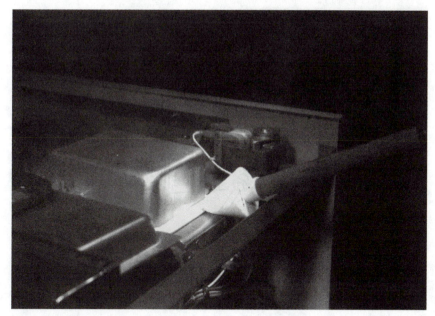

11-5 Check for leakage around the waveguide assembly.

choke area. If the door is warped, you might try pulling out on the door handle to un-cover a leakage spot at the corner. Usually a defective door will leak at the top and bottom corner areas.

Another method for checking the door is to slightly push down on the door latch and take a leakage test. Do not push down too hard or the latch will release the door, stopping the oven. Just press down slightly before latch interruption. Now take a leakage test around the end and corner door areas while holding down the door latch handle (figure 11-6).

Check for leakage around the door viewing window after testing around the door area. Double-check all exhaust and air inlet openings. You might find them on the top or on both sides of the microwave oven. Make sure the screws inside the oven front door are not loose or missing. Microwave energy might leak if the screws are not properly tightened. Slowly approach these areas for possible high leakage. When high leakage is suspected, do not move the probe horizontally along the oven surface or you could cause probe damage.

Norelco RR7000 door adjustment

Door adjustment on a microwave oven is very important to prevent excessive leakage. RF leakage levels should be checked after all door, switches, and associated repairs are made. Care should be exercised for proper door adjustments.

1. Close the door and depress switch.
2. Place a ¼-inch shim between bottom of door and bottom trim (figure 11-7). The door should align with control panel trim.

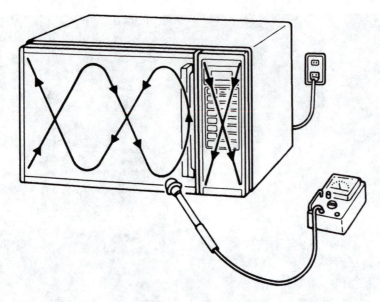

11-6 Pull on the door without unlatching it and test for leakage at the door seams.

3. Loosen the four screws on the hinge side until the door can easily be moved back and forth.
4. Squeeze the door against front frame and at the same time tighten the four screws. Tighten center screws first, then outer screws.

Tips on taking leakage tests

Do not exceed full-scale meter deflection. Slowly approach the tested area for signs of high-leakage radiation. Go slowly. Take your time in taking leakage tests around the door. Proceed at no faster than one inch per second (2.5 cm/sec) along any suspected area.

Make sure the plastic cone is slipped into place and keep the probe perpendicular to the testing surface. Keep your hands on the probe handle to prevent a false reading. Slightly pull out on the oven handle and then take a leakage test.

Always handle the leakage tester with extreme care. Keep the delicate test instrument in its case when not in operation. Most manufacturers recommend the leakage tester be sent back in for calibration at least once a year. If in doubt about a leakage reading, go back over and check it once again.

Be careful when taking leakage measurements after replacing the magnetron, when there is cabinet damage, or when replacing the front door. You might encounter excessive RF radiation. If the instrument indication goes to full scale at any time, the tester might be damaged due to the presence of excessive microwave power.

Always approach the oven slowly with the probe while observing the meter. The oven should be set at maximum power output. Hold the probe two or three feet from the oven door. While watching the meter, move slowly toward the door, oven surface,

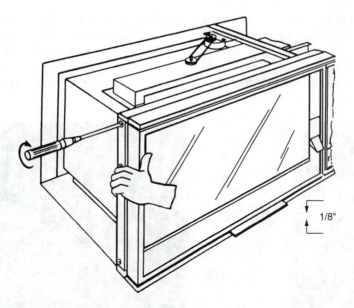

1/8"

11-7 Realignment and adjustment of the door on a Norelco RR7000 oven. Courtesy Norelco Corp.

or gap between the door. When high leakage is noticeable, do not move the probe along the oven surface; you might burn out the probe. Often the greatest leakage is at the corners of the door. Proceed with the leakage test when excessive leakage is not found on the oven.

Typical leakage tests

The test probe must be held by the handle to avoid the false reading of the test instrument. Hold the probe perpendicular to the cabinet door or component to be checked for radiation. To check the door, run the probe cone on the door and cabinet seam. While watching the meter, move the probe slowly (not faster than one inch per second) along the gap area. Check for the highest or maximum indication on the meter.

Keep the probe perpendicular to door corners or access area so the probe will not be closer than the cone space (two inches). Always use the two-inch spacer with the probe. Often leakage occurs around the door corners. Double-check the door area. When any leakage is noted in the top center or end areas of the door, push tightly on the door to determine if the door has some play between it and the oven. If the leakage reading decreases by pushing on the door, check for door adjustment. The door must be tight against the oven seal.

After the door is adjusted, take another reading. Usually the long door latches do not seal tightly in the oven or microswitch area. Adjust the door latch head at a position where it smoothly catches the latch receptacle through the latch hole. You might have to reset or adjust the monitor and interlock switch assembly. Adjust the door to the manufacturer's specifications. Check for leakage at the door screen and sheet metal seams.

Take leakage tests around the magnetron tube after installing a new tube. Check for possible leakage between the magnetron and the waveguide assembly. Run the cone over the exhaust and inlet air openings for possible leakage (figure 11-8). The door and all cabinet openings should be checked for leakage after the back cover has been replaced. You will find that most ovens have no or very little leakage (less than 0.1 mW/cm^2).

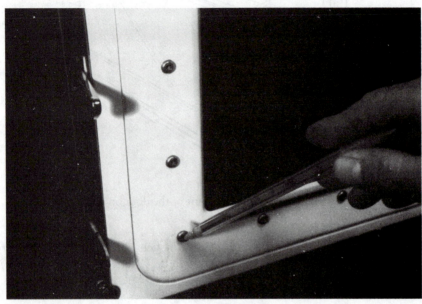

11-8 Check loose screws on the door and shelves for arcing.

Record all leakage readings of any oven on the repair ticket for future references. Although customers do not always understand the leakage reading, they want to know if the radiation will cause damage. Most manufacturers want the submitted warranty report to include the oven leakage, while others require the make and model number of the leakage tester to be recorded. If the leakage is over 5 mW/cm^2, report the leakage reading to the manufacturer at once.

Customer leakage tester

There are two readily available microwave leakage detectors on the market. The GC-20-224 model is a small, round instrument at a cost of $14.95 (figure 11-9). Radio Shack has a microwave leakage detector Model 22-2001 for less than $15.00. Although these two detectors are not certified leakage instruments, they might spot potentially dangerous RF leakage from a faulty microwave oven and satisfy the customer who is afraid of possible microwave radiation.

The GC micrometer radiation leakage detector can be operated like any leakage tester. Measure one cup of cold water and place it in the microwave oven. Close the door and set the cook selector timer on high for 60 seconds or longer. You can also do the leakage test while cooking food. Hold the meter perpendicular to the oven door

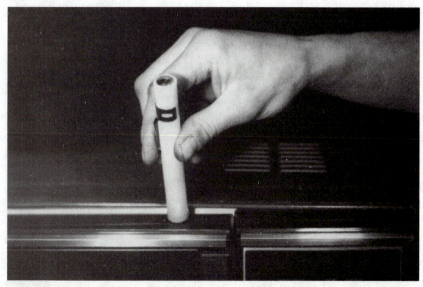

11-9 Check door for radiation leakage with a customer-type leakage tester.

edges, like the cone probe of any leakage tester. Move the meter at approximately two inches per second. If the meter stays in the green zone, the emission level is considered safe. Any time the needle enters into the red zone, the oven should be checked with an authorized microwave leakage tester or at a microwave service center.

There are several low-priced microwave oven leakage detectors on the market to check for leakage around door gaskets, latches, and door edges. Usually these instruments are less than $20 and are used for detection of RF leakage. They are ideal for the operators who are afraid of microwave radiation and can check it themselves.

The Tenma microwave oven leak detector from MCM Electronics of Centerville, Ohio, is easy to use (figure 11-10). Simply run the front end (arrow) of the test instrument along the seams of the door and along the intake and outlet openings and watch the meter movement. The meter has a red and green scale. If the meter reads in the green area, no relative amount of microwave energy is escaping. When the meter hand goes into the red scale, call the local microwave repair technician.

Just remember: These two low-cost microwave leakage instruments are not certified, but they might prevent possible serious and even permanent health problems. The microwave oven should be checked with a certified microwave leakage detector after each repair.

Leakage record keeping

After adjustment and repair of any microwave energy or microwave-energy-blocking device, record the leakage reading on the service invoice. Also, don't forget to place the leakage reading in the service literature. When making out a warranty repair report, most manufacturers ask to have the leakage test recorded on the warranty report. This report might take the repair agency and manufacturers off the hook, if for some unknown reason a customer tries a lawsuit for radiation burns.

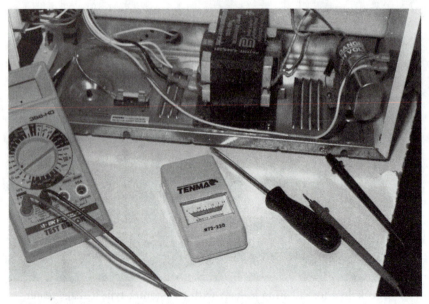

11-10 A low-cost microwave leakage tester by Temma. Courtesy MCM Electronics, 650 Congress Park, Centerville, Ohio 45459-4072

Always report to the manufacturer if the microwave leakage is more than 2 mW/cm² (1 mW/cm² for some ovens in Canada). Don't let the customer have the oven until the manufacturer releases it. Follow the manufacturer's instructions in repairing or replacing a microwave oven. The factory might want the oven shipped directly to them for possible repairs.

Replacing the door

Check the door for warped or damaged areas when excessive radiation leakage is noted at the top or corners of the door. In most cases, the door is opened with the handle or an extra hand at the top of the door. Eventually, the door becomes warped or might be damaged while it is open. A broken or worn hinge might let the door sag, creating possible leakage. Inside-oven explosion of food might cause the door to be blown open, causing damage to the door. In some ovens, you might find that a broken latch-head assembly is not holding the door snug against the open face plate. Definitely replace the door assembly when the front glass is cracked and broken or the door cannot be properly adjusted.

Inspect the choke and door seals at the inside of the oven door. Check for damaged areas. Make sure that all the inside door screws are not loose or missing. In some doors, the glass can be removed and replaced. Usually the whole door assembly should be replaced when damaged or warped out of line.

After determining that the door assembly should be replaced, check the parts list for the correct part number. Always list the correct part and model number when ordering any oven component. All oven door components must be ordered directly from the oven manufacturer or the manufacturer's distributor.

A typical oven door might be hinged at the top and bottom or have a long piano-type hinge down the entire backside of the door (figure 11-11). A hinge plate over the piano-type hinge assembly must be removed before the door is free. The new door should be replaced in the reverse order of removal. When only two pivot area hinges are found, the top hinge holds the door towards the oven, while the bottom hinge supports the door load.

11-11 Adjust the door by moving the hex nuts at the top door hinge.

Make sure the new door is tight and level with the front piece of the oven. Check the latch lever or door hook alignment pin for smooth operation. Some doors are held tight to the front piece by correct adjustment of the latch interlock assemblies. Others are adjusted at the door hinge areas. It's best to follow the manufacturer's literature for correct alignment.

Although most oven door hinges are held in place with hinge-mounting nuts or Phillips screws, you might encounter a new Torx screw in some ovens. These screws resemble the Phillips head, except the Torx screw has a star-type indentation. Do not try to remove or tighten these Torx screws with an ordinary screwdriver. Pick up a Torx screwdriver, size T-20 or T-28, at your local hardware store.

Remove the outer cover of all ovens before attempting to adjust or remove the door. Then discharge the HV capacitor. Here are several typical microwave oven door replacements and adjustments.

Amana model RR-40

Open the door wide open. Remove the two counter-balance mounting screws from each side. Lift off the door. In this model the door handle and door glass might be separate, replaceable items. Adjust the level and tighten the two mounting screws after door replacement.

Hardwick model EN-228-0

Remove the cabinet trim from the main front by removing several screws on top, bottom, and sides. Lay the oven on the backside and remove four hex-head nuts that secure the door hinge to the main cabinet. Lift the door handle and door from the base unit.

To install a new door, reverse the above procedure. Level the door and keep $\frac{1}{16}$-inch spacing between the control panel and door edge. Tighten the hex nuts starting at the top and ending at the bottom area. Open the door and notice proper door alignment. Loosen hex nuts and readjust the door, if needed.

Norelco Model MCS 7100

Turn the oven on its backside. Loosen the upper hinge-mounting nuts and re-move the door assembly. Be careful not to lose the small nylon spacer at the top and bottom hinge. Make sure that both nylon spacers are in place when installing a new door. Loosen lower hinge-mounting nuts (figure 11-12).

Press the door against the oven face plate near the hinge and retighten the hinge-mounting nuts. Set the oven dial for normal operation. Check the play in the door. Readjust the door interlock switches if there is too much play. Loosen the two side mounting screws of the metal mounting bracket of the latch switch assembly. Press in on the door and pull back on the latch assembly. Tighten the latch bracket mounting screws. Check the door for proper clearance. Make sure the upper and lower door switches are functioning properly.

Panasonic NN-9807 door assembly

1. Remove the four screws holding the door.
2. Remove the door by carefully pulling outward, starting at the upper right-hand corner.
3. Remove the door key spring and remove the three screws holding door handle (figure 11-13).
4. Separate door A from door E by freeing the six catch hooks on door A (figure 11-14).
5. Remove the door keys, door key lever, door key spring, and door key pins.
6. Reverse order to assemble the new door.

Quasar model MQSS-207W

Remove the two hex nuts holding the upper hinge to the oven base. Open the door and pull the door towards you. Be careful. The top door arm must come through the slot in the oven assembly. Leave the bottom hinge intact. Lift the door up and off the bottom hinge pin.

Before installing the new door, check for a washer spacer on the bottom hinge. Adjust the door parallel to the cabinet. The door should be tight against the cabinet base, with no clearance. The top and bottom hinge can be adjusted to align the door correctly with the control panel. Tighten all four hex nuts to secure the hinge and door assembly. Open and close the door for proper operation.

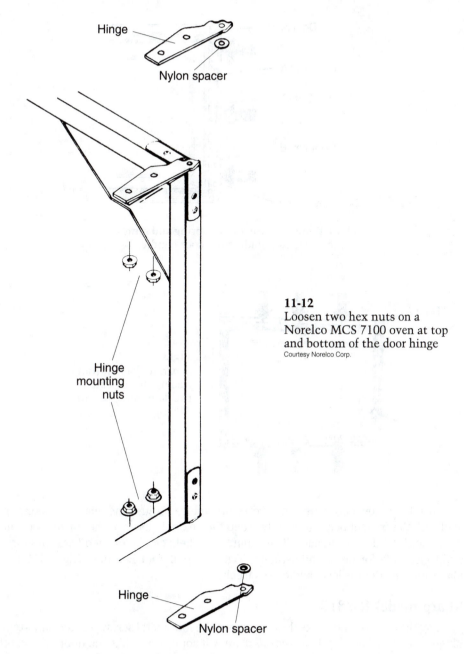

Hinge

Nylon spacer

11-12
Loosen two hex nuts on a
Norelco MCS 7100 oven at top
and bottom of the door hinge
Courtesy Norelco Corp.

Hinge
mounting
nuts

Hinge

Nylon spacer

Samsung model RE-705 TC

Remove the small spring attached to the door arm CAM. Remove the rod pin that prevents the door from flying open. You will find two metal hex screws at the top and bottom hinge. Remove them and pull the door out. Be careful when pulling the door arm CAM through the slotted area.

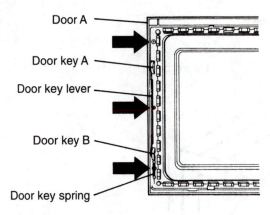

11-13 Remove the door key spring and remove the three screws holding the door handle. Courtesy Matsushita Electric Corp. of America

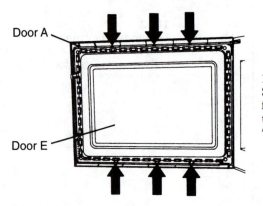

11-14 Separate door A from door E by freeing the six catch hooks on the door. Courtesy Matsushita Electric Corp. of America

Install the new door with the reverse procedure. Adjust and align the door parallel with the control-board assembly. Keep the hex nuts loose at the top and bottom hinge until the door is in place. These hinge brackets can be pushed back and forth and to the side for proper adjustment. Tighten up all four hex nuts (figure 11-15). Make sure the door is level and closes freely.

Sharp model R-7810

Lay the oven on the backside and remove the four metal screws of the piano-type hinge assembly. The metal back-strap plate will come loose. Lift the door up and off the base assembly. Be careful not to damage the control panel assembly.

Install the new door in the reverse procedure. Make sure the door is parallel with the bottom line of the oven face plate. The latch heads should pass through the latch holes without binding. Keep the door tight against the oven face plate. Insert all four metal screws. The door can be aligned before the screws are tightened in the long hinge assembly. Check the door for proper clearance and operation.

11-15 Remove the hex nut screws to take the door off this early microwave oven.

After each door replacement or adjustment, double-check around the door for poor alignment and leakage. Try the door several times in opening and closing operation. Go around the door and check for possible radiation leakage with a radiation leakage meter. If any leakage is noticed, realign and adjust the door.

The door on a microwave oven is designed to act as an electronic seal to prevent microwave energy leakage from the oven cavity. If light can be seen on some point around the door, it's still normal, provided the radiation meter does not show any leakage. A normal door might not be airtight or moisture tight. When light movement of air or moisture appears around the oven door, this does not mean the oven is leaking radiation energy. When these conditions occur, double-check for leakage with the radiation monitor survey meter.

Some electronics technicians use a dollar bill between the door and oven to test the snugness of the door. If the door is proper and tight, the bill cannot be pulled easily out of the door. Realign the door if the dollar bill is easily pulled out.

Replacing waveguide assemblies

If the waveguide has been removed or replaced with a damaged waveguide assembly, or when replacing the stirrer motor or antenna, make sure the waveguide bolts and nuts are tight (figure 11-16). Visually inspect any high places in the waveguide assembly. Remove and check for particles under the metal waveguide. If the waveguide is not tight, RF hot spots might occur, with possible RF leakage. Double-check the waveguide assembly with the radiation monitor probe.

11-16 Check the waveguide for hot spots. Inspect around the mounting bolts.

Last-minute checks

1. Do not allow an oven to be operated with the door open or with improper interlock operation.
2. Check all interlocks, proper door closing, wear and tear of rubber or choke seals, damaged door hinges or latches, and damaged door windows.
3. After completing all repairs, check the magnetron, waveguide, and oven cavity for proper alignment, snug bolts and nuts, and good connections.
4. Repair and replace all components with exact oven part numbers.
5. Check for proper fuse. Do not clip across the fuse terminals to make the oven operate. Replace defective fuses with those of proper amperage and type.
6. After all repairs are made, make microwave leakage tests with approved radiation monitor meters required by federal laws. Record the measurement on the service invoice.

GE JEBC200 microwave leakage test

Place 275 ml of water in a 600 ml beaker (WB64X5010) in the center of the oven. Set the microwave leakage tester at the 2450-MHz scale. Turn oven on for a five-minute test. Hold the probe perpendicular to surface being tested and scan surfaces at a rate of one inch per second. Check the entire door and control panel, the viewing surface of the door window, and the exhaust vents. The maximum leakage should not be more than 4 mW/cm^2. Record the date on the service invoice and on the microwave leakage report.

Samsung leakage procedures

The equipment needed to perform leakage tests is a certified microwave energy survey meter, 600 cc glass beaker, and mercurial or digital 100 degrees Celsius or 212 degrees Fahrenheit thermometer. Pour 275 ±15 cc of 20 ±5°C (68 ±9°F) of water in a beaker which is graduated to 600 cc, and place the beaker in the center of oven.

Start the oven and measure leakage with a survey meter. Set survey meter to 2450 MHz. When measuring the leakage, always use a 2-inch spacer cone with the probe. Hold the probe perpendicular to the cabinet door. Place the spacer cone of the probe on the door/or cabinet door seam and move along the seam, the door viewing area window and the exhaust openings moving the probe in a clockwise direction at a rate of 1 inch/sec. If the leakage testing of the cabinet door seam is taken near a corner of the door, keep the probe perpendicular to the areas making sure that the probe end at the base of the cone does not get closer than 2 inches to any metal. If it gets closer than 2 inches, erroneous readings may result.

Check the microwave leakage with the cabinet or back cover removed with the same water test (figure 11-17). Operate the oven at the highest power level. By using the survey meter and spacer cone as described before, measure around the opening area of magnetron, the surface of the air guide and the surface of the wave-guide. Avoid the high voltage components.

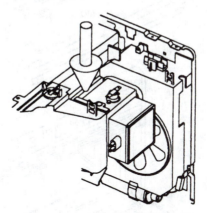

11-17 Check for radiation leakage around the magnetron in a Samsung oven. Courtesy Samsung Electronics America, Inc.

Do not exceed the limited scale of meter. The test probe must be held on the grip of the handle, otherwise a false reading can result if the technician's hand is between the handle and the probe. When high leakage is suspected, do not move the probe horizontally along the oven surface; this may cause probe damage.

After repairs are made on any microwave oven, take a leakage test. The measured leakage should not be more than 5 mW/cm^2 after repair or adjustment. Make a repair record for the measured values, and keep the data. If the radiation leakage is more than 5 mW/cm^2 after determining all parts are in good condition, function properly and the identical parts are replaced as listed by the manufacturer, call the manufacturer service department. Do not let anyone operate this oven with high radiation.

GE JEBC200 door removal

The oven door in this microwave oven tilts downward flat, instead of swinging outward. The oven door consists of a handle, outer glass, sub-door assembly, choke cover, hinge assembly, and latch pawls. The glass, handle, and latch pawls can be replaced individually. To remove the door (figure 11-18):

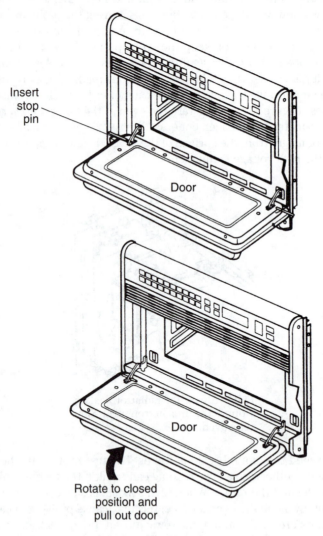

11-18 Remove the oven door in a GE JEBC200 oven by inserting a small Allen wrench/drill bit.

1. Open the door and insert small Allen wrench/drill bit (holding pin) into the hole of the hinge body. Place tape on each side of hinge hole to protect door from being scratched.
2. Close the door against the inserted material (holding pin) and lift. The door will come off its hinge.

To service bottom hinge springs (figure 11-19):

1. Remove the door.
2. Remove the glass.
3. Loosen the bottom trim on affected side.
4. Remove the inserted material (holding pin) from hinge body on affected side. The hinge hook must be pulled away from the door to remove holding pin.
5. Remove three screws holding the old hinge assembly.
6. Pull the hinge hook out and reinsert holding pin.
7. Replace the door.

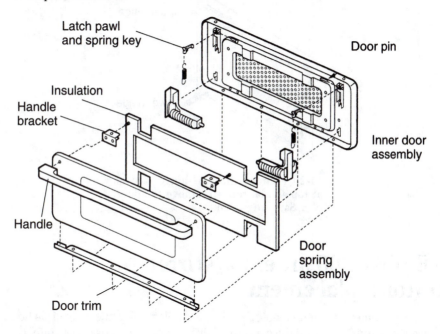

Latch pawl and spring key

Door pin

Insulation

Handle bracket

Inner door assembly

Handle

Door spring assembly

Door trim

11-19 Remove the door glass and loosen the bottom trim to service a GE JEBC200 oven.

Goldstar ER-505M door removal

Remove the outer case of microwave oven and control panel. Remove 2 screws holding the oven hinge assembly to the oven unit. Be careful and not lose the hinge washers at the lower hinge and on the upper hinge. Leave the oven "L" hinge upon oven.

When replacing and mounting the door assembly to the microwave oven, adjust the door assembly parallel to the oven chassis by moving the upper and lower hinge. Adjust the door so there is no play between the inner door surface and oven frame assembly. If the door is not mounted properly, leakage can occur between door and oven. Take radiation leakage test all around the door after replacement.

Samsung MW2500U door replacement

To replace the door assembly, remove 4 hex bolts securing the upper hinge and lower hinge. Remove the door assembly (figure 11-20). After replacing the door, check if the primary interlock switch, the secondary interlock switch and the interlock monitor switch operate normally. Check for leakage around the door. Microwave emission should not exceed the limit of 5 mW/cm^2.

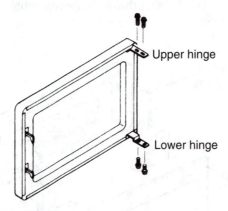

11-20 Remove the two hinge screws at top and bottom to remove the door of a Samsung oven. Courtesy Samsung Electronics America, Inc.

GE JEBC200 antenna/stirrer motor replacement

The antenna and antenna motor are located below the removable glass floor. To service the antenna, remove the glass shelf with a suction cup or putty knife. Lift the antenna off the antenna motor shaft. To service the antenna motor:

1. Disconnect the power and remove oven from the wall.
2. Remove the antenna.
3. Carefully put the oven on its back on a pad or blanket.
4. Remove two screws to access the panel on bottom of shaft.
5. Disconnect the motor leads and 2 screws that secure the motor (figure 11-21).

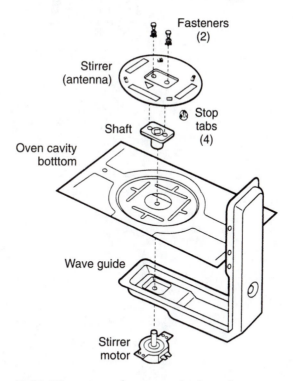

11-21 Disconnect the motor leads and remove two screws to take out the stirrer motor. Courtesy General Electric Co.

12
CHAPTER

Microwave oven case histories

In this chapter, I will describe actual microwave oven troubles that I've encountered during my years of servicing microwave ovens. The service problems in this chapter might be very similar to those you are now experiencing. Many manufacturers make microwave ovens for other firms, so different models can be identical in operation. I chose these case histories to help you locate the various problems found in microwave ovens. You can compare the problems you encounter to those solved in this chapter (figure 12-1).

12-1 The defective oven relay might have an open solenoid winding or dirty contact points.

Nothing works

K-Mart model SKR-6705/dead oven

The fuse was replaced in this oven and the light came on. When opening and closing the door, the fuse blew again (figure 12-2). Replaced the monitor switch, QSW-M0046YBEO. This oven is manufactured by the Sharp Corporation.

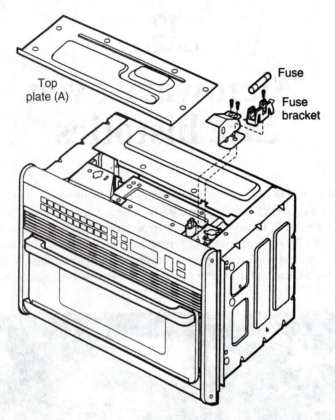

12-2 The 15-amp fuse prevents an overload in the circuit of a GE JEBC200 oven. Courtesy General Electric Co.

K-Mart SKR-7805/dead/no light

Replaced 15-amp fuse. Checked monitor and interlock switches for correct operation. Readjusted oven door.

Litton model 70/05 830/dead

Checked the door for too much play after replacing the 12-amp fuse. Adjusted the latch switch and door for proper closing. Sometimes, if there is too much play between door and base area, the monitor switch might hang up, causing the fuse to blow once again.

Litton model 70-05/dead

Replaced fuse and installed a new latch switch. While inside the oven, checked for dark or dead light bulbs. This model happens to be a commercial microwave oven. Checked for defective on/off switch. This switch can be located on the front of the oven, at the top, or on the side (figure 12-3).

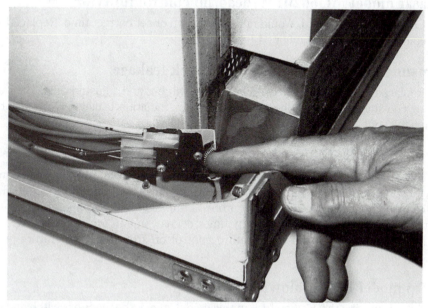

12-3 The door sensing interlock is at the top of this microwave oven.

Norelco model RK-7000/dead

After checking the fuse and voltage to and from the control board, a new board was ordered. Replacing the entire control board solved the dead condition. I replaced only the control board, not the push-button assembly (part number 6000-000-00200).

Sanyo EM375TS/dead

A Sanyo EM375TS model had no oven light or oven operation. The 20-amp fuse was blown. Still no oven operation after replacing the line fuse. The thermal thermostat of the oven cavity was open and replaced with part number 617-124-1235.

Norelco model MCS 8100/dead/blown fuse

After replacing the fuse, the oven began to operate—except for fan rotation. I found that the fan assembly screws had come loose, letting the fan blade jam. The fan motor would overheat and blow the 15-amp fuse. Replaced the bolt and nuts. (You might find them in the bottom oven area.)

Norelco model MCS 9100/dead

Control board assembly would not count down—no fan operation. Replaced control board assembly 8100-000-00008.

Norelco model MCS 8100/dead/nothing

Removed outside cover and replaced blown 15-amp fuse. Checked the fuse continuity with the low-ohmmeter range of the VOM. Found a loose plug going to the power control board. Repaired plug and terminals before starting up the oven.

Quasar model MQ6620TW/dead/no control function

Checked control board. Would not count down or set correct time. Replaced entire control board (part number 84-90 564A41).

Samsung model RE-705TC/dead/check leakage

Replaced 15-amp fuse. Checked interlock switches and adjusted the door. Ran a 4-hour cooking-water test. Checked the door for radiation leakage. Customer had complained of a loose door and possible leakage.

Sharp model R6770/dead

In some Sharp models that have a rotating temperature control, the dead condition might be caused by improper control setting. Check the temperature control for the off condition. If the temperature probe is unplugged, the temperature control should be set in the off position. Many times ovens come in for repair because the operator forgot to turn the temperature control off from the temperature-probe cooking operations.

Sharp model R7650/blown fuse

After replacing the 15-amp fuse in this model, check the monitor switch and primary interlock switch. Replace the monitor switch if it keeps blowing fuse when you open the door.

Sharp model R7650/dead/lights on

The fuse was good (the oven lights were on), but there was no cooking operation. Check each interlock switch after discharging the capacitor. Notice if the switches have black or burned marks at the terminal connections. Clip leads across each switch terminal, and monitor with a voltmeter. If the switch is open, you will have the power line voltage (120 Vac) across these terminals. Make sure you have the right switch or interlock. In this case, the secondary (upper) interlock switch was open (QSW-M0055YBEO).

Sharp model R7650/oven lights/no operation

The oven lights came on, which means the fuse and timer were operating. Checked interlock and cook switch; they were normal. While checking the door, found a broken latch spring. The plastic latch hook was down and would not trigger or press on the interlock switch. These latch springs must be replaced in the oven door (figure 12-4).

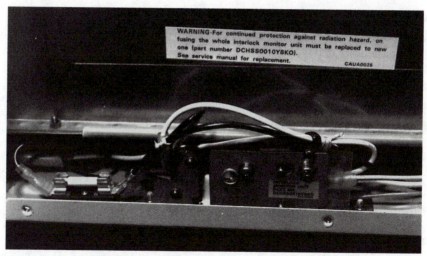

12-4 Some early microwave oven's interlock switches are mounted along the side.

Sharp model R7650/dead/won't turn on

Replaced the 15-amp fuse. Checked the upper interlock and monitor. Replaced both for erratic operation (figure 12-5).

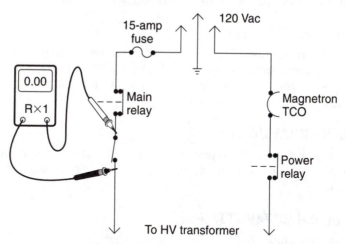

12-5 Check the interlock switch by the R×1 ohmmeter scale of a DMM or VOM.

Sharp model R7650/oven light on/dead

Nothing happened when the cook button was pushed. Replaced the interlock switch.

Sharp model R7650/dead

Replaced the open 15-amp fuse. Checked all interlock switches for proper operation. Replaced the defective interlock switch QSW-M0036YBEO.

Sharp model R7650/dead/won't come on

Replaced the spring in door latch and also replaced the top interlock switch (part number MSPRD0013YBEO).

Sharp model R7710/dead

After replacing the blown 15-amp fuse, checked all interlock switches. Found that the monitor switch was sometimes hanging up. Replaced both the monitor and the secondary interlock switch.

Sharp model R9200/dead

In this model, both latch springs were broken inside the oven door. The latch hooks tied down and did not engage the microswitch of the interlock switches. Replaced both with part number LSTPP00179BFO.

Sharp model R9500/oven won't come on

The fuse checked normal. Voltage was going into the control board, but not the control function. Replaced the control board number BUNTK085-DE00.

Sharp model R9600/dead

Replaced the blown fuse. Found out the door had too much play. The oven door should be snug against front piece of oven. Pulled back the door latch assembly and tightened up assembly. Sometimes a drop of cement on the screws will keep them from loosening up during oven operation.

Sharp model R9750/dead

Fuse was okay. Found a loose plug in a connection on the readout control board. Repaired and rechecked the oven. Bench-check time took four hours of an intermittent water test.

Toshiba model ER749BTI/dead

Replaced blown fuse. Replaced 5-amp monitor switch.

Oven lights out

Samsung MW8600 oven/no light/no cook

The oven light in a Samsung oven would not come on or there was no cooking. Upon checking the schematic after replacing the 15-amp fuse, checked the ac voltage across LV transformer, which was normal. When the power relay contacts were shorted with a clip wire, the oven operated. The power relay solenoid was open and replaced.

Sharp model R7704A/light out and noisy

In this model, I replaced both light bulbs. One was completely out and the other was very dark—indicating only a few more hours of operation. Found loose bracket on the cooling fan. Replaced metal screw and ran cooking test.

Sharp model R8200/no oven light/no operation

Straightened top oven arm. The arm was bent out of line and would not trigger primary interlock switch. Replaced dead oven bulb.

Toshiba model ER749BT/flashing oven light

This oven operated correctly with an intermittent oven light. Removed both covers. Found the light bulb was loose in the socket. Tightened the bulb and checked all interlock switches.

Timer rotated but no operation

Litton model 37000/timer on/no operation

The defrost or timer motor would not run in this microwave oven. Found that the bottom door interlock switch was not making contact. The switch assembly had slipped back and was not making contact. A new interlock switch was installed. Secured the switch assembly with a dab of glue on each nut to keep the assembly in place.

Magic Chef M5-2/no timer operation

The timer was rotated in oven operation, but would not count down or shut off. It would not move. The high voltage was discharged, and timer motor switch terminals were normal—except that the winding of the timer was open. A continuity measurement across the timer coil was infinite. The normal timer coil resistance should be between 6 and 7 kilohms in this model.

Amana RS415T/dead blower motor

Disconnect power from oven and discharge high-voltage capacitor. Place ohmmeter to the R×1 scale. Remove fan blower terminals if it will not rotate. Check the motor field coil resistance of fan motor terminals (5 to 10 ohms); with no meter reading, replace fan motor assembly. The blower motors in this model have an internal fuse; if fuse opens, the fan motor must be replaced.

Another method is to disconnect motor terminals. Connect another ac cord to the motor terminals. Monitor the line voltage. If motor is dead with 120 Vac at motor terminals, replace blower motor assembly.

Litton model 370.00/no fan or defrost motor operation

This oven light came on, but there was no fan operation and no cooking mode. Suspected a bad timer. Found a defective latch or interlock switch had slipped out of position and latch arm would not engage switch. Replaced and remounted the interlock switch. Cemented switch into position.

Norelco model MCS 6100/no fan or timer action

Replaced upper primary interlock switch assembly. (This switch is mounted on a metal strike switch assembly and is held in place with metal screws.) Checked all interlock switches. Opened and closed the door several times while in the cooking mode to make sure all interlock switches were working.

Norelco model MCS 7100/bad latch switch

With no fan and no cooking action, the upper strike switch assembly was re-placed (part number 6100-000-00055). The upper primary and lower secondary door switches are mounted on a latch slide-assembly that moves up when the door is closed. As the slide moves up, the switches are activated by latch hooks mounted in the door. Be sure to line up the adjustable strike assemblies with the door hook by simply loosening the screws on the strike switch assembly and tightening the screws into position.

Sharp model R6750/intermittent/now dead

In this model, the timer was not dependable. Sometimes the timer was too slow, but in other cooking modes it was too fast. Finally, the oven went dead. The 15-amp fuse was replaced. The sluggish timer assembly was replaced (figure 12-6). After re-placing the new timer, a four-hour cooking and leakage test was made before releas-ing the oven.

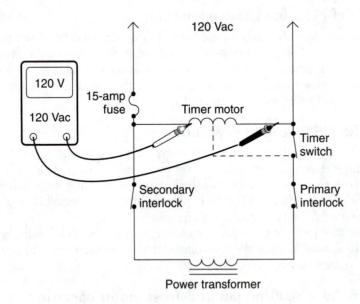

12-6 Check the timer motor winding and switch contacts with the R×1 ohmmeter range.

Sharp model R7704/intermittent timer

This oven was timed with a watch and appeared intermittent. A whole new timer unit was installed. In some of these bell-type timers, the end of the cooking cycle sounds like a thud instead of a clear bell. Adjustment of the bell clapper can be made by bending arm in or out.

Sharp model R7600/light/no timer action

When the timer was rotated, the blower motor or cook light did not come on. Found a defective interlock switch. All interlocks were checked for proper operation before cooking tests.

Sharp model R7600/intermittent

Front door hook would not trigger the microswitch. Bent latch hook in place and ended up replacing interlock switch assembly. If door is too loose, pull the interlock switch assembly back and tighten securely. Make sure that all latch hook assemblies will trigger the interlocks.

Sharp model R8200/timer off

The timer ran very slow in this model and sometimes would almost stop. Several cooking tests were made before a new timer assembly was installed. This timer assembly must be ordered directly from the manufacturer or the manufacturer's distributor (figure 12-7).

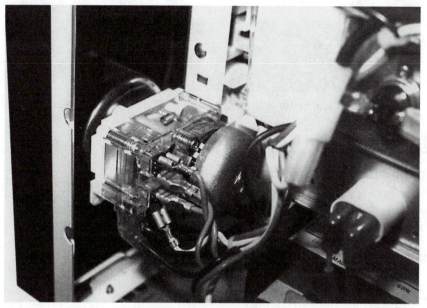

12-7 Inspect the timer connection and check with the low ohm scale of a DMM.

Lights on but no fan operation

Norelco model MCS 6100/no fan operation

Checked for a broken lead connection of the fan motor assembly (figure 12-8). With some ovens, the fan motor might be plugged in and the plug will vibrate, causing the plug to work loose and disconnect the ac to the fan motor. Measure the ac voltage across the field terminal of the fan motor (120 Vac). If voltage is normal with no fan motion, check the motor winding with the ohmmeter. Replace fan motor assembly.

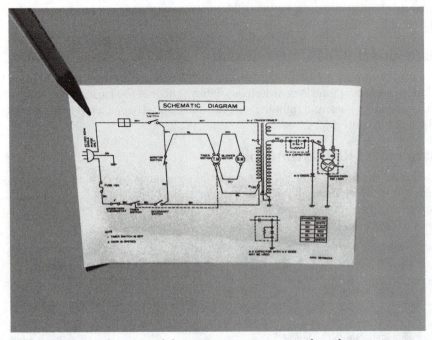

12-8 Look for a schematic of the microwave oven pasted on the oven cover.

Flickering lights

Amana model RS470P/house lights flicker

The oven was suspected of drawing heavy current until it was discovered that it was being operated from an extension cord. The technician requested that an electrician put in another grounded circuit from the fuse box to the microwave outlet.

Litton model 402.000/flickering cook light

The customer complained that when the oven was operating, the cook light constantly flickered. Most ovens with neon bulb cook lamps will flicker to some extent. These neon bulbs have a voltage-dropping resistor and operate from the 120-volt power line (figure 12-9). When the bulb becomes very dim, check the bulb for dark areas of the glass envelope. Replace the neon bulb. Sometimes cleaning off the grease and dirt will help the brightness of these small neon bulbs.

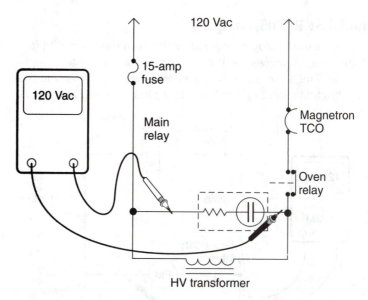

12-9 Check the oven neon light with an ac voltage test.

Sharp model R7650/light flickering, door is open

Check for a dirty or defective secondary interlock switch if the oven light continues to flicker with the door open. First, tighten all light bulbs in their respective sockets. Then replace the upper secondary switch with part number QSWMC055Y-BEO.

Norelco model MCS 8100/lights on/no fan

No fan or cooking operation was noted in this oven. When the door was moved, the oven began to operate. The door was adjusted and both interlock switches were adjusted for correct microwave cooking operation.

Sharp model R6770/lights only/no operation

The only problem with this oven was that the operator had turned the temperature probe on but was not cooking with the probe. When set in the on position, the oven will not cook with probe out. I showed the operator how to run oven in the off temperature-probe position with normal cooking procedures.

Sharp model R7650/lights/dead operation

Checked all interlock door switches. The secondary upper interlock switch was defective. Adjusted the interlock switches so the door fit tight against the oven base area.

Sharp model R7704A/no fan operation

A badly soldered connection was found on one of the motor terminals. If arcing has occurred, remove the wire, scrape with a pocketknife, and tin with rosin solder paste. Reconnect the terminal wire and solder it.

Sharp model SKR7705/fan quits

The fan motor would stop rotating after the oven was operating for several minutes. With the oven unplugged and the HV capacitor discharged, the fan blade was rotated by hand. The blade would barely move. Washing out the fan motor bearings and proper lubrication solved the intermittent fan rotation (figure 12-10).

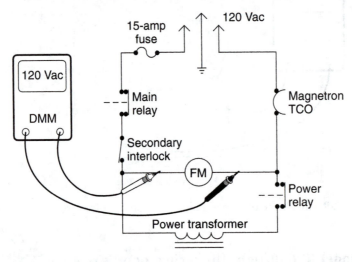

12-10 Measure the ac voltage across the motor winding if the fan motor will not operate.

Cook light on but no turntable rotation

K-Mart model SKR90505A/turntable won't turn

The turntable within the oven cavity would not turn in this model. Turned the oven on one side to get at the turntable motor assembly. The turntable motor was removed. Glancing at the motor field coil indicated the motor had become very warm.

The motor field coil was charred and a new turntable assembly was ordered (ROMTE0061YBEO). A broken plastic coupling from motor to turntable had bound the motor so it could not rotate. The turntable motion was restored with a new motor and coupling assembly FCPL0018WRKO (figure 12-11).

K-Mart model SKR7805A/no turntable rotation

All oven functions were perfect—except for no turntable rotation. Removed the motor turntable and found a broken plastic coupling. Replacement of the plastic bushing (FCL-0018WRKO) cured the no turntable rotation.

Panasonic NN-9507/no turntable

This microwave oven cooked normally, but no turntable rotation. The drive or turntable motor was removed and terminals removed from the motor. A quick resistance measurement indicated an open drive motor. The motor was replaced with the original part number A63264250AP.

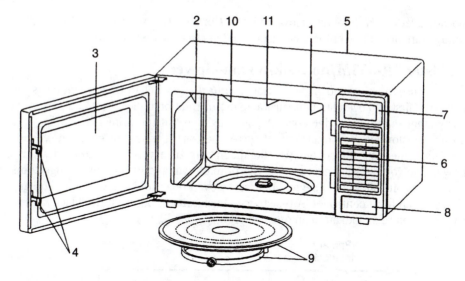

12-11 Remove the tray before you turn the oven over to check for a defective turntable motor.

Sharp model R7600/fan motor on/no oven light

Here the turntable motor was used as a trouble indicator. The turntable did not rotate, nor was there a cooking mode. The fan motor was rotating, but there was no oven light. The defective component turned out to be a defective interlock on the left-hand side of the cabinet.

This switch assembly can be pushed back and forth for adjustment after loosening two metal screws. The interlock switch must be triggered with a long arm from the bottom of the door with the door closed. When the door is open, the ac switch contacts are open. After correct adjustment, tighten the two screws securely. Drop cement on the metal screws to hold them in place.

Sharp model R7700/intermittent turntable rotation

With this model, the turntable might operate for several cooking modes and then stop rotating in the middle of a long cooking session. A broken plastic coupling was replaced after dropping the turntable motor down for observation. Check the turntable motor and gear box for correct lubrication.

Sharp model R7704A/no turntable rotation

All the other oven functions were okay, but the turntable would not operate. Found a broken and jammed plastic bushing. Replaced the plastic bushing with part number PCPL 0018 WRKO. This same coupling part number is used in all turntable assemblies.

Amana RS471P oven/cooks unevenly

Check the customer's symptoms that apply to the oven. Make sure that the rotating antenna is rotating. Check for dry or gummed-up bearings and rough edges.

Do not overlook a defective blower motor. Connect the Magnameter to the high-voltage circuits. If current is low, suspect a defective magnetron tube.

Goldstar ER-653M/no cook/no fan blower

Because there was no blower action, an interlock switch or thermal unit was suspected at first. No oven lamp or turntable motor rotation indicated that a part in the low-voltage circuits was defective. A pigtail light test at the primary winding of the power transformer and a quick ac voltage test across the triac showed no line voltage.

Upon checking the schematic, the Micon controller was found to be defective. Power line voltage was measured at the primary winding of the low-voltage transformer on controller. After several hours, the oven relay solenoid was found open at pins 1 and 3 from controller. A resistance measurement between the black and red wire on oven relay indicated an open winding (figure 12-12).

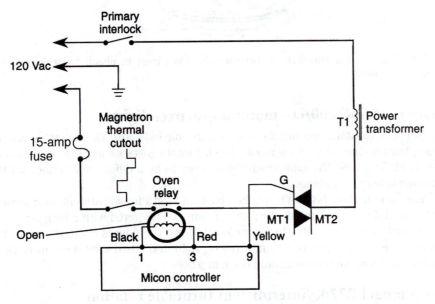

12-12 The oven relay solenoid was found open in a Goldstar ER-653M oven.

Hardwick M-250 oven problems/no cook

A pigtail light was connected across the primary winding of power transformer and ac meter across terminals MT1 and MT2 of the triac. When the oven was fired up and programmed for cooking, there was no indicator light, but the power line voltage was measured on the meter and across triac connections. Either no gate voltage at triac or a defective triac was indicated. Replacing triac solved the no-cook problem.

Hardwick M-250/no cooking/magnetron

A very loud pop was heard when the high-voltage capacitor was discharged, indicating very high voltage. After connecting the Magnameter, higher than normal HV was found at the magnetron (3.9 kV). No plate current was measured. The next two sentences reflect a defective magnetron. A short test across heaters was good. But, when the heater leads were removed, the continuity test showed open. Replacing the defective magnetron solved the no-cook problem.

Sharp R-9550/no turntable operation

The customer complained that after using the oven for cooking a roast the day before, the next time the oven was fired up, the turntable would not rotate. A visual inspection inside the oven gave no indication as to what was wrong. The oven power cord was pulled, the oven was turned over, and turntable motor was inspected.

A quick continuity test across the motor terminals indicated continuity. The wrapping paper on motor coil was burned with heavy brown area. Holding a screwdriver blade at the motor laminations caused vibrations, indicating that ac voltage was applied to the motor field coil. Both motor and gear box were replaced with a jammed gear box (RMTE0061YBEO).

Sharp model R9700/no turntable rotation/dead oven

The turntable motor was used as a trouble indicator. Besides no turntable rotation, the whole oven was dead. The 15-amp fuse and interlock switches were normal. Replacing the defective control unit BUNTK085LDE00 solved the no-turntable-rotation problem.

Normal oven operation but no heat or cooking

Goldstar ER-410M/low-voltage circuit normal/no cook

The low-voltage circuits were normal with the monitor pigtail light across the primary winding of transformer. The Magnameter was attached to the high-voltage circuits, and the oven was fired up again. High voltage was noted without any plate current of magnetron. The high-voltage capacitor was discharged and a resistance test run across filaments for continuity. After removing the filament wires, the continuity across heaters was open. Replacing the magnetron (2M140A) solved the no-cook problem (4B71537A).

Goldstar ER-411M/no cooking

The blower and oven lamp was on but no cooking in the Goldstar oven. Upon checking with the schematic, the oven relay and all interlocks were working except for maybe the power transformer. The Magnameter was connected to the magnetron and showed no signs of HV or current. The voltage measured across the triac indicated 120 Vac. By placing a clip lead from T1 to T2, the oven began to operate. Replacing the triac solved the no cook problem.

K-Mart model SKR9S05/lamp out/no cooking

Normal fan motion. Replaced burned-out oven lamp. Found high voltage at the magnetron, but no current. The high voltage was higher than normal, indicating an open magnetron. Installing an FV-MZOY11WRKO magnetron solved the no-heat problem (figure 12-13).

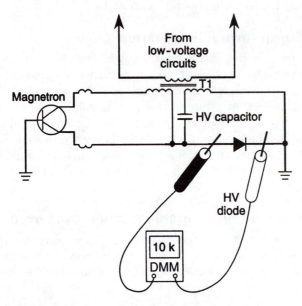

12-13 A 2-kilohm resistance was measured across a leaky HV diode in a K-Mart STR-7705 oven.

Litton model 402.000/blown fuse/no heat

After replacing the 15-amp fuse the oven was normal, except no heat or cooking. No high voltage was found on the heater terminal of the HV magnetron. Always discharge the HV capacitor to measure continuity between magnetron and chassis ground.

Only 1.5 ohms was found between heater terminals and ground. The short was still present when the heater terminals were removed. The magnetron must be shorted internally. In this model, check the HV diode, which is mounted down inside the magnetron shield area (figure 12-14). Removed one end of the HV diode and tested it—found that it was shorted. One could have easily ordered a new magnetron when only a shorted HV diode needed to be replaced.

Litton model MD I0D5I/no heat/defective capacitor

No heat or cooking in the oven—everything else operated. The HV capacitor was discharged and no arc was noticed between the two capacitor terminals. No high voltage or current readings were noticed on the magnetron. Undoubtedly, the HV circuits were defective.

Resistance readings were made between the HV diode and ground with no results. The resistance reading of the HV and heater windings were normal. A 10-megohm reading between heater and chassis ground was above normal. Power line voltage (120 Vac) was found at the HV transformer.

12-14 You can replace a defective HV diode with a universal HV replacement diode manufactured by another company.

Because no high voltage was measured at the magnetron, the HV capacitor was suspected. Two alligator lead clips were clipped across another HV capacitor and then placed across the suspected capacitor leads. The high voltage came up with the correct magnetron current. Replacing the open HV capacitor provided adequate heat in the oven.

Magic Chef M15-4/no cooking

Although the timer and blower motor operated, no cooking occurred in the high-voltage circuits. To determine if the high-voltage circuits were performing, a Magnameter was connected to the high-voltage circuits and pigtail light across the primary winding of transformer. No high voltage was found on the meter, and no low voltage at the pigtail light. S6, temperature limit switch in series with the primary winding of transformer lead, was open. Replacing defective temperature limit switch solved the no-cooking problem.

Magic Chef model MW207-4/no cooking

No high voltage was measured at the anode side of the HV diode. Resistance measurements of both transformer windings were normal. Power line voltage was measured on the primary winding of the HV transformer. The resistance measurement between anode of the HV diode and ground was about 10 megohms. Either the HV diode or capacitor was defective.

A new HV capacitor was clipped across the capacitor terminals with no high voltage. The cathode end of the HV diode was disconnected from chassis ground. Just clipping a new diode into the circuit solved the HV problem. Be careful while working around the HV components. Discharge that capacitor each time the ac plug is pulled (figure 12-15).

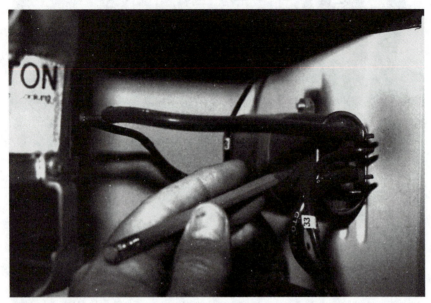

12-15 Discharge the HV capacitor before testing it for leakage.

Magic Chef model MW207-4/no heat/no high voltage

In another Magic Chef microwave oven, the results were the same—there was no high voltage. Of course, the HV problem could easily be seen. The cathode lead of the HV diode was broken right in the body of the diode. A new HV diode solved the no-heat/no-HV problem. In some cases, when the diode lead is broken off and is long enough, another wire can be soldered to it instead of installing a new replacement.

Montgomery Ward model KSA-8037A/no cooking

Low high voltage readings were noted on the heater terminals of the magnetron, with very little current drain. The HV capacitor was discharged and a resistance reading was taken between the heater terminals and chassis ground. A 5-kilohm measurement indicated a leaky magnetron or HV diode.

The cathode end of the diode mounted to the chassis was removed. A resistance reading across the diode could not be measured. No doubt, the magnetron was leaky. Sure enough, a 5-kilohm reading was found from the heater terminal to the metal shield of the magnetron (figure 12-16). Installing a new magnetron FV-MZ0033YBKO solved the no-cooking problem.

Montgomery Ward model L65-10/47X/no heat/no cooking

An HV measurement was above normal in this Ward's microwave oven with no current reading. The HV capacitor was discharged. Both heater terminals were removed. The heater terminal transformer winding showed continuity. No resistance reading

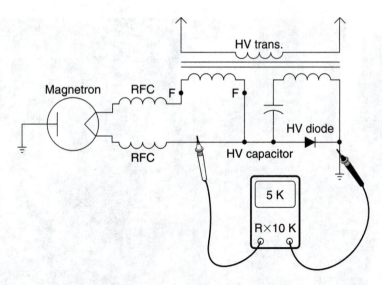

12-16 A 5-kilohm measurement from heater to common ground indicates a leaky magnetron or HV diode.

was noted between the heater terminals of the magnetron. Here, less than 1 ohm should be measured. A new magnetron was installed.

Montgomery Ward model 67X175/no cooking

No high voltage or current was found on the magnetron tube. Discharged the HV capacitor before taking resistance measurements. When taking resistance measurement across the HV diode, a bubble-type bulge was found on one side of the diode. Sure enough, the HV diode measured less than 1 kilohm. If the HV diode or capacitor becomes leaky, the 15-amp fuse might open. It depends on how long the oven is on with a leakage across the HV transformer.

Samsung MW5700/no heat

The blower motor and oven lamp was on in this oven with no heat or cooking of food. A pigtail light was clipped across the primary winding of transformer and indicated power line voltage. The oven was shut down and a continuity ohmmeter measurement indicated a 1.5-kilohm leakage. Either the HV diode, capacitor, or magnetron were defective. When removing the HV diode ground from the chassis, the HV diode showed leakage of 1.5 kilohms.

Norelco model MCS 6100/no heat/no cooking

High voltage was normal with no current reading in the microwave oven. Continuity checks on the HV and heater windings were quite close. When high voltage is present with no current reading, you can assume that the magnetron is open. Sure enough, the heater terminals were open on the magnetron (figure 12-17). A new magnetron was installed (6100 00 00029).

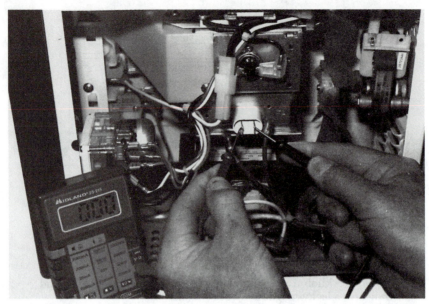

12-17 The normal heater (or filament) of the magnetron should have less than 1 ohm of resistance.

Norelco model R6008/normal, except no heat

No HV or current reading was found on the magnetron. Regular power line voltage was found on the primary winding of the HV transformer. Both the HV and heater windings were normal. A resistance measurement of 1.5 kilohms between heater and chassis ground indicated either a leaky magnetron or HV diode.

One end of the diode was removed to check for leakage across the diode terminals. The resistance was about 3700 ohms. The HV diode was replaced and another resistance check was made; now the resistance was 4125 ohms. Perhaps the new diode was leaky. The ground end was removed and the diode was normal. A resistance reading from heater to ground measured a little over 4 kilohms. Both the HV diode and magnetron were replaced. Of course, this is a very unusual case. You might find one component defective, but usually not both. The defective diode part number is 6000 000 0079 and 6000 000 00085 for the magnetron.

Norelco model RR 7000/dead/no heat

After replacing the 15-amp fuse, the fan rotated—but there was no cooking. Poor line voltage was monitored at the primary winding of the HV transformer. High voltage was present with no current from the magnetron. Cooking was resumed with a new magnetron (part number 6000 000 00085).

Norelco model MCS 7100/no heat/no cooking

No high voltage or current was found on the magnetron. The HV capacitor was discharged and resistance measurements were made. A resistance measurement of 71 ohms was found across the HV diode. The magnetron was suspected of a dead short. Of course, the low-ohm reading was still present with the heater leads removed.

One lead of the HV diode was removed and the diode was good. A quick resistance measurement of only 1.5 ohms was found across the HV capacitor. The 71-ohm reading across the diode to ground was actually the total reading of resistance of the HV power transformer winding and shorted capacitor (figure 12-18).

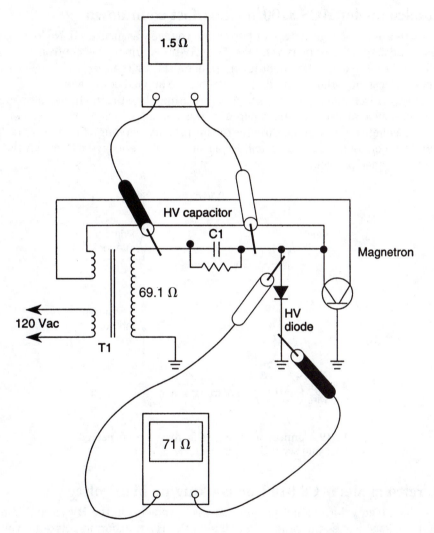

12-18 A leaky C1 had only 1.5 ohms across its terminal with no high voltage or heat.

Norelco model MCS 7100/slow cooking

The complaint with this oven is that it took longer to cook food than usual. The low voltage across the power transformer primary winding was monitored with a VOM. A current and HV meter monitored the magnetron. When the oven first came on, all voltage was fairly normal with about half the normal current reading.

After the oven was on for several minutes, the current would almost come up to where it should. Anytime high voltage is present with low or no current pulled by the magnetron, suspect a low-emission magnetron tube. Replacement of the magnetron (number 6100 000 00029) solved the slow-cooking problem.

Norelco model MCS 8100/no heat/fast countdown

In this oven, no high voltage or power line voltage was measured. No low voltage was found across the primary winding of the HV transformer. The controller seemed to be working, except the countdown appeared faster than usual. The VOM was clipped across the primary winding to monitor the applied ac voltage.

Always discharge the HV capacitor before attempting to attach the voltmeter or take resistance measurements. A clip wire was connected across the triac connections, feeding the ac voltage directly to the primary winding (figure 12-19). The oven came on and worked as usual. Replacement of the leaky triac solved the HV and low-voltage problem.

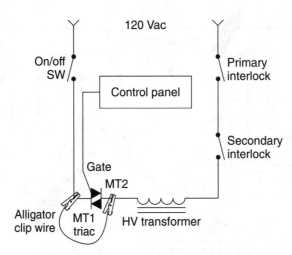

12-19 Connect a wire clip across the suspected triac and see if the oven operates.

Norelco model MCS 8100/no cooking/no high voltage

When high voltage is not present, suspect problems in the HV circuits. Check the ac voltage across the primary winding of the HV transformer. Here no voltage was measured. A quick glance at the magnetron tube showed an overheated thermal cutout. The wires to the thermal switch were burned and the cutout showed signs of excessive heat. A clip wire across the thermal switch applied ac to the HV circuits. The defective thermal switch was replaced with part number 8100 000 00098.

Samsung MW-5500 fan rotates/no cooking

Line voltage was monitored at the primary winding of the power transformer, indicating normal voltage (118 Vac). The Magnameter was connected to the high-voltage circuits. No high voltage or plate current was found on the tester. A leakage test across the high-voltage diode indicated 1075 ohms. The stacked diode was replaced with a universal oven replacement.

Samsung 4530U/no triac action

Although the oven light and fan blower operated when the oven was turned on, no cooking resulted. A pigtail light was connected across the primary winding of the power transformer, with no indication. Suspected triac or control board circuits. Power line voltage was found across terminals T1 and T2 of the triac (figure 12-20). Another triac was installed and solved the no-cook problem.

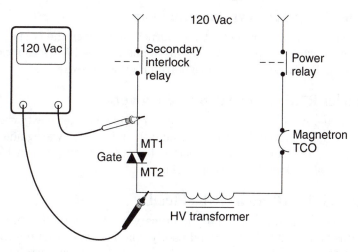

12-20 If 120 Vac is measured across triac with the oven turned on, the triac is defective, or there is no gate voltage.

Sharp model R6600/dead/no cooking

Replaced blown fuse. High voltage normal. Replaced open magnetron FVMZ00IY BKO.

Sharp model R7600/turntable rotates/no heat

Both the turntable and fan were rotating. No low voltage (120 Vac) on the primary winding of the HV transformer. Found a defective thermal cutout. The resistance of 5.1 ohms was measured across the switch terminals. No resistance should be measured across the thermal switch.

Sharp model R7704A/everything operates except cooking

Measured only 1.5 Vdc from heater terminal to chassis ground. Practically a dead short was measured from heater terminals to ground. Replaced shorted magnetron.

Sharp model R7704/will not cook

Installed a new magnetron tube FV-MXO 111 WRKO.

Sharp model R7710/not enough heat

The oven was checked out thoroughly at the shop. Cooking tests were perfect. When checked in the customer's home, only 105 Vac was measured at the ac outlet. Recommended electrician run a separate line from the switch box.

Sharp model R7710/no heat/defective diode

No high voltage was found after replacing the 15-amp fuse. Power line voltage (120 Vac) was found at the primary winding of the HV transformer. Resistance measurements turned up a shorted diode RH-0Z0039W REO.

Sharp model R7804/no voltage/no heat

In this model, the thermal cutout was burned open; even the connecting wires showed signs of overheating. Ac voltage was applied to the HV transformer, but no heat or cooking. Replaced defective magnetron FVMZ0091-WRKO.

Sharp model R7810/no heat/burned cover

The customer had complained that the cover inside the oven cavity was burning. The old cover was removed. Wiped out around the microwave waveguide outlet for excessive grease that might have collected there. A new magnetron FV-MZ0102 MBKO was installed to provide normal cooking.

Sharp model R8310/no heat/bad lead

Power line voltage was applied to the primary winding of the HV transformer— but there was no output voltage. Found poorly crimped leads from the transformer winding. Scraped around clip and crimped area. Applied rosin paste and soldered each transformer wire connection.

Sharp model R9310A/no cooking/bad leads

The complaint was poor heat and then no cooking at all. High voltage was present across the HV diode, and no current measurement was found on the magnetron. Found excessive burned heater cables at the heater connections. Cleaned off connections and flat heater terminals. Soldered both connections.

Sharp model R9310A/lights up/no cooking

Found one bad heater connection on the magnetron. Cleaned off and soldered each connection. Gave the oven the four-hour cooking test.

Sharp R-9700/oven operates/little heat

This oven came in with poor cooking symptoms and a water test was performed. The water was fairly warm after three minutes of operation. The oven might have a defective control unit, magnetron, high-voltage diode, high-voltage capacitor, defective power transformer, or poor wire contacts. The browning element was normal, indicating problems within the high-voltage circuits.

The Magnameter was connected to the high-voltage circuits to determine if high voltage was present and if the magnetron was drawing plate current. The high voltage was higher than normal, with only 97 mils of current. After discharging the high-voltage capacitor, the filament leads were checked and inspected and appeared normal. Sometimes burned or charred heater terminals can cause improper heating. The magnetron had low emission and was replaced.

Tappan model S6-1026-1/no heat

Found bad connections of crimped leads from the power transformer winding. Cleaned and applied solder paste. Resoldered each crimped transformer connection.

Thermador model MC17/no heat/voltage normal

Replaced defective magnetron L5261A.

Oven goes into cooking cycle, but then shuts down

Norelco model R6008/starts but no cooking

The oven quit operating after several minutes of cooking. All other functions were normal except no heat. Monitoring the voltage at the primary winding of the HV transformer indicated no voltage when the oven quit cooking. A new thermal cutout solved the shutdown problem. Clip a VOM or 100-watt bulb across the thermal switch to see if it opens up (figure 12-21). The entire ac line voltage will be measured or the bulb will become bright when the switch opens.

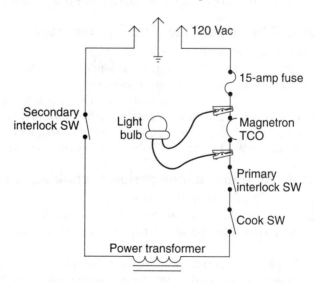

12-21 Clip a 100-watt bulb across the thermal switch and see if it lights when the oven shuts down.

Norelco model MCS 7100/popped and quit

The complaint with this oven was that it was cooking beautifully, then made a popping sound and quit. No high voltage was measured. Resistance measurement of the power transformer was good. A leaky HV capacitor (6100 000 00005) popped and the oven ceased cooking.

Norelco model 57500/shuts off after several minutes

At first, a triac was suspected of shutdown after no heat was noticed. A clip wire across the suspected triac indicated the triac was normal. The oven still quit after a few minutes of operation. Very little dc voltage was measured at the gate of the triac, indicating a possibly defective control board. Before ordering a new control board, a new triac was installed—just in case it was shorting out the gate voltage. But a new control board brought the oven back to life.

Norelco model MCS 8100/just went dead after several minutes

Replaced the 15-amp fuse. No high voltage was measured on the magnetron. After several service checks, a leaky HV diode was replaced (6000 000 00079). Still, the oven was inoperative. Replacing the control board (8100 000 00080) solved the "just-went-dead" symptom.

Norelco model MCS 8100/ran one minute and quit

The ac voltage was monitored with a VOM at the primary winding of the HV transformer. Found no voltage when the oven was not operating. Everything was working except no heat. Clipped a lead across the triac after discharging the HV capacitor. The oven continued to operate. The intermittent triac was replaced.

Sharp model R6770/no cooking after five minutes

You would never suspect anything wrong with the oven if only a few minutes of cooking was required, but after five minutes or longer the oven would not cook. Some overheated component was shutting the oven down.

The magnetron was too warm with the thermal switch opening up. When in the no-cook operation, the vari-motor was not rotating. After several hours of checking, the contact points of the temperature control assembly were opening up. Replacement of the assembly (QPWB F0002W REO) cured the shutdown problem.

Sharp model R-7804/after a long period of cooking, quits

After 12 minutes of operation, this microwave oven shut down the cooking process. The fans and all other functions seemed to be operating. A lot of service time can be involved when attempting to locate the defective component with long cooking periods.

The oven would start to operate after it cooled down for a few minutes. Usually heat shutdown problems are caused by the magnetron tube. If possible, monitor the voltage on the primary winding of the HV transformer. Place a 100-watt bulb clipped across the thermal switch. Monitor the high voltage and current if a special meter is handy.

In this case, the light bulb came on when the oven quit cooking. A new thermal switch did not solve the problem. Then a call was made to the factory service, and a thermal kit was provided. A piece of insulation was placed behind the thermal switch, solving the long breakdown period.

Erratic or intermittent cooking

Goldstar ER-505M/intermittent cooking

This oven might operate for several days and then suddenly quit operating. The intermittent oven was checked with a pigtail light monitored across the primary winding of HV transformer. The light went out after the oven had been operated off and on for a total of 4 hours. (Note: Some components take time warming after the oven is turned on.) The magnetron thermostat was monitored across with a pigtail light and the triac with an ac meter. When the oven quit, the magnetron thermo unit went open. The pigtail light came on. The thermostat was replaced with exact part number 3870069C.

Hardwick M-240/poor cooking

The customer complained the oven took too long to cook meat in the microwave oven. A Magnameter was connected to the high-voltage circuits to determine if high voltage and plate current were normal. The magnetron current was fairly stable, with normal high voltage.

A close inspection of the blower motor indicated normal operation, but the radiating antenna was not normal. The radiating fan belt was off and broken in two. Replacing the drive belt solved the slow cooking process.

Litton model 70/65.730/normal high voltage

The HV and low-voltage circuits were monitored for a change in voltage. A current meter was tied to the leg of the magnetron. When intermittent, the current would drop to 0. The high voltage increased with no change in the low-voltage meter reading. Replaced defective magnetron.

Norelco model MCS 6100/intermittent operation

Found poor ac power in the home and improper line voltage. Requested electrician to install new outlet.

Norelco model MCS 7100/slow and erratic

An HV and current meter showed erratic current reading. The current would remain about half scale and then go up. Replaced magnetron 6100 000 00029.

Norelco model MCS 7500/erratic operation

An erratic low-voltage reading at the primary winding indicated a defective triac or control board. Shorting across the triac turned up a defective control board 7500 000 00/00.

Norelco model MCS 8100/opened door/dead

Sometimes, when the door was opened and then closed, the oven would go dead. Found an intermittent relay 8100 000 00011 (figure 12-22).

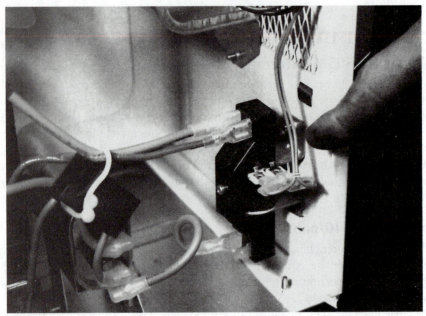

12-22 Sometimes you can hear the relay energize when power is applied.

Norelco model MCS 8100/intermittent interlock

This oven would operate normally for several days, then become intermittent. Replaced erratic upper interlock switch assembly 6100 000 00055.

Quasar model MQ6620 TW/intermittent operation

Replaced intermittent latch switch 40-90344A79.

Samsung model MW8600/improper operation

The microwave oven did not cook as it should. Sometimes cooking took longer, and at other times the oven lights seemed to blink. Even a closer inspection of the oven at the shop showed nothing unusual. Checking the home outlet and power line voltage solved the improper operation. The line voltage measured 100 volts. Recommended that a new outlet be installed by the local electrician.

Sanyo model EM-8205/erratic/poor leads

Found poorly soldered or crimped-on leads of the power transformer. Scraped back the enameled wire. Tinned and soldered each crimped-on connection.

Sharp model R6770/erratic cooking

Replaced defective magnetron FV-MZ0078WRKO.

Sharp model R7704A/erratic heat

Incorrect time in cooking cycle. Replaced erratic timer QSWTE 0101WREO.

Sharp model R6770/erratic cooking/erratic current

Monitored both high voltage and current of the magnetron. Replaced defective magnetron FV-MZ0078WRKO. In some ovens, you can tell when the magnetron is heating up by seeing if heat is coming out of the vented blower area. However, this does not indicate the oven is cooking. Correct high voltage and current readings will determine if the magnetron is functioning properly.

Sharp model R7710/intermittent operation

Right side of the filament terminals were all black. Replaced defective clip and wire. Soldered both heater connections.

Sharp model R7704/intermittent/magnetron

Monitored high voltage and current. Erratic current reading. Replaced erratic magnetron FV-MZ0011WRKO.

Sharp model R7704/dead, then intermittent

Replaced blown 15-amp fuse. Installed both monitor and interlock switch (QSW-M0100WREO and QSW-M00 64YBEO, respectively).

Sharp model R9310A/intermittent cooking

Burned heater wires. Replaced clips going over heater terminals. Cleaned off heater terminals. Soldered both connections.

Sharp model R9310/sometimes cooks

Found a defective interlock switch QSW-M0100WREO. Readjusted loose door.

Tappan model 56-1026-1/intermittent cooking

Sometimes the oven would cook and at other times not. Everything was normal, except no heat. The filament of the magnetron tube was intermittent. Replaced magnetron. You can quickly tell if a magnetron is oscillating or cooking by setting a portable TV near the oven, while the oven is operating. A large noise band is generated across the TV screen on the lower TV channels (channels 2 to 5) when the magnetron is operating.

Oven light stays on

Norelco model MCS 8100/lights and oven stay on

The oven lights and cooking operation would stay on and not shut off when time was up. Replaced leaky triac assembly.

Magic Chef model MW3117Z-SP/light on

The light stayed on. Replaced latch switch. Readjusted door.

Sharp model R9315/oven light stays on

Replaced erratic secondary interlock switch QSW-M0096WREO.

Keeps blowing fuses

Hardwick M-250/oven keeps blowing 20-amp fuse

The fuse blew when the oven was not being used. The Circuit Saver was installed instead of the fuse. Right away, the circuit breaker in the fuse saver kicked out. The overload would kick out when the oven was plugged into the wall. A quick inspection found a charred varistor across the primary winding of the low-voltage transformer (figure 12-23). Cutting the varistor loose solved the problem. No doubt, lightning or power line outage had caused the varistor to arc over and remain shorted.

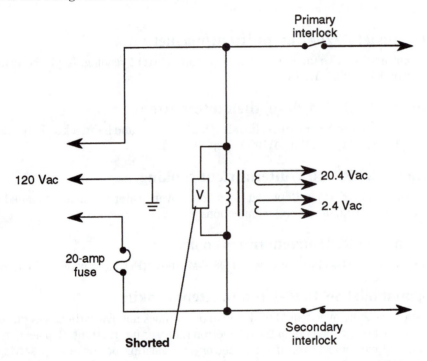

12-23 The shorted varistor in a Hardwick M250 oven kept opening the fuse.

Hardwick M-240/fuse opens once a month

Once every three or four weeks, the fuse would blow inside the oven. The house fuse was normal. After operating the oven off and on, on the service bench for three weeks, the oven appeared normal. No doubt, trouble existed at the customer's residence.

When the oven was returned to the customer, the power line voltage was checked and was fairly normal (117 Vac). The oven worked fine after several cook tests. Finally, a ground test was made between oven and power outlet box. A poor ground of

4.75 kilohms was found. The customer was advised to have a local electrician install a new grounded outlet wired directly from the fuse box for the microwave oven.

Amana model RR-4D/fuse blows when the door is opened

When opening the door, the fuse would open. Suspected defective monitor switch. Found metal lever at top of oven would not let the microswitch release. This safety switch would blow the fuse each time. Also replaced safety interlock switch.

Frigidaire 56-5991/blows fuse

After operating for a few seconds the fuse blew and the oven quit operating. Since the fuse did not blow at once, the secondary circuits of the magnetron were checked. The HV capacitor was discharged and a continuity ohmmeter test was made across the HV diode. A resistance measurement of 1.7 kilohms indicated a leaky capacitor, diode or magnetron. Last but not least, the magnetron was checked and found leaky.

Litton model 465/oven shuts off, blows fuse

When the oven shut off, the 15-amp fuse sometimes opened. Only after cooking for 10 minutes or longer would this condition occur. Replaced an erratic triac.

Quasar model MQ-6620TW/blows fuse when plugged in

Each time the oven was plugged into the wall outlet, the fuse blew. Found that the varistor in the ac power line was arcing. This component is located in a clear plastic sleeve of the wire cable assembly (figure 12-24). Replaced.

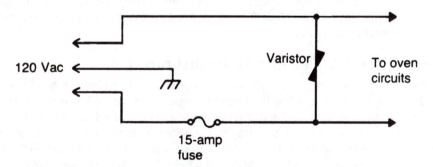

12-24 A lightning strike or power outage can damage the varistor across the ac power line.

Litton model 370.00/blown fuse, diode

The fuse continually opened in this microwave oven. A shorted HV diode was found and replaced.

Sharp model R7710/dead/keeps blowing fuses

Replaced shorted magnetron tube FV-MZ0115MRKO.

Sharp model R9315/fuse opens/hot operation

Found an interlock switch hangup causing the monitor switch to open. Replaced both monitor and interlock switch assembly. The fuse opened when the oven was operated for more than 12 minutes. Discovered that the magnetron was shorting out. In this case, two different troubles caused the fuse to blow.

Whirlpool model REM7400-I/blows fuse when plugged in

As the ac cord was inserted into the power outlet, the fuse opened. Replaced a frozen monitor switch. This switch is located in the middle of three separate switches.

Hum or noisy operation

Amana model RS 458P/loud buzzing

The oven, when first turned on to cook, made a loud buzzing and vibrating noise. With the cover off, the noise was greater. Removed and replaced power transformer.

Hardwick model M-250/hum

Most noises caused in the microwave oven are the blower motor, radiating antenna, triac, and power transformer. Here, the small controller transformer was causing the audible hum. Replaced the transformer.

Hardwick M250/sparking in oven cavity

Look for metal ties or objects that might have metal on them, such as a cup or bowl with metallic paint. Make sure the temperature probe is not in the oven. Check the radiating-stirrer antenna blade for touching or being very close to the oven metal cavity, producing sparks or arc-over.

Hardwick M-250/loud hum when first turned on

Sometimes a clicking noise in the oven can be caused by a blower fan blade striking the metal framework. A low hum can be caused by a controller or power transformer. But this low hum noise was heard as the oven started up. After checking both transformers and motors, the triac was checked. The triac had been changed at one time or another. Wires on MT1 and MT2 were reversed, placing triac 180° out of phase. The white wire should go to MT1 and black wire to MT2 (figure 12-25). Make sure the controller is not damaged.

K-Mart SKR7705A/noisy turntable

When the turntable rotated, you could hear a grating noise. Removed turntable assembly and replaced broken plastic coupling FCPL-001YWRKO.

Norelco model MCS 8100/noisy fan

Found loose fan-mounting bolts on this one.

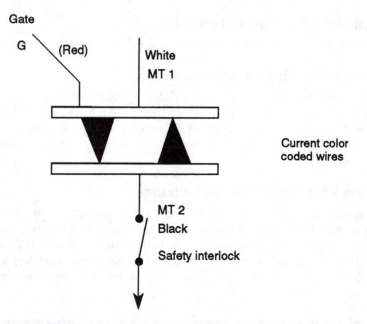

12-25 A loud hum was heard in the Hardwick M250 oven with reversed MT1 and MT2 on the triac.

Norelco model MCS 8100/hum with closed oven

The oven would begin to hum when the door was closed. Removed and replaced leaky triac assembly 8100 000 00005.

Quasar model MQ6600/hot spot and noisy fan

The customer complained of a possible hot spot. The oven was normal. The fan was loose. Repaired and remounted the fan assembly.

Sharp model R6770/relay chatter

Sometimes the relay started to chatter when the oven switch was pushed. Erratic fan operation. Replaced erratic on/off switch.

Sharp model R7704/hum and buzz

When the cook button was pushed, you could hear a loud hum. Removed the outside cover and now could hear a loud hum and buzz from the power transformer. Replaced noisy transformer RTN-00144WREO.

Sharp model R7705A/shock and noisy

Found a noisy timer. Replaced timer. Checked oven for poor grounding. Was normal in the shop. Found the oven was not grounded in the home.

Sharp model R7804/noisy turntable

Removed noisy turntable assembly. Lubricated and greased gearbox assembly.

Sharp model R8320/fan noise at the top

In the convection oven, a fan blade is located at the top of the oven. After removing several layers of metal ducts, I saw that the fan blade was striking a metal bracket of the heating element. This fan is located in the round center circle of the heating element. Removed heating element and repaired mounting bracket. Replaced all components.

Sharp model R9510/constant buzzing

In the cook cycle, the power transformer was buzzing very loudly (figure 12-26). Particles or iron vanes inside the transformer were making the noise, which would decrease when the load was taken off it. Discharged the HV capacitor. Took one lead off the heater connection of the magnetron. Replaced transformer RTRN-0160MKEO. These noisy transformers must be replaced because they are dipped and cannot be repaired.

12-26 A noisy power transformer must be replaced.

Cannot turn oven off

Norelco model MCS 8100/hums/can't be turned off

After running for 10 minutes, the oven began to hum and would keep on operating when shut off. Replaced leaky triac 8100 000 00005.

Norelco model MCS 8100/fan fast operation/no shutoff

The fan motor seemed to run fast in this oven after shutting off. The oven would keep on operating in the cooking mode. Replaced leaky triac.

Whirlpool REM7400-1/cannot turn oven off

Replaced defective on/off switch (figure 12-27).

12-27 Check the on/off switch if the oven cannot be shut off with a relay instead of a triac.

No fan or blower rotation

Hardwick M-240/blower motor stopped

The blower motor stopped running after the oven was on a few minutes. The motor was suspected of gummed-up or dry bearings, but the fan blade would spin fairly easily by hand. The ac meter was clipped across the blower terminals to determine if fan or voltage was the problem.

Because a thermal protector was found in the fan circuit, a pigtail light was connected across its terminals. When the blower quit, the pigtail light came on, indicating an overheated magnetron. Because the magnetron felt fairly cool, a new thermal unit was installed. This thermal unit should open at 250°F to 260°F and was opening before a 90°F temperature was reached. A new thermal cutout solved the problem.

K-Mart model SKR-7715/no fan rush

The fan was not rotating in this microwave oven. Discharged the HV capacitor. Measured the ac voltage across the motor leads. Found a bad fan connection when attaching the meter. Replaced and resoldered fan connection.

Litton model 401/slow fan blower

The oven fan would not come on for at least three to five minutes after the magnetron started up. The fan blade did not spin freely by hand. Spraying oil into the end and front bearings took care of the fan problem (figure 12-28).

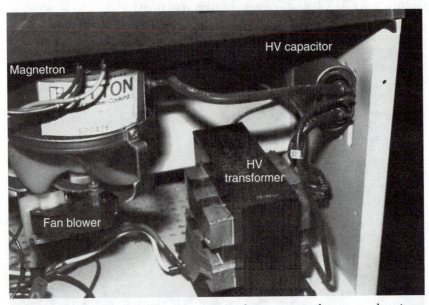

12-28 If the fan is slow to come on, suspect dry or gummed up motor bearings.

Norelco model MCS 8100/no fan rotation

Sometimes fan would start out slow and not rotate. Removed fan assembly. Cleaned up and lubricated fan bearings.

Norelco model MCS 8100-1/intermittent fan operation

Sometimes the fan rotated and other times not. Repaired bad fan cable plug connection.

Sharp model R7704A/intermittent fan and oven

After five minutes of cooking, the oven shut down. When in the shutdown condition, the cooling fan was not rotating. When the fan stopped, the magnetron overheated, kicking out the thermal switch. Found poor motor field connection. Repaired and soldered.

Oven shock and leakage
K-Mart SKR-7705/clicks and shocks

Checked ac leakage from metal chassis to ac power cord. Was normal. Customer suspected the oven of door leakage. Checked for leakage around the door and readjusted the door. Suspected static electricity because the kitchen floor area was carpeted.

Norelco model MCS 7100/intermittent operation/shocks

The intermittent operation was caused by a two-wire extension cord. Low ac voltage was measured at the oven in the home. The oven was given a four-hour cook test in the shop. Installed a new three-prong outlet and grounded the oven.

Sharp model R5480/customer afraid of shock

The oven had a burned waveguide cover and sparks were flying out of the waveguide area. Replaced both waveguide covers and magnetron. The oven was checked for possible poor grounding. A thorough door leakage check was made.

Sharp model R5600/customer shocked by oven

The power cord and internal components were checked for ac leakage. Proper grounding of the oven in the home solved the possible shock hazard. Also checked for possible door leakage.

Sharp model R7704/customer suspects leakage

Radiation leakage was checked around the door and all vent openings. After the oven door was adjusted, the oven was double-checked for leakage. Before door adjustment, only 0.1 mW/cm2 leakage was noted at the top of the door.

Sharp model R7705/checked for shock and leakage

The oven and power cord were checked for ac leakage, and the door and openings were checked for radiation leakage. Requested that customer properly ground the oven.

Sharp model R7710/sparks flying all over oven

The customer complained that the inside of the oven had sparks flying all over. A new magnetron and waveguide cover were replaced. Magnetron part number FV-MZ0102 WRKO. Waveguide cover part number PC0VP0Z04WREO.

Sharp model R7800/sparks from oven

Replaced arcing magnetron tube FV-MZ00912 RKO.

Too hot to handle

K-Mart model SKR-7705/sparks in oven

Flashing and sparks flying in the oven. Replaced the defective magnetron and waveguide cover.

Montgomery Ward model 67X175/burning in oven

One plastic shelf was burning at one metal screw area. Excessive grease over the years had run down behind the plastic shelves. Each time the oven was turned on, the plastic started to burn. Washed out all grease with soap and water. Removed shelves. Installed two new ones.

Norelco MCS 6100/warped plastic cover

Popcorn was placed in a brown paper bag. The bag exploded and started a fire in the oven. The plastic shield and front door plastic was replaced. Some brown paper bags are made of recycled paper. If the bag is made of recycled paper, sometimes pieces of metal are introduced into the paper during the recycling process. The metal can cause the bag to start a fire, which might end up destroying tile in the microwave oven.

Norelco model MCS 8100/burns food

The owner complained that sometimes the oven burned the food. Checked timing and control board. Cooking tests were normal in the shop. Found a very high power line voltage at the farmhouse (127 Vac). Referred the customer to the power company.

Sharp model R5380/oven smokes, won't cook

Replaced magnetron tube.

Sharp model R5480/something burning in oven

Removed burning waveguide cover. Washed out all grease. Installed a new cover PCOVP0039YBPO.

Sharp model R6770/too hot, burns food

This microwave oven was cooking too fast. A current meter revealed the magnetron was drawing excessive current. The magnetron got extremely hot. Replaced defective magnetron FVMZ0078 WRKO (figure 12-29).

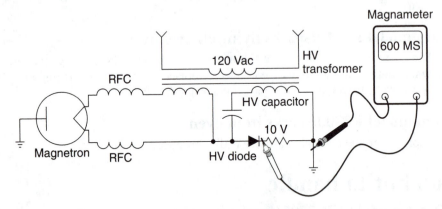

12-29 Do a Magnameter test if food burns or cooks too fast.

Sharp model R7704A/something smoking

After the 15-amp fuse was replaced, the oven began to smoke. Before the cord could be pulled, the fuse opened once again. The suspected magnetron showed a leakage between heater and ground of 48.7 ohms with a digital VOM (figure 12-30). The magnetron FV-MZ0091 WRKO was replaced.

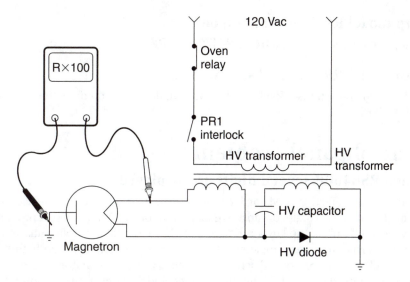

12-30 A leaky magnetron in a Sharp R7704A oven caused the transformer to smoke.

The oven was turned on and something appeared extremely hot. The HV capacitor was discharged. The secondary winding of the power transformer RTRN-KL44WREO was extremely hot. Undoubtedly, the shorted magnetron broke down the HV winding before the fuse was blown. Both the transformer and magnetron were replaced.

Sharp model R7404/smokes and quits

Replaced 15-amp fuse. Replaced magnetron and waveguide cover.

Sharp model R7650/arcing in oven

Found that eggshells went up through the open waveguide assembly. Excessive arcing from the magnetron to ground. Cleaned up the magnetron antenna area. Installed new metal gasket PGSK-0004YBEO.

Sharp model R7800/smoking in oven cavity

Replaced the waveguide cover. Installed a new magnetron. Original part number FV-MZ0091 WRKO was subbed at the factory for a FV-MZ0 111 WRKO.

Sharp model R7844/firing inside, now dead

Replaced 15-amp fuse. Replaced defective magnetron FY-MZ0111WRKO.

Sharp model R9S00/sharp arcing

Found the fuse blown. The fuse immediately opened when a loud arcing was heard. The magnetron tube was arcing internally. Replaced the magnetron FV-MZ0091WRKO.

Sharp model R9504/burning food

Replaced defective magnetron FV-MZ0091WRKO.

Sharp model R9750/sparks, now dead

Replaced 15-amp fuse. Replaced the waveguide cover and defective magnetron FV-Z0091WRKO.

Control panel problems

Amana RSB460P/several digits are displayed

Suspect a defective keyboard assembly if a certain pad produces no programmed results, two digits or more are displayed, or touching any pad provides no programming. Replace the keyboard assembly. If the digital controller misses a pre-program shortly after warmup; suspect a defective component in the high-voltage section. Pull the power cord, discharge the high-voltage capacitor, and connect a Magnameter into the circuit. If the high voltage and plate current are normal when the digital controller provides misprogram, replace controller.

GE JEBC100/no cook/no display lights

The oven lamp or blower motor did not operate in this oven and the display did not light up. Ac was found across the LVT transformer, but no secondary voltages. The primary winding socket was disconnected and showed no continuity of winding. After replacing the LVT transformer, the oven cooked and operated.

Goldstar model ER-505M/does not count down

When the start button was pushed, the setting of time did not start to count down. Interlocks were normal. Suspected the control panel board (PCB). When the oven was fired up again, it worked okay. When the PCB was pushed the next time, the operation was intermittent. Repairing the loose connection at pin 10 solved the countdown operation (figure 12-31).

Goldstar ER-505M/display okay/no cook

The timer setting would not count down when the start pad was pressed. The triac and low-voltage transformer were monitored with a pigtail light and ac voltmeter. No light or meter indication pointed to low-voltage circuits or controller. Power line voltage and low transformer secondary voltage was found at controller. All connections were checked and pushed down on controller. Replaced micon controller.

Goldstar ER-711M/oven cannot be programmed

At times, no programs could be placed into operation—and when programmed once again, only a few pads operated. No doubt, the control board was defective. The harness between push buttons and control were checked and seemed normal. The low-voltage transformer voltage was normal at 2.4 V, 24 V, and 48 V. A new control board was ordered and it solved the intermittent symptom (2Q10064A).

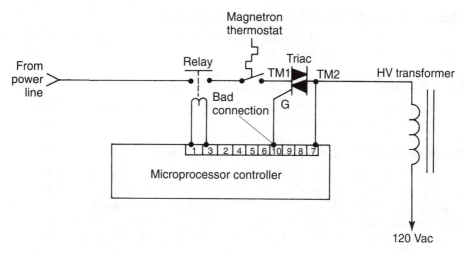

12-31 A bad connection on terminal 10 caused intermittent gate voltage in the Goldstar ER-505M oven.

Goldstar model ER-410M/no cooking

The blower motor operated with no cooking. No line voltage was found across the primary winding of the HV transformer. Line voltage was found across the triac, indicating a defective triac or no gate voltage. No gate voltage was measured at the G terminal of the triac. Replacing the control board solved the no-cook problem.

Hardwick M-250/no program function

At first, the control panel was suspected of being defective. Ac power line voltage was measured across the low-voltage transformer. No secondary voltage (20 Vac) was found on the red leads of transformer. Both primary and secondary leads were checked for continuity. Replacing the low-voltage transformer solved the no-program function with the open primary winding.

Hardwick M-270/oven stops during programmed cycle

First, I check for a blown fuse. Install the Circuit Saver if fuse continues to open. Check the power transformer and magnetron. Monitor triac with the pigtail light. If lights up, replace triac. Suspect a magnetron or oven thermal cutout for overheating. Test with a pilot light across thermal terminals. If lights up, replace thermal unit. Notice if the blower motor is operating. The magnetron might overheat if blower motor will not rotate.

Hardwick MM-270/display numbers incorrect

Make sure that each touch pad is working. Check the nonfunctioning ones with the R×1 scale of the ohmmeter. Pull the power cord and discharge the high-voltage capacitor. If the touch pads are normal, replace microprocessor control assembly.

Hardwick model M-270/incorrect numbers

The numbers appeared incorrectly in the display tube. Replaced the control board assembly with microprocessor.

Norelco model MCS 8100/opened door/dead

Replaced 15-amp fuse. Control board would not count down properly. Replaced control board 8100 000 00008.

Norelco model MCS 8100/intermittent operation

Monitored voltage at the primary winding of HV transformer. Suspected a defective triac when I found it was intermittent when no voltage was applied to transformer. Shunted triac with a clip lead. The oven began to operate. Monitored voltage at the gate terminal of triac with the clip lead removed. Found that gate voltage was very low when the oven was intermittent. Replaced the control circuit board—8100 000 00008.

Norelco model MCS 8100/behind 8 ball

The right-hand side number 8 was always shown behind any number. When the oven was turned on, you could see a faint number 8 in the background. A new control board was ordered and installed.

Sharp model R9750/time card error

When the time card was pushed in on this model, the cooking timer showed up as an error. Replaced entire card reader assembly. DPNC-S0004PAZZ.

Door problems

Hardwick model M-240/steam at door

In this model, the steam seemed to flow into the door while cooking. The door seals were checked with the air vents. Improper venting and air flow allowed steam to enter the door area.

Litton model 70-05/dead/leakage

Replaced 15-amp fuse. Checked cooking and leakage test. Found 0.25 mW/cm^2 leakage at top of door. Readjusted door and made another leakage test.

Norelco model MCS 6100/door open

The owner said the oven blew up. Found both door latches broken. The door would not catch or lock. Replaced both latch switches—5100 000 00055. Gave oven four-hour cooking test. Checked for leakage after adjusting the door assembly.

Norelco model MCS 7100/door doesn't close

This door appeared warped, or someone had tried to adjust it. Removed door entirely and checked for warped door. Door assembly okay. Installed and adjusted door (figure 12-32). Checked for leakage.

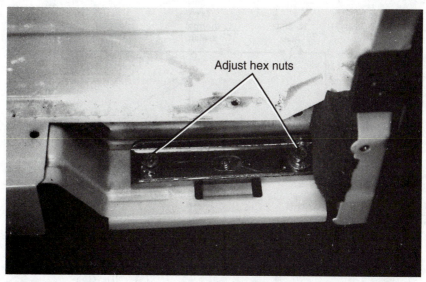

12-32 Adjust the hex nuts for proper closing and test for any sign of leakage around the door area.

Norelco model MCS 8100/check door and shield

The customer complained of a warped shield—6100 000 00020. Ordered and replaced. Repaired and adjusted the door so there was no play between door and base cabinet.

Norelco model MCS 8100/stuck door

Removed door. Repaired door guide assembly. Installed door assembly and adjusted for too much play. Performed leakage and cook tests.

Norelco model MCS 8100/door will not shut

Replaced lost screw in the switch housing. The switch assembly was down, so the hook would not lock in. Adjusted door. Make leakage test.

Norelco model MCS 8100/binding door

Readjusted the door latch assembly. Readjusted the door assembly. Checked the other interlock switch assemblies. Made cook and leakage tests.

Samsung MW5700/no touch control

The fuse was found open and replaced. But there was still no operation from the control panel. Monitored the low voltage at the primary winding of the power transformer and found no ac voltage. Suspected triac, but ohmmeter tests showed it was normal. After checking the circuit diagram, the ac voltage was measured across the low-voltage transformer on the control panel with no ac voltage (figure 12-33). Found an open thermal unit. Replaced thermal switch with an exact replacement part—3589-001-0183.

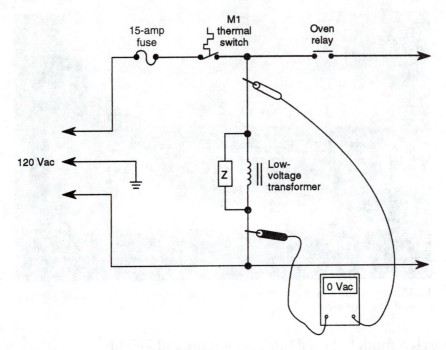

12-33 A defective thermal TCO (M1) in a Samsung MW5700 oven caused no ac voltage at the LVT or fan motor.

Sharp model R3700/loose door latch

A loose rivet and the door assembly was found to have too much play. Repaired door latch assembly with a new pop rivet.

Sharp model RS480/door won't open

The door was stuck on this microwave oven. Pushing down on the latch would not release the door. After removing the back cover and discharging the HV capacitor, one of the door latch assemblies helped to release the door. Replaced the broken door latch assembly.

Sharp model R6770/intermittent operation

Sometimes the oven would operate and at other times, it quit. Found a broken door spring in the door assembly. Replaced spring and readjusted the door. Checked for leakage (figure 12-34).

Sharp model R7600/blows fuse

Every once in a while the fuse would blow for no known reason, until the operator said the oven quit when the door was opened. Checked door for correct closing. Replaced left-side interlock assembly and also readjusted door. Performed cooking and leakage tests.

12-34 Check for a broken spring on the plastic latch if it does not engage the interlock assembly.

Sharp model R7600/must raise up door to operate

The owner had to raise up on the door to make the oven operate. Replaced door interlock monitor switch assembly.

Sharp model R7650/oven light/no heat

Fuse was good with oven lights on. All interlocks were checked and found okay. Located a broken spring in the door latch assembly. Replaced, checked leakage, and gave a cooking test.

Sharp model R7650/intermittent/now dead

Replaced 15-amp fuse. Realigned the front door. Checked for door leakage.

Sharp model R7650/door loose

The oven began to operate when the customer pushed against the door. A new interlock monitor switch was installed. To keep the switch assembly in place so the latch head will contact the microswitch, slide the assembly tight against the front piece. Hold the unit in place by placing a thin screwdriver blade in the mounting slot, then push the assembly forward as you tighten the nuts holding the switch assembly.

Sharp model R7710/intermittent cooking

When in the intermittent state, the whole oven would shut down. Readjusted door and interlock switches.

Sharp model R7804/no heat

No heat and no cooking were the oven symptoms. When loading oven for a water cook test and when closing the door, the oven sometimes came on. Readjusted door and checked interlock assemblies.

Sharp model R9200/poor door closing

Adjusted door. Replaced damaged control panel assembly HPNLC1017C-CZZ. Checked for leakage.

Sharp model R9310A/broken door

The glass was broken out of the door. Undoubtedly, something was thrown at the microwave oven. Replaced entire door assembly DD0RF0129WRKO. Adjusted door and made leakage tests.

Sharp model 9310/loose door

Found too much play between door and oven base plate. Sometimes the oven quit operating when door was pulled outward. Adjusted the door and interlock switch assemblies. Made cooking and leakage tests.

Lightning damage

Norelco model MCS 8100/control board

Had to replace the entire control board in this oven, which had lightning damage. Check the oven for a damaged thermistor. Sometimes only the thermistor and maybe some wiring needs to be replaced. If any doubts exist about the control board operation, replace it. Control board part number 6000 000 00010.

Sharp model R6460/lightning damage/dead

Replaced the 15-amp fuse. Replaced blow-apart interlock switch QSW-M0039YBEO.

Sharp model R8310/temperature control board

Replaced blown fuse. Oven was still dead. Checked and found oven temperature control board was damaged. Removed and replaced the entire assembly—2KBBK0106WRVA.

Miscellaneous oven problems

Amana CRS470P/no temperature control

Suspect a defective temperature probe when it does not operate or control the cooking temperature in the oven cavity. Check the resistance of temperature probe at room temperature. Set ohmmeter at the R×1 kilohm scale. Place ohmmeter on probe tip and ground area. If the resistance is real low or higher than 50 K, replace the defective temperature probe.

Hardwick model MM-2270/longer cooking

The complaint on this oven was it appeared to take a little longer than normal to cook food. Perhaps the magnetron was getting weak, but I found that the radiating antenna was not rotating. The antenna belt was off.

Norelco model MCS 6100/warped cover

The top waveguide cover was warped and part of it was hanging down. Replaced cover 6100 000 00020. Checked cooking and leakage tests.

Norelco model MCS 8100/keeps running

Replaced shorted triac assembly 8100 000 00005.

Norelco model S-7500/power outage

After a lightning and wind storm, replaced control board assembly. Varistor and wiring on control board damaged (figure 12-35).

12-35 Inspect the control board if the oven was struck by lightning.

Norelco model MCS 8100/oven won't shut off after 15 minutes

This oven operated okay on small cooking warm-up of less than 15 minutes. When food was cooked more than that time, the oven would not shut off and the power plug had to be removed. Replaced leaky triac assembly.

Samsung model 70STC/firing in door

Found two loose screws in the door. Checked door screws for leakage and arcing. Sometimes arcing will occur around metal screws inside the oven cavity.

Samsung 4630U/no temperature cooking

The oven operated normally, except when the temperature probe was used. With the temperature probe inserted or removed, no control of temperature was noted. A resistance measurement across the temperature probe jack was infinite. Using the hot water test and thermometer, the resistance was checked when the water was 100°F. The resistance should be around 7 or 8 kilohms, but it measured 22.12 kilohms. A new temperature probe was ordered.

Samsung MW2500U/oven does not operate at all

Nothing operated in this oven with a normal fuse. The flame sensor and magnetron TCO was checked and showed continuity. Since there were no lights the LVT transformer was checked without a normal measurement. The transformer primary wires were removed and tested again. The primary winding should have 275 ohms, but it was found open. Replaced transformer with exact part—79193-219-020.

Sharp model R7650/flickering lights

When the door was opened, the oven lights begin to flicker. The oven lights were normal with the door closed. Replaced the secondary interlock switch.

Sharp model R7710/slow cooking

The customer complained of dead spots and too much time for the cooking process. Oven was okay on several water-cooking tests. Checked oven in the home. Had only 112 Vac on extension cord when oven was operating. Suggested electrician install power outlet from the fuse box.

Sharp model R7810/temperature probe shuts off

After cooking two minutes with the temperature probe, the cooking probe shut off. Oven works okay without probe. The dealer exchanged probes; the same trouble occurred. The oven would shut off after two or three minutes of cooking with the probe. Replaced entire temperature probe electronic assembly FPWBF-0007WRKO.

Sharp model R8010/no convection cooking

Microwave cooking normal. No convection cooking. In this oven the heating element is on one side and consists of a round element. The select switch was turned to convection cooking. The fan motor turned with no heat. Removed the cover and discharged the capacitor. Checked the resistance across the heating element. Resistance should be less than 10 ohms. No continuity measured. Ordered and replaced heating element.

Sharp model R8310/poor convection cooking

The customer complained of poor heat from the heating element. Also, a hot smell was noted. The microwave oven cooking was normal. Found fan blade was not rotating. Replaced broken belt going from magnetron fan motor to heating element. The heating element fan is located right in the center of the heating element, blowing heat down from top of oven.

Sharp R-9105/no convection cooking

In this oven the convection heating element is turned on with a heater relay, controlled with a control panel and relay unit. The microwave oven cooking was normal. Either the control, oven heater relay, or heating element was defective. A pigtail light was connected across the heating element terminals and another tell-tale light connected across relay heater contacts (figure 12-36).

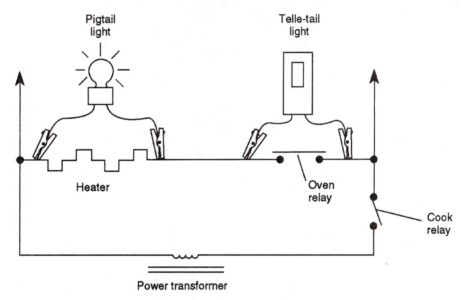

12-36 Place a pigtail or neon light across heater and relay terminals to check for open heater element in a Sharp R-91RP oven.

When the oven was turned on, the heater light was bright, indicating a defective heating element or connecting wires. The high-voltage capacitor was discharged and a continuity across the heating element indicated the element was open. These heating elements should normally measure from 5 to 25 ohms. Replacing heating element (RHET-0037WREO) solved the convection cooking problem.

Sharp model R9210/dead/power voltage

After a storm and power outage, the microwave oven was dead. Replaced the 15-amp fuse. The fuse would keep blowing. Replaced the burned line varistor with a

GE-750 line protector. Repaired two wires burned on the circuit board. Checked cooking and ran a leakage test.

Samsung MG5920T/no convection cooking

In this model the browner relay comes on and is controlled by the control board to operate the convection oven. The microwave oven worked perfectly—except there was no convection cooking. The HV capacitor was discharged and the pigtail light was clipped across the heater control sensor. (When the pigtail lights, the defective component is the one clipped across or parallel with the pigtail light.) No light was noted across the browner relay terminals. When the pigtail light was placed across the 1000-watt heater the light came on, indicating a defective heater number 76403-0001-00.

13
CHAPTER

Important do's and don'ts

For safety

Don't operate the oven when the oven cavity is empty. Do use at least a liter of water or a paper cup full of water in the oven cavity when making tests.

Don't overload the line-voltage circuits. This can cause the oven to operate erratically or intermittently in the home. Do check the line voltage.

Don't let the owner use a light extension cord for oven operation. Do suggest a separate outlet and always check for proper grounding of the microwave oven.

Don't operate or allow the oven to be operated with the door open. Do warn the owner of possible injury when the oven operates with the door open. If you find a defective monitor switch, be sure to replace it.

Do check the following components before servicing the oven: interlock operation, proper door closing, seal and choke surfaces, loose or damaged hinges, and evidence of cabinet damage (figure 13-1). Don't forget to replace all damaged oven door components.

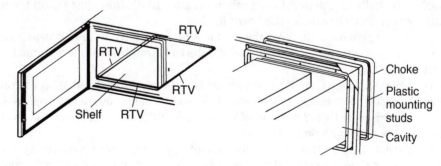

13-1 Check the door for leakage with a certified leakage meter.

Don't, under any circumstances, place your hands in the oven while the oven is operating (figure 13-2). Do pull the power plug before testing or removing the suspected component. You might receive a terrible burn or shock that could be fatal.

13-2 Do not place your hands or tools in the oven while it is operating.

Don't forget to remove the glass turntable tray or cooking test equipment from the oven before doing repair work. Do remove everything from the oven cavity. Don't work around the microwave oven when the owner or several people are hovered around asking questions. Do keep a cool head and always think what should be done for quick repair and safety.

Do keep your hands away from the ICs and critical components when replacing an electronic control board. Handle the board by the edges, keeping your body at ground potential with the metal oven cabinet (figure 13-3). Don't just rip off the insulated protection foil and slap the board in place.

Don't ever remove a clip over the contacts of a defective interlock switch and leave it that way as a repair. Do replace all interlock switches with the original replacement part when available.

Do several leakage tests if the customer is concerned about radiation or complains of possible oven leakage. Don't forget to report a dangerous, leaky oven to the manufacturer if more than 2 mW/cm^2 of leakage is found in the oven. Warn the owner not to use the oven until properly serviced.

Do check the magnetron, waveguide assembly, and microwave generating components for possible damage and leakage. Don't let the oven go out without proper leakage and cooking tests.

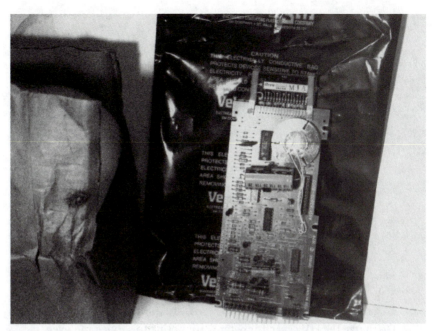

13-3 Handle the panel control board by the edges and wear a wrist grounding strap.

For proper oven repairs

Don't forget to discharge the high-voltage (HV) capacitor after pulling the ac cord. Do pull the plug and discharge the two capacitor terminals with a metal screwdriver (figure 13-4). Make sure the capacitor terminals are discharged, not just from one side to the metal case.

Don't forget to pull the power cord and discharge the HV capacitor before taking continuity and resistance measurements with the VOM. Do double-check the discharging of the HV capacitor. Look for an arc or spark when the two terminals are shorted.

Do use alligator clip leads when connecting test instruments to the various components. Don't hold the test probes in your hand when taking actual voltage measurements (figure 13-5). Insulate the test instrument with a book or rug for possible ground with the test instrument set on the metal cabinet.

Don't ever attempt to use an ordinary VOM in making HV measurements. Do use a correct HV probe or HV test instrument.

Don't stick any metal tools such as screwdrivers or nut drivers in the oven while it is operating. Do be careful when the oven is cooking. Use only a piece of wood or plastic probe tools.

Don't leave the meat probe in the metal oven cavity with the oven operating. Do remove the probe and cable when checking or making tests on the oven.

Don't forget to open and close the oven door several times to check for a possible defective interlock switch. Do make sure the oven door fits snugly, with no play.

13-4 Discharge the HV capacitor before attempting to service the oven.

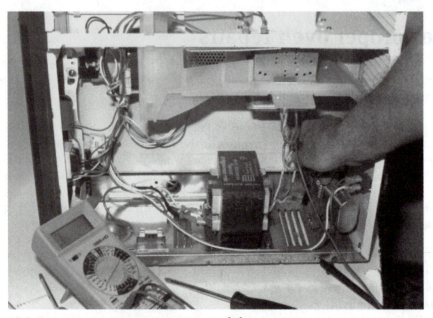

13-5 Do not connect a test instrument while oven is operating.

Do use a 100-watt lamp bulb with connecting cable clips across the 15-amp fuse terminal when the oven keeps blowing fuses. Don't forget the lamp is only an indicating device and will be very bright when a shorted component is found in the oven.

Don't forget to mark down where each wire goes when removing a defective component (figure 13-6). Always mark the wiring leads in the service manual with all components having more than two connecting wires. Do be careful. Correct component replacement saves time and money.

13-6 Write down the lead wire connections to the control panel, transformer, triac, magnetron, and relays before removing them.

Don't substitute oven parts if they do not have the same rating or physical size. Do remember the HV capacitor and diode, interlock switch, and magnetron tubes can be interchanged with the same or other manufacturers—providing they are the same part number and physical size.

Do follow the manufacturer's instructions in replacing the control board. Don't forget to read the instructions for checking off the various components before replacing the control board.

Don't forget to replace all mounting screws in components or the back cover of the metal cabinet. Do check the front of the oven cabinet for loose or outside tab mounts before the cover is locked in place.

Don't forget to run a cooking test on the oven after all repairs are made. Do keep the oven operating for a two-hour test on the service bench. Now is the time to take that final leakage test.

Don't forget to make a leakage test after each repair. Do make a careful leakage test around the door area. After replacing the magnetron, always make a leakage test around the waveguide area.

Don't forget to clean out the air vents and intake areas (figure 13-7). Do brush out all vents and clean out all pieces of food collected at the top vent area.

Do clean up the outside cabinet and controls with window spray or cleaning detergent before returning the oven to the customer. Don't forget to take out the cooking test equipment. You will need it for the next oven repair.

Don't forget to remove your wristwatch while working around the oven. Do take off the watch when replacing the magnetron tube.

Don't attempt to repair microwave ovens if you wear a pacemaker. Do consult your physician.

13-7 Clean out the air vents when an oven is in for service.

14
CHAPTER

Where and how to obtain parts

Even after you locate a defective component, securing the replacement part could take a while. Even though some oven components are interchangeable, they should be purchased from the correct oven manufacturer (figure 14-1). Most oven manufacturers issue the parts, or they have distribution centers around the country who can get them for you. Go directly to local chain stores, such as Sears and Wards if you need to order their oven components.

14-1 Magnetrons, reburnished control panels, HV diodes, and capacitors are available from the manufacturer and mail-order firms.

Many microwave oven manufacturers have service depots and personnel located in various states for warranty repair service. Just give them a call. You might find a list of warranty stations enclosed in the oven operation literature. Call or write to the following manufacturers to obtain service manuals and oven parts.

Amana Refrigeration, Inc.
Amana, IA 52204
(319) 622-5511

Caloric Corp.
Topton, PA 19562
(215) 682-4211

Crosley Corp.
P.O. Box 1959
Winston-Salem, NC 27102
(919) 761-1212

Emerson Radio
One Emerson Lane
North Bergen, NJ 07047

Frigidaire Div. WC1
300 Phillips Road
Columbus, OH 43228
(614) 272-4100

General Electric Co.
Appliance Park
Louisville, KY 40221
(502) 452-4311

Goldstar Electronic Inc.
1050 Wall Street West
Lindhurst, NJ 07071
(201) 480-8870

Hardwick Stove Co.
740 King Edward Avenue
Cleveland, TN 37311
(615) 478-4610

Hitachi Sales Corp. of America
401 West Artesia Boulevard
Compton, CA 90220
(213) 537-8383

Hotpoint
General Electric Co.
Appliance Park
Louisville, KY 40225
(502) 452-4311

J. C. Penney Co.
(Check at local store)

Jenn-Air Corp.
3025 Shadeland Avenue
Indianapolis, IN 46226
(317) 545-2271

Kitchen Aid
701 Main Street
St. Joseph, MI 64501
(616) 982-4500

Litton Microwave Cooking Products
4450 Mendenhall Road South
Memphis, TN 38101
(901) 366-3000

Magic Chef
740 King Edward Avenue
Cleveland, TN 37311
(615) 472-3371

Maycor (Division of Maytag)
240 Edwards Street
Cleveland, TN 37311

Maytag
One Dependability Square
Newton, IA 50208
(515) 792-7000

Modern Maid
Topton, PA 19562
(215) 682-4211

Montgomery Wards
(Check with local stores)

Norelco American Phillips
High Ridge Park
Stanford, CT 06903
(203) 329-5700

Panasonic Co.
One Panasonic Way
Secaucus, NJ 07094
(201) 348-7185

Quasar
1325 Pratt Boulevard
Elk Grove Village, IL 60007
(312) 228-6366

Roper Sales Co.
1507 Broomtown Road
Lafayette, GA 30728
(404) 638-5100

Royal Chef
Gary & Dudley
2300 Clinton Road
Nashville, TN 37209

Sampo of America
5550 Peachstreet Ind. Boulevard
Norcross, GA 30728
(404) 449-6220

Samsung Electronics Corp. America
One Samsung Place
Ledgewood, NJ 07852
(201) 691-6200

Samsung Distributor
J&J International Inc.
211 Parsippany Rd.
Parsippany, NJ 07054
(800) 627-4368

Samsung Distributor
Fox International LTD, Inc.
23600 Aurora Road
Bedford Heights, OH 44146
(216) 439-8500

Samsung Distributor
Hermans Electronics
1365 NW 23rd St.
Miami, FL 33142
(305) 634-6591

Samsung Distributor
Amkotron Electronics
14821 Spring Ave.
Sante Fe Springs, CA 90670
(310) 802-0120

Samsung Electronics
301 Mayhill Street
Saddlebrook, NJ 07662
(201) 587-9600

Sanyo-Fisher, Inc.
21350 Lassen Street
Chatsworth, CA 91311
(818) 996-7322

Sharp Electronics Corp.
Sharp Plaza
Mahwah, NJ 07430
(201) 265-5600

Tappan Appliance Division
300 Phillips Road
Columbus, OH 43228

Tatung of America
2850 El Presido Street
Long Beach, CA 90810
(213) 637-2105

Thermador/Waste King
5119 District Boulevard
Los Angeles, CA 90040
(213) 562-1133

Toshiba America, Inc.
Home Appliance Division
82 Otawa Road
Wayne, NJ 07470
(201) 628-8000

Whirlpool Corp.
2000 M-63 North
Benton Harbor, MI 49022
(616) 926-5000

White-Westinghouse
Div. WC1
300 Phillips Road
Columbus, OH 43228
(614) 272-4100

List of universal mail-order oven parts

AIM (Authorized Microwave of Iowa)
P.O. Box 171
99 W. Broadway
Eagle Grove, IA 50533-0171

MCM Electronics
650 Congress Park Drive
Dayton, OH 45459-9955

Service literature

The manufacturer's service literature is one of the most important pieces of information in servicing microwave ovens. Besides several circuit diagrams and parts lists, you might find trouble charts on how to service a certain model. Invaluable tips and charts are found in the oven service literature (figure 14-2).

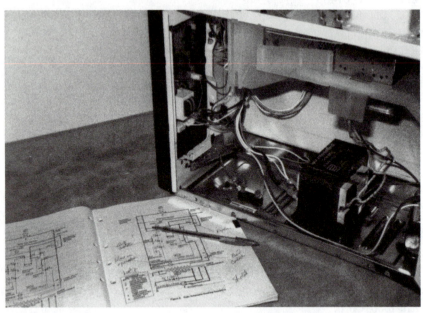

14-2 The manufacturer's schematic is a must in tough oven problems.

Very few manufacturers include voltage or current measurements on the circuit diagram. When any given oven has been repaired and is operating correctly, it's wise to take voltage and current measurements. Just write the measurement right on the circuit diagram for future reference. Remember that high voltage (HV) and current readings must be taken with an instrument manufactured especially for microwave oven repair.

When unusual or difficult problems are found in the oven, they should be marked on the schematic for future reference. Often these same problems crop up and it's difficult to remember all of them (figure 14-3). These difficult problems can be compared with problems with other ovens, saving valuable service time. If the service manual is not available, you can order one directly from the manufacturer.

Warranty parts

Most oven manufacturers' components are warranted for one or two years. The magnetron might be guaranteed from five to seven years. Periodically, a warranty service shop is issued a copy of the warranty description for each oven. You might find the warranty period listed in the owner's manual. Oven parts that are found to be in the warranty period must be treated a little differently than those out of warranty.

With some manufacturers, only the larger components are returned in the warranty period. Others request that only the magnetron tube be returned. For instance, Sharp Electronics Corporation requests that the warranty station return all

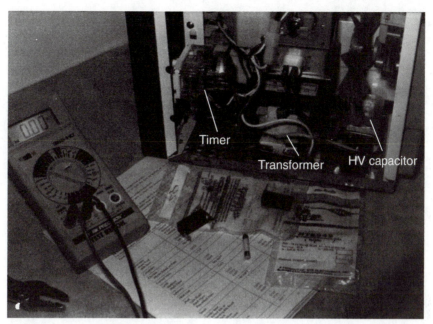

14-3 Before moving any transformer or oven component, write down the location of the terminal connectors and color-coded wires.

components except small items such as switches, fuses, and HV diodes. Norelco might request the return of warranty components such as control boards and magnetrons. Each oven manufacturer has a different warranty replacement procedure. All warranty claims should be followed correctly to receive credit and payment for services rendered.

How to order parts

When ordering a replacement part for a defective component, always include the part number, model number of the oven, and component description. The defective part should be listed in the parts list of the service manual. Some ovens have both a physical and part drawing number for locating each component. Listing the reference number, if available, might also help secure the correct component. Simply find the part in the layout drawing, then check the number in the part number.

You must include all numbers or letters in the part number. Often, the oven part number is rather long. For instance, in a Sharp microwave oven, there are a total of 13 numbers and letters in the part number. The waveguide cover part number for a Sharp R9330 model is PC0VP0204WRE0. A turntable coupling assembly for the same oven is FCPL-0018WRK0. Notice that the dash mark is included in the total of 13 letters and numbers.

In a Norelco oven, the part number has a total of 12 numbers. The first four numbers are the model number of the oven. A controller board in an MSC 8100 Norelco oven has the part number 8100 000 0000 8. The part number begins with a 6100 or 7100 number, which shows that the component is common to both ovens (figure 14-4).

WARNING

THERE EXISTS HIGH-VOLTAGE ELECTRICITY WITH HIGH-CURRENT CAPABILITIES IN THE CIRCUITS OF THE HGIH VOLTAGE TRANSFORMER SECONDARY AND FILAMENT TRANSFORMER SECONDARY. IT IS EXTREMELY DANGEROUS TO WORK ON OR NEAR THESE CIRCUITS WITH THE OVEN ENERGIZED.

WARNING

DO NOT TOUCH ANY PART OF THE CIRCUITRY ON THE TOUCH CONTROL CIRCUIT, SINCE A STATIC ELECTRIC DISCHARGE MAY DAMAGE THIS CONTROL PANEL.

14-4 Discharge the HV capacitor so you won't get hurt or damage valuable test equipment.

If a service manual is not available, the component can be ordered by giving the model number of the oven and number marked on the part. You might find that a magnetron tube number stamped directly on the body is the same as the part number. But if possible, it's best to order the oven service manual and then order the component from the parts list. You are ensured of obtaining the correct component with this method.

Returning critical components

All critical parts such as the magnetron, transformer, and control board should be sent back in the original cartons. Often these components are packed in two different boxes. Special magnetron boxes have additional packing to prevent breakage to the glass portion of the magnetron. Inspect the vent area of the magnetron for possible damage before installing the new tube.

Controller or circuit boards are packed in a special nonstatic sleeve and are taped up. Replace the circuit board in the same enclosure, insert in the part box, and then place the box inside a larger one for protection. Never send a defective power transformer and a circuit board back in the same carton. The heavy transformer might break or otherwise damage the circuit board. You should have no problems if the defective component is returned in its same shipping carton (figure 14-5).

Before sealing the final box, enclose the warranty service report. Most oven manufacturers require that the service dates be included with the defective component. When several different warranty components are returned, tape the service

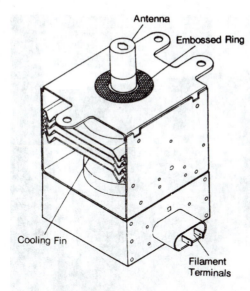

Antenna

Embossed Ring

Cooling Fin

Filament
Terminals

14-5
Place defective parts that are in
warranty back in the container
they came in.

form to each defective part. Every oven manufacturer has their own service form
and it's best to follow their warranty part procedures.

Substitution of parts

Although the manufacturer requests that both the oven technician and the repair person use exact part numbers, there are a few components that can be purchased from electronics distributors, warehouses, or mail-order firms. The biggest
problem in substituting parts is the physical size (figure 14-6). Replacement components should have the same operating voltage or higher and the correct wattage.

You should always replace defective parts with the original part numbers when
possible. But sometimes, substitution is unavoidable. The manufacturer might be
out of business or critical parts orders could be tied up for a while, resulting in parts
substitution.

Warranty replacement parts sometimes pile up in the dealers stock, and they
should be used up as soon as possible. Here are a few components that can be substituted:

Fuses

The microwave oven employs a ceramic-type fuse that is about $1\frac{1}{4} \times \frac{1}{4}$ inch in
size. These fuses come in 10, 15, 20, 25, and 30 amps. The common ceramic fuses are
15 and 20 amp. These fuses can be purchased almost anywhere ovens are sold. Replace with the exact amperage as the original.

Interlocks

A lot of the interlock switches can be exchanged from one oven to another. Make
sure that the switch is the same. Check and compare the size, closed or open state,
two switches in one, or single switch.

14-6 Replace the magnetron tube with one of the same wattage, voltage, and physical size.

HV capacitor

Replace the capacitor with one with the exact same working voltage or higher, and the same capacitance or higher. These capacitors are easily interchanged between ovens.

HV rectifier

Microwave oven diodes can be subbed with ECG 541 and 548 replacements. RCA 9306 and 9305 are 1000 and 500 mA forward current types. Pick these up at wholesale parts houses, distributors, and mail-order firms.

Magnetrons

Sometimes magnetron tubes can be interchanged because several manufacturers make ovens for many different firms. Make sure the tube will fit physically and has the correct wattage and fan blower application. Make sure the HV and heater terminals are in the same direction. Match up the correct mounting and direction of air flow from the fan blower. Remember, some of these magnetrons have a fan blower mounted on the bottom side.

Do not install a magnetron from a small 250–400-watt oven into a 600–700-watt oven; nor should you install a 600–700-watt magnetron tube in a 400-watt oven. Check the oven for an external HV rectifier. If the diode is inside the tube cavity, you cannot replace it with a regular magnetron. Most magnetron replacements do not contain an internal rectifier. If the magnetron replacement has the same wattage, physical size, and correct direction of air flow, no doubt it will work satisfactorily.

Service forms

Although some oven manufacturers issue their own warranty forms for service and parts reports, many prefer the Narda service form number 317-515. The Narda service report can be used for television, radio, stereo, and oven service reports. This service form consists of the customer's name, address, and where the oven was purchased (figure 14-7). Don't forget to include the model and serial number of the oven (also include the date purchased, received, and repaired). Some manufacturers request that a microwave leakage test be shown. This leakage reading should be marked on all microwave oven reports for the service technician and manufacturer's protection. All Narda warranty business forms can be obtained from:

Narda, Inc.
2N Riverside Plaza
Suite 222
Chicago, IL 60606
(312) 454-0944

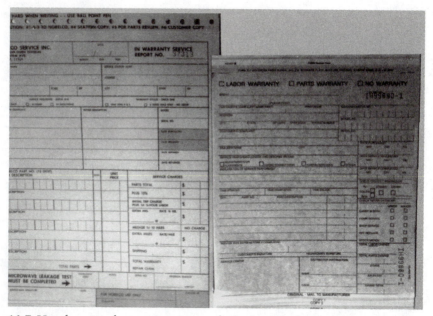

14-7 Use the manufacturer's warranty forms or a Narda Service Form for warranty repairs.

Appendix:
Selected schematics

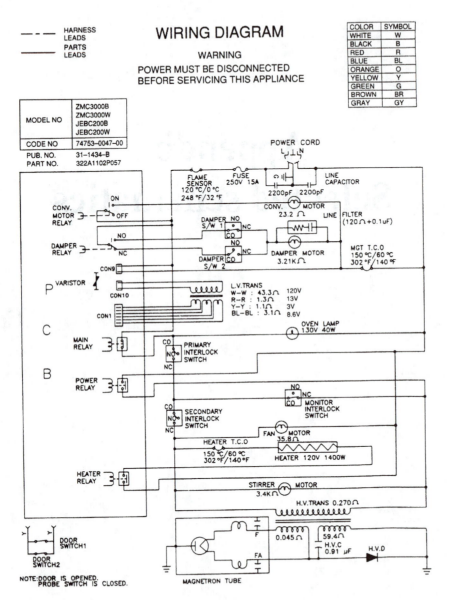

A-1 General Electric schematic diagram of model JEBC200 oven. Courtesy General Electric Co.

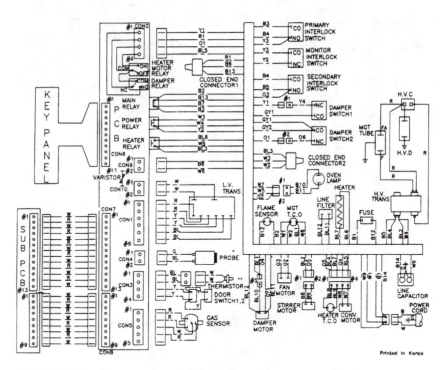

A-2 GE wiring diagram of model JEBC200 oven. Courtesy General Electric Co.

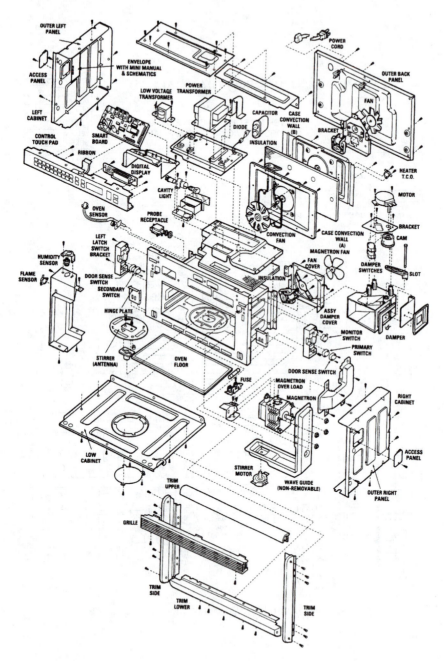

A-3 GE JEBC200 tilt-down door parts assembly. Courtesy General Electric Co.

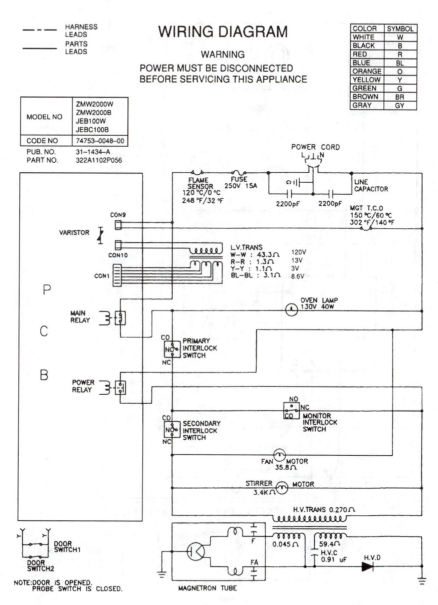

A-4 GE model JEBC100 microwave oven diagram. Courtesy General Electric Co.

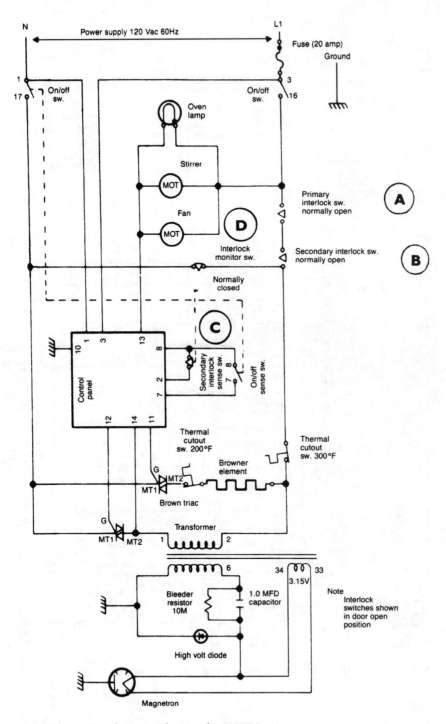

A-5a The wiring diagram of a Norelco R7700 microwave oven. Courtesy Norelco Corp.

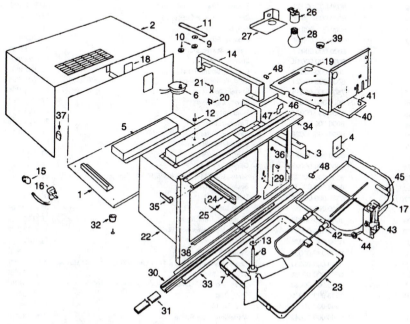

Item No.	N.S.I. Part Number	Description	
1	6000-000-00001	Wrapper, Back and Bottom	805T015F15
2	6000-000-00003	Wrapper, Top and Sides (Walnut)	809T011F04
3	7500-000-00013	Angle, Air Baffle	008T378F05
4	7500-000-00014	Angle, Probe Receptacle	008T435F01
5	6000-000-00006	Baffle, External Vent	024T123F03
6	6000-000-00007	Motor, Stirrer	463T070P07
*	6000-000-00008	Screw, Stirrer Motor (10Ax3/8" Ph. Tr.Hd) (2)	636T091P01
7	6000-000-00009	Fan, Stirrer	239T028F01
8	6000-000-00010	Shaft, Stirrer	652T005P01
*	6000-000-00011	Screw, Stirrer Shaft to Fan (6-32x3/8" Ph. Pan Hd) (2)	636T095P31
*	6000-000-00012	Nut, Stirrer Shaft to Fan (6-32 W/Ex. T Washer) (2)	479T069P01

* Not illustrated

A-5b The oven body layout of Norelco S7500 oven with important oven parts.
Courtesy Norelco Corp.

Item No.	N.S.I. Part Number	Description	
9	6000-000-00013	Fastener, Shaft Retainer	241T017P151
10	6000-000-00014	Grommet, Stirrer Pulley (2)	287T026P01
11	6000-000-00015	Belt, Stirrer Drive	044T002P02
12	6000-000-00016	Bushing, Stirrer Shaft	102T027P01
13	6000-000-00017	Washer, Stirrer Shaft	789T022P03
14	6000-000-00018	Channel, Air Duct	136T258F04
15	6000-000-00020	Bushing, Strain Relief	102T024P03
16	6000-000-00021	Cord, Power Supply	171T019P04
17	7500-000-00015	Stirrer Cover	174T165P04
18	6000-000-00023	Cover, Light Access	174T169F01
19	6000-000-00024	Duct Bottom	211T032F05
20	6000-000-00025	Fastener, Stand Off	241T017P153
21	6000-000-00026	Fastener, Purse Lock (Ground Wires)	241T017P154
22	N/A	Front Main Wave Guide	N/A
23	6000-000-00027	Tray, Oven (Glass)	736T005P01
24	6000-000-00028	Glide, Tray (4)	280T046P01
25	6000-000-00029	Screw, Glide Mtg. (10AB x 5/8'' Ph. Fillister) (8)	636T091P19
26	6000-000-00030	Socket, Light	662T027P02
27	6000-000-00031	Shield, Oven Light	655T302P02
28	6000-000-00032	Light Bulb	374T018P01
29	6000-000-00033	Grommet, Door Strike	287T003P08
30	6000-000-00034	Cover, Microwave Seal (2)	174T166P02
31	6000-000-00035	Seal, Microwave (2)	643T008P09
32	6000-000-00036	Spacer, Foot (4)	665T132P02
*	6000-000-00037	Screw, Foot and Bottom Trim Mtg. (8Ax3/4'' Ph.Tr.Hd) (4)	636T090P10
33	6000-000-00038	Trim, Bottom	740T385P05
*	6000-000-00039	Screw, Bottom Trim Mtg. (8A x ½'' Ph. Tr. Hd)	636T069P07
34	6000-000-00040	Trim, Top	740T386P05
35	6000-000-00041	End Cap (4)	123T079P01
36	6000-000-00042	Medallion, Door Strike	445T004P01
37	7500-000-00016	Fastener, Wrapper (3)	241T017P152
38	6000-000-00044	Trim, Front Main	740T392P02
39	6000-000-00045	Bushing, Wire (2)	102T028P01
40	7500-000-00017	Triac, Heat Sink Assembly	742T001P02
41	7500-000-00018	Fastener, Triac (3)	241T017P198
*	7500-000-00019	Screw, Wrapper (8A x ½'' W/Int. Washer) (23)	636T090P46
42	7500-000-00020	Probe, Roast Control	563T008P01
43	7500-000-00021	Receptacle, Probe	597T029P01
44	7500-000-00022	Nut, Receptacle Mtg.	479T082P01
45	7500-000-00023	Insulator	331T117P01
46	7500-000-00024	Insulator, Front Frame (Cap to Touch Panel)	331T118P01
*	7500-000-00025	Kit, Touch-up Paint (Almond)	508T020P02
47	7500-000-00026	Cover, Corner (2)	174T213P01
48	7500-000-00027	Spacer, Eyelet (4)	665T155P01
*	7500-000-00028	Grounding Washer, Wave Guide (Int. Tooth) (4)	789T053P30
*	7500-000-00029	Screw (10A x ½'' Ph.Tr.Hd.) Back Wrapper	636T091P02
*	7500-000-00030	Screw (8A x ½'' W/Int. Washer) Wrapper	636T090P46

* Not illustrated

A-5c Continuation of parts list for Norelco S7500 oven.

NN-8507/NN-8807/NN-8907 APH

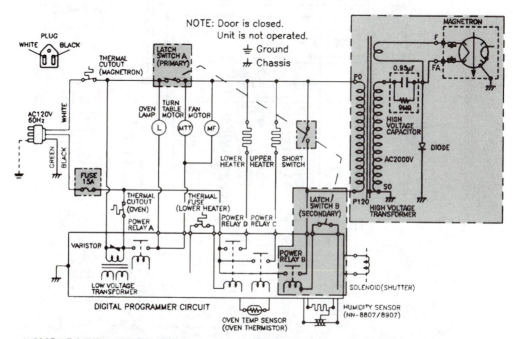

A-6 The schematic diagram of the Panasonic NN8507/NN8807 microwave oven. Courtesy Matsushita Electric Corp. of America

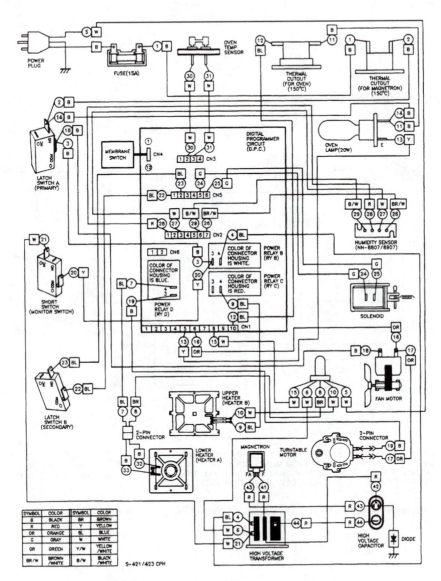

A-7 The wiring diagram of the Panasonic ovens NN8507/NN8807 CPH.
Courtesy Matsushita Electric Corp. of America

EXPLODED VIEW

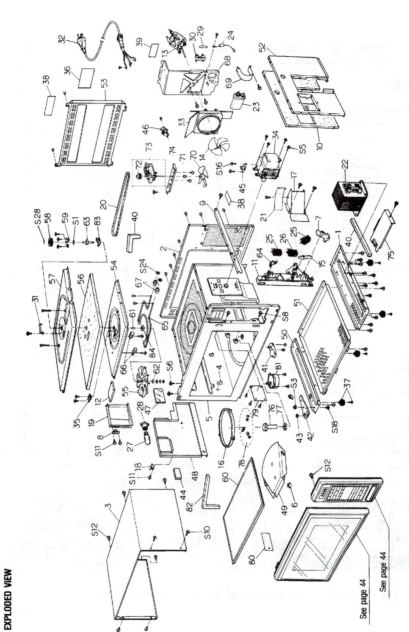

A-8 An exploded view of parts in the Panasonic NN9807, NN9507, MQ8897 BW. (Page numbers refer to oven manual.) Courtesy Matsushita Electric Corp. of America

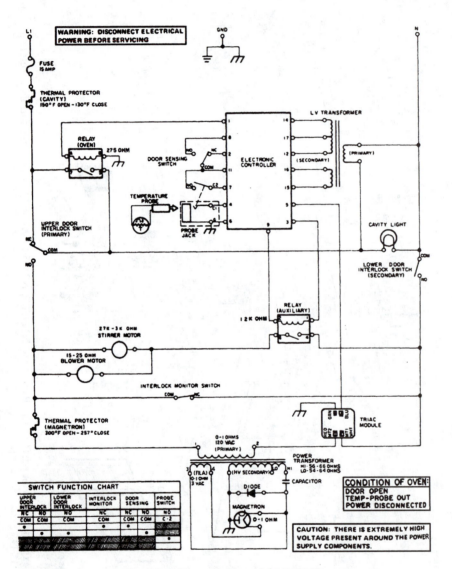

A-9 The schematic diagram of Panasonic model NN9807 oven. Courtesy Matsushita Electric Corp of America

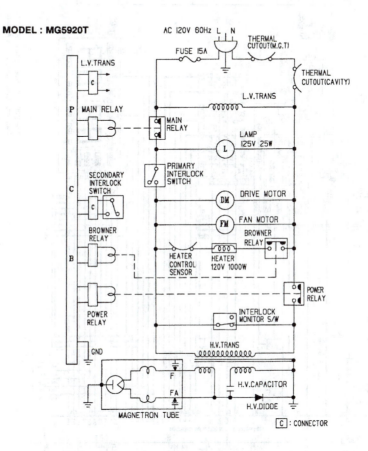

MODEL : MG5920T

WARNING: POWER MUST BE DISCONNECTED BEFORE SERVICING THIS APPLIANCE.

NOTE: DOOR IS OPEN.
FOR SERVICING REPLACEMENT, USE 18GA. 105°C THERMOPLASTIC COVERED WIRE EXCEPT FOR HIGH VOLTAGE LEADS OR THE IDENTICAL PART AS NOTED IN THE PARTS LIST.

A-10 Schematic diagram of Samsung oven/browner MG5920T. Courtesy Samsung Electronics America, Inc.

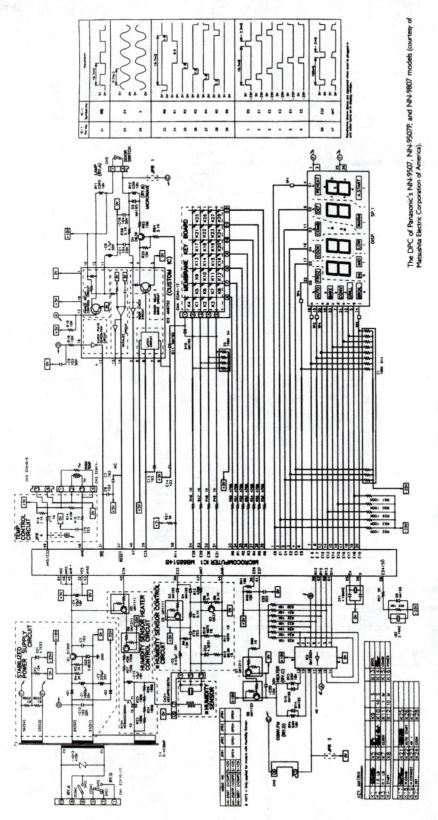

A-11 The DPC of the Panasonic NN9507, NN9507P, and NN9807 models. Courtesy Matsushita Electric Corp. of America

The DPC of Panasonic's NN-9507, NN-9507P, and NN-9807 models (courtesy of Matsushita Electric Corporation of America).

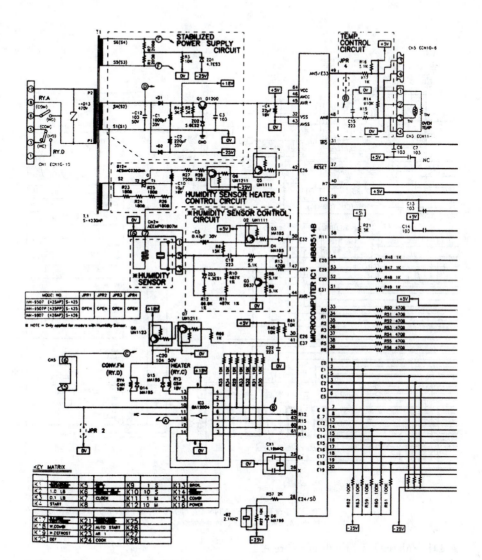

A-11a Enlarged left half of figure A-11.

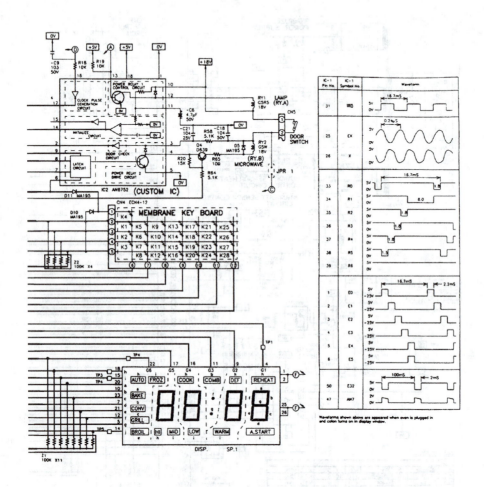

The DPC of Panasonic's NN-9507, NN-9507P, and NN-9807 models (courtesy of Matsushita Electric Corporation of America).

A-11b Enlarged right half of figure A-11.

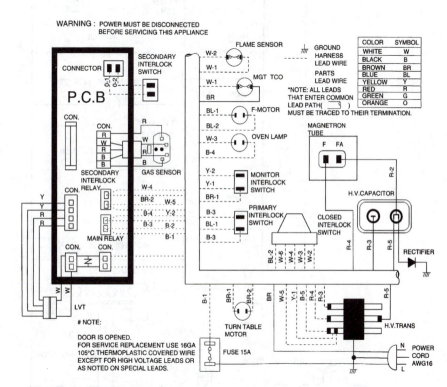

A-12 Schematic diagram of the Samsung auto/sensor MW5820T microwave oven. Courtesy Samsung Electronics America, Inc.

MODEL : MW2500U

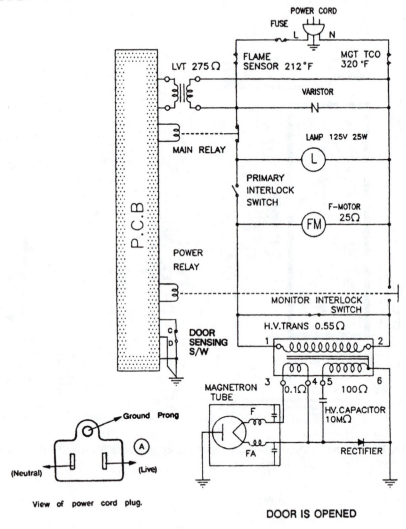

A-13 Schematic diagram of the Samsung Compact oven MW2500U. Courtesy Samsung Electronics America, Inc.

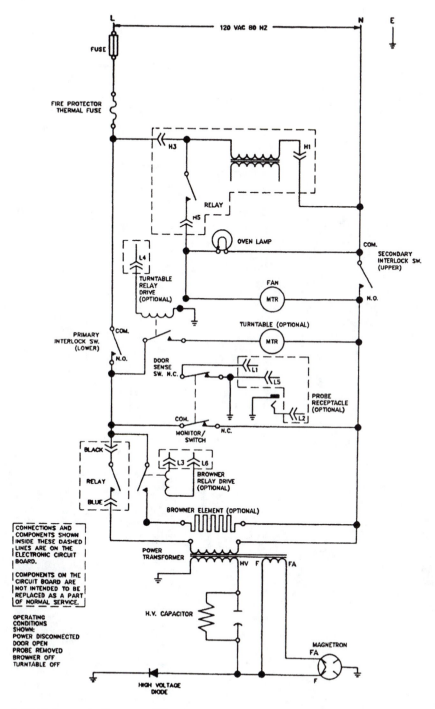

A-14 Schematic diagram of Tappan oven. Courtesy Tappan, brand of Frigidaire Co.

Glossary

ac voltage Alternating current.

anode The positive electrode in the magnetron receives the electrons. In this case, the outer shell of the magnetron is the anode and is always grounded.

bell or buzzer An electronic or mechanical sounding device to indicate that the cooking process is finished. The bell might have a mechanical striker on the timer assembly. The electronic buzzer or speaker sounds a loud tone when the oven cooking cycle is over.

blower motor The blower motor keeps the magnetron cool. The blower motor circulates air and helps remove moisture from the oven cavity.

capacitance The electric size of a capacitor. Sometimes called capacity. The basic unit of capacitance is the farad. A capacitor permits the storage of electricity between two insulated conductors.

capacitor A capacitor stores electric energy, blocks the flow of direct current, and permits the flow of alternating current. The large capacitors found in the microwave ovens are in a metal container. Most HV capacitors in the oven are oil-filled types.

cathode The cathode element provides the source of electrons. In the magnetron tube, the cathode is centered within the anode cavity and has a highly negative voltage potential.

choke An inductance used in a circuit to present high impedance to frequencies without limiting the flow of direct current. In the microwave oven, a groove or shaped area to reflect guided waves within a limited frequency range.

circuit A path in which electrical current can flow.

control board An electronic board used to control the time and operation of the oven cooking process. The controller or control board can be replaced separately or repaired.

current The flow of electricity. The current found in the magnetron circuit is measured in milliamperes.

defrost To intermittently remove ice from frozen food for quick microwave cooking. The defrost circuit might consist of a motor, switches, control board, and CAM assembly in the oven.

dielectric A material that serves as an insulator. A dielectric material is used between the metal foil layers of the HV capacitor.

diode Passes current in one direction and blocks current in the reverse direction. The HV diode has a cathode and anode terminal. The cathode terminal is always at chassis ground in the microwave oven circuits.

door assembly The front door assembly opens so food can be inserted in the oven for cooking. The door assembly has hinges at one end and latch or hook switch assemblies to hold the door in place.

electron The smallest electric charge. A negatively charged particle. Electrons flow from the cathode to the anode terminal at a high rate of speed in the magnetron.

fan blower A fan motor to provide cooling of the magnetron and circulate the air in the oven areas.

ferrite A compressed, powdered iron material.

filament or heater The heater or filament element heats up the cathode terminal to emit free electrons. The filament or heater terminals are at a high negative voltage in the oven circuit.

filament transformer In some microwave ovens, a separate transformer is used to provide voltage to the filament or heater terminals of the magnetron.

fuse A chemical 15-amp fuse is found in all microwave ovens to protect the oven circuit.

fuse resistor A service that senses an increase in current in the transformer secondary and opens the transformer primary circuit.

impedance A combination of resistance and reactance, expressed in ohms.

inductance The property of a circuit that causes a magnetic field to be induced.

interlock switch The interlock switch prevents the operator from using the oven without having the door closed. Interlock switches might be called monitor and safety switches.

latch switch An interlock switch located and controlled by the oven door.

leakage tester A government-approved survey instrument to measure radiation leakage of the front door, vents, and waveguide areas.

light switch Controls the oven light when the door is closed or opened.

magnetron A vacuum tube in which the flow of electrons from the heated cathode flow to the anode element. The speed of the electrons is controlled by a magnet and an electrical field. The magnetron produces very short electrical waves (microwaves). The frequency of these waves is 2450 MHz.

oven light Furnishes light to the oven cavity. Some ovens have two lights.

power cord Provides power from the ac outlet to the oven circuits.

power transformer A large power transformer provides filament or heater voltage for the magnetron and high voltage to the HV circuits. A small power transformer provides low ac voltage to the control board.

radiation leakage RF microwave leakage that might occur around the door and vent areas.

relay An oven relay provides power to the HV transformer and magnetron. The auxiliary relay energizes when the oven is placed in the cook cycle. The power oven relay might be controlled by a digital programmer circuit.

resistor A device that limits the flow of current, providing a voltage drop. A high-megohm resistor can be found in the HV circuits. A 10-ohm resistor can be used to measure current in the HV diode circuit.

stirrer motor A motor or blade that circulates the microwaves within the oven cavity. The stirrer blade might be driven by the fan or blower motor by a long drive belt.

temperature probe When inserted into meat or other food in the oven cavity, the probe determines the temperature or cooking time.

thermal switch When overheated, the thermal switch opens up the power line voltage, preventing the magnetron from overheating.

timer A device to control the amount of cooking time.

triac An electronic switch to apply ac voltage to the HV transformer. Usually, the gate voltage of the triac is controlled from the control board.

voltage-doubler circuit In a microwave oven, the voltage-doubler circuit consists of a silicon rectifier and HV capacitor. The high ac voltage from the secondary winding of the power transformer is applied to the voltage-doubler circuit, where it is rectified and is always less than the doubled input voltage.

watt The unit of electric power. A microwave oven might pull more than 200 watts.

waveguide A metal enclosure for the conduction or transmission of microwaves. The waveguide in the microwave oven is found between the magnetron and the oven cavity.

wavelength The distance between corresponding points of two successive ac waves.

Index

About the author

Homer L. Davidson worked as an electrician and small appliance technician before entering World War II and taught Radar while in the service. After the war, he opened his own Radio & TV repair shop for 38 years. He is the author of more than 30 books for McGraw-Hill. His first magazine article was printed in *Radio Craft* in 1940. Since that time, Davidson has had over 1000 articles printed in 48 different magazines. He is at present TV Servicing Consultant for *Electronic Servicing & Technology* and Contributing Editor for *Electronic Handbook* magazines.